NOTES ON
COBORDISM THEORY

BY

ROBERT E. STONG

PRINCETON UNIVERSITY PRESS
AND THE
UNIVERSITY OF TOKYO PRESS

PRINCETON, NEW JERSEY
1968

Published in Japan exclusively
by the University of Tokyo Press;
in other parts of the world by
Princeton University Press

Printed in the United States of America

Preface

These notes represent the outgrowth of an offer by Princeton University to let me teach a graduate level course in cobordism theory. Despite the fact that cobordism notions appear in the earliest literature of algebraic topology, it has only been since the work of Thom in 1954 that more than isolated results have been available. Since that time the growth of this area has been phenomonal, but has largely taken the form of individual research papers. To a certain extent, the nature of cobordism as a classificational tool has led to the study of many individual applications rather than the development of a central theory. In particular, there is no complete exposition of the fundamental results of cobordism theory, and it is hoped that these notes may help to fill this gap.

Being intended for graduate and research level work, no attempt is made here to use only elementary ideas. In particular, it is assumed that the reader knows algebraic topology fairly thoroughly, with cobordism being treated here as an application of topology. In many cases this is not the fashion in which development took place, for ideas from cobordism have frequently led to new methods in topology itself.

An attempt has been made to provide references to the sources of most of the ideas used. Although the main ideas of these sources are followed closely, the details have frequently been modified considerably. Thus the reader may find it helpful to refer to the original papers to find other methods which are useful. For example, the Adams spectral sequence gives a powerful computational tool which has been used in determining some theories and which facilitates low dimensional calculations, but is never used here. Many of the ideas which appear are of the "well known to workers in the field - but totally unavailable" type and a few ideas are my own.

The pattern of exposition follows my own prejudices, and may be roughly described as follows. There are three central ideas in cobordism theory:

1) Definition of the cobordism groups,

2) Reduction to a homotopy problem, and

3) Establishing cobordism invariants.

This material is covered in the first three chapters. Beyond that point, one must become involved with the peculiarities of the individual cobordism problem. This is begun in the fourth chapter with a survey of the literature, followed by detailed discussion of specific cobordism theories in the later chapters. Finally, two appendices covering advanced calculus and differential topology are added, this material being central to the 'reduction to a homotopy problem' but being of such a nature as to overly delay any attempt to get rapidly to the ideas of cobordism.

I am indebted to many people for leading me to this work and developing my ideas in this direction. Especially, I am indebted to Greg Brumfiel, Peter Landweber, and Larry Smith for numerous suggestions in preparing these notes, and to Mrs. Barbara Duld for typing. I am indebted to Princeton University and the National Science Foundation for financial support. Finally, I am indebted to my wife for putting up with the foul moods I developed during this work.

CONTENTS

CHAPTER I

Introduction - Cobordism Categories

In order to place the general notion of cobordism theory in mathematical perspective recall that differential topology is the study of the category of differentiable manifolds and differentiable maps, primarily in relation to the category of topological spaces and continuous maps. From a slightly less theoretical point of view, it is the study of differentiable manifolds by topologists using any methods they can find. The guiding principle is that one does not study imposed structures such as Riemannian metrics or connections and this distinguishes differential topology from differential geometry.

As in any subject, the primary problem is classification of the objects within isomorphism and determination of effective and computable invariants to distinguish the isomorphism classes. In the case of differentiable manifolds this problem is not solvable, since for any finitely presented group S one can construct a four dimensional manifold $M(S)$ with fundamental group S in such a way that $M(S)$ and $M(T)$ will be homeomorphic if and only if S and T are isomorphic, but one cannot solve the word problem to determine whether two finitely presented groups are isomorphic (Markov [76]). In special cases one can solve the problem, but cobordism theory works in another way - by introducing an equivalence relation much weaker than isomorphism.

Briefly, two manifolds without boundary are called 'cobordant' if their disjoint union is the boundary of some manifold. It is worthwhile to note that every manifold M with empty boundary is the boundary of $M \times [0,\infty)$. To get a nontrivial theory it is standard to restrict attention to compact manifolds.

The first description of this equivalence relation was by H. Poincaré: Analysis Situs, Journal de l'École Polytechnique, 1 (1895), 1-121 (section 5, Homologies). His concept of homology is basically the same as the concept of cobordism used today.

The next development of cobordism theory was by L. S. Pontrjagin: Characteristic cycles on differentiable manifolds, Math. Sbor. (N.S.), 21 (63) (1947), 233-284 (Amer. Math. Soc. translations, series 1, no. 32). This paper shows that the characteristic numbers of a closed manifold vanish if the manifold is a boundary (providing the invariants for classification).

The cobordism classification of manifolds is reasonably elementary in dimensions 0,1, and 2, since manifolds are themselves classified in these dimensions. Using geometric methods the cobordism classification problem in dimension 3 was solved by V. A. Rohlin: A 3-dimensional manifold is the boundary of a 4-dimensional manifold, Doklady Akad. Nauk. S.S.S.R., 81 (1951), 355.

The first application of cobordism was by L. S. Pontrjagin: Smooth manifolds and their applications in homotopy theory, Trudy Mat. Inst. im Steklov no. 45, Izdat. Akad. Nauk. S.S.S.R. Moscow, 1955 (Amer. Math. Soc. translations, series 2, vol. 11, 1959). Pontrjagin attempted to study the stable homotopy groups of spheres as the groups of cobordism classes of 'framed' manifolds. This amounts to the equivalence of a homotopy problem and a cobordism problem. The lack of knowledge of manifolds has prevented this from being of use in solving the homotopy problem.

The major development of cobordism theory is the paper of R. Thom: Quelques propriétés globales des variétés differentiables, Comm. Math. Helv., 28 (1954), 17-86. This paper showed that the problem of cobordism is equivalent to a homotopy problem. For many of the interesting manifold classification questions the resulting homotopy problem turns out to be solvable. Thus, Thom brought the Pontrjagin technique to the study of manifolds, largely reversing the original idea.

For a brief sketch of cobordism theory there are three survey articles of considerable interest. For an insight into the early development of the theory (up through Thom's work) see V. A. Rohlin: Intrinsic homology theories, Uspekhi Mat. Nauk., 14 (1959), 3-20 (Amer. Math. Soc. translations, series 2, 30 (1963), 255-271). A short article which covers many of the examples of cobordism classification problems is J. Milnor: A survey of cobordism theory, Enseignement Mathematique, 8 (1962), 16-23. Contained in the survey of differential topology by C. T. C. Wall: Topology of smooth manifolds, Journal London Math. Soc., 40 (1965), 1-20, is a discussion of representative cobordism theories, with outlines of the methods by which these problems are solved.

Cobordism Categories

In order to formalize the notion of cobordism theory, it seems useful to set up a 'general nonsense' situation. As motivation, one may consider the properties of differentiable manifolds.

Let $\mathcal{D}$ denote the category whose objects are compact differentiable manifolds with boundary (of class C^∞) and whose maps are the differentiable maps (again C^∞) which take boundary into boundary. This category has finite sums given by the disjoint union and has an initial object given by the empty manifold. For each object of $\mathcal{D}$ one has its boundary, again an object of $\mathcal{D}$, and for each map the restriction of it to the boundary.

Further, the boundary of the boundary is always empty. This defines an additive functor $\partial : \mathcal{D} \longrightarrow \mathcal{D}$. For any manifold M, the boundary of M is a subset whose inclusion is a differentiable map $i(M) : \partial M \longrightarrow M$. This inclusion gives a natural transformation $i : \partial \longrightarrow I$ of additive functors, $I : \mathcal{D} \longrightarrow \mathcal{D}$ being the identity functor. Finally, the Whitney imbedding theorem shows that each differentiable manifold is isomorphic to a submanifold of countable dimensional Euclidean space. Thus $\mathcal{D}$ has a small subcategory $\mathcal{D}_0$ (submanifolds of R^∞) such that each object of $\mathcal{D}$ is isomorphic to an object of $\mathcal{D}_0$.

Abstracting these properties, one has:

<u>Definition</u>: A <u>cobordism category</u> $(\mathcal{C},\partial,i)$ is a triple in which:

1) $\mathcal{C}$ is a category having finite sums and an initial object;
2) $\partial : \mathcal{C} \longrightarrow \mathcal{C}$ is an additive functor such that for each object X of $\mathcal{C}$, $\partial\partial(X)$ is an initial object;
3) $i : \partial \longrightarrow I$ is a natural transformation of additive functors from ∂ to the identity functor I; and
4) There is a small subcategory $\mathcal{C}_0$ of $\mathcal{C}$ such that each object of $\mathcal{C}$ is isomorphic to an object of $\mathcal{C}_0$.

As noted in motivating this definition, $(\mathcal{D},\partial,i)$ is a cobordism category. There are many more examples, and in fact the purpose of cobordism theory is to study the interesting examples. The precise choice of this formulation is based, somewhat vaguely, on the definition of 'adjoint functors'.

The purpose of this definition is not to establish a general nonsense structure; rather the definition will be used to follow the framework of previously developed theory and to try to unify the ideas. To begin, one has in any cobordism category the idea of a 'cobordism relation'.

<u>Definition</u>: If $(\mathcal{C},\partial,i)$ is a cobordism category, one says that the objects X and Y of $\mathcal{C}$ are <u>cobordant</u> if there exist objects U and V of $\mathcal{C}$ such that the sum of X and ∂U is isomorphic to the sum of Y and ∂V. This will be written $X \equiv Y$.

One has easily:

<u>Proposition</u>:

a) $\equiv$ is an equivalence relation on the objects of $\mathcal{C}$.

b) $X \equiv Y$ implies $\partial X \cong \partial Y$.

c) For all X, $\partial X \equiv \emptyset$ where $\emptyset$ is an initial object.

d) If $X \equiv X'$, $Y \equiv Y'$ and Z and Z' are sums of the pairs (X,Y) and (X',Y') respectively, then $Z \equiv Z'$.

<u>Proof</u>:

a) $X + \partial\emptyset \cong X + \partial\emptyset$;

$X + \partial U \cong Y + \partial V \Longrightarrow Y + \partial V = X + \partial U$; and

$X + \partial U \cong Y + \partial V$, $Y + \partial W \cong Z + \partial T$ implies $X + \partial(U+W) \cong X + \partial U + \partial W \cong Y + \partial V + \partial W \cong Z + \partial V + \partial T \cong Z + \partial(V+T)$.

b) $X + \partial U \cong Y + \partial V$ implies $\partial X \cong \partial X + \emptyset \cong \partial X + \partial\partial U \cong \partial Y + \partial\partial V \cong \partial Y + \emptyset \cong \partial Y$.

c) $\partial X + \partial\emptyset \cong \emptyset + \partial X$ since $\partial\emptyset$ is initial.

d) $X + \partial U \cong X' + \partial U'$, $Y + \partial V \cong Y' + \partial V'$ gives $Z + \partial(U+V) \cong Z' + \partial(U'+V')$

<u>NOTE</u>: In all of the above $A + B$ denotes an object which is a sum for A and B. **

<u>Note</u>: If one is unhappy with equivalence relations on a category, one may reduce to considering $\equiv$ as an equivalence relation on the <u>set</u> of isomorphism classes of objects of $\mathcal{C}$. This is the reason for the assumption about existence of $\mathcal{C}_0$.

<u>Definition</u>: An object X of $\mathcal{C}$ is <u>closed</u> if ∂X is an initial object. An object X of $\mathcal{C}$ <u>bounds</u> if $X \equiv \emptyset$ where $\emptyset$ is an initial object.

<u>Proposition</u>:

a) X closed and $Y \equiv X$ implies Y closed.

b) X and X' closed implies their sum is closed.

c) X bounds implies X is closed.

d) X and Y bound implies their sum bounds.

e) X bounds and $Y \equiv X$ implies Y bounds.

<u>Proof</u>:

a) follows directly from b) above.

b) $\partial X \cong \emptyset$, $\partial X' \cong \emptyset$ implies $\partial(X+X') \cong \emptyset + \emptyset \cong \emptyset$.

c) $X \equiv \emptyset$ implies $\partial X \cong \partial\emptyset \cong \emptyset$.

d) $X \equiv \emptyset$, $Y \equiv \emptyset$ implies $X + Y \equiv \emptyset + \emptyset \cong \emptyset$.

e) is immediate since $\equiv$ is an equivalence relation. **

<u>Proposition</u>: The set of equivalence classes of closed objects of $\mathcal{C}$ (under $\equiv$) has an operation induced by the sum in $\mathcal{C}$. This operation is associative, commutative, and has a unit (the class of any object which bounds).

<u>Proof</u>: The existence of $\mathcal{C}_0$ implies that the equivalence classes form a set. That the sum in $\mathcal{C}$ gives rise to an operation on this set follows immediately from the propositions. Associativity and commutativity hold for isomorphism classes of objects, hence also here. **

Definition: The cobordism semigroup of the cobordism category $(\mathcal{C},\partial,i)$ is the set of equivalence classes of closed objects of $\mathcal{C}$ with the operation induced by the sum in $\mathcal{C}$. This semigroup will be denoted $\Omega(\mathcal{C},\partial,i)$.

Remarks: 1) $\Omega(\mathcal{C},\partial,i)$ may also be described as the semigroup of isomorphism classes of closed objects of $\mathcal{C}$ modulo the sub-semigroup of isomorphism classes of objects which bound.

2) The semigroup $\Omega(\mathcal{O},\partial,i)$ is quite easily identifiable as Thom's cobordism group $\mathfrak{N}_*$ of unoriented cobordism classes of closed manifolds. In order to clarify this slightly, in the usual expression for equivalence one has X equivalent to Y if there is a V with $\partial V = X \cup Y$. Then $X \cup \partial V \cong Y \cup \partial(X\times I)$ giving $X \equiv Y$. The implication $X \cup \partial U \cong Y \cup \partial V$ implies $X \cup Y = \partial T$ is an easy geometric argument by looking at components and piecing together manifolds with boundary by means of tubular neighborhoods of their boundary components.

Within the literature of cobordism there are a few standard constructions performed. These may be generalized to the categorical situation as will now be shown.

Construction I: Let $(\mathcal{C},\partial,i)$ be a cobordism category, $\mathcal{X}$ a category with finite sums and an initial object, and $F : \mathcal{C} \longrightarrow \mathcal{X}$ an additive functor. For any object X of $\mathcal{X}$, form a category $\mathcal{C}/X$ whose objects are pairs (C,f) with C an object of $\mathcal{C}$ and $f \in \mathrm{Map}(F(C),X)$ and whose maps are given by letting $\mathrm{Map}((C,f),(C',f'))$ be the set of maps $\phi \in \mathrm{Map}(C,C')$ such that the diagram

$$\begin{array}{ccc} F(C) & \xrightarrow{F(\phi)} & F(C') \\ & {}_{f}\searrow \quad \swarrow_{f'} & \\ & X & \end{array}$$

commutes.

If $\emptyset$ is an initial object of $\mathcal{C}$ and $\phi : F(\emptyset) \longrightarrow X$ is the unique map, then $(\emptyset,\phi)$ is an initial object of $\mathcal{C}/X$. If (D,g) and (D',g') are objects of $\mathcal{C}/X$ and $D+D'$ is a sum for D and D' in $\mathcal{C}$, then $F(D+D')$ is a sum for $F(D)$ and $F(D')$ in $\mathfrak{X}$. The maps g and g' give a well defined map $g+g' : F(D+D') \longrightarrow X$, and $(D+D', g+g')$ is the sum of (D,g) and (D',g') in $\mathcal{C}/X$.

Let $\tilde{\partial}(C,f) = (\partial C, f \circ F(i_C))$ and $\tilde{\partial}(\phi) = \phi \circ i_C$ to define the functor $\tilde{\partial} : \mathcal{C}/X \longrightarrow \mathcal{C}/X$. Define the natural transformation $\tilde{i} : \tilde{\partial} \longrightarrow I$ by $\tilde{i}_{(C,f)} = i_C : \partial C \longrightarrow C$.

Then $(\mathcal{C}/X, \tilde{\partial}, \tilde{i})$ is a cobordism category.

Remarks: 1) This is the algebraic-geometric (Grothendieck style) notion of the category of objects over a given object.

2) If one begins with the category $(\mathcal{B}, \partial, i)$ and takes $F : \mathcal{B} \longrightarrow \mathfrak{X}$ to be the forgetful functor to the category of topological spaces and continuous maps, then $\Omega(\mathcal{B}/X, \tilde{\partial}, \tilde{i})$ is the unoriented bordism group $\mathfrak{N}_*(X)$ as originally formulated by M. F. Atiyah: Bordism and cobordism, Proc. Camb. Phil. Soc., 57 (1961), 200-208.

Construction II: Let π be a small category, $(\mathcal{C}, \partial, i)$ a cobordism category, and let $\mathrm{Fun}(\pi, \mathcal{C})$ be the category whose objects are functors $\Phi : \pi \longrightarrow \mathcal{C}$ and whose maps are the natural transformations.

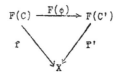

If $\emptyset$ is an initial object of $\mathcal{C}$, the constant functor $\Phi : \mathcal{A} \longrightarrow \mathcal{C} : A \longrightarrow \emptyset$ is an initial object of $\mathrm{Fun}(\mathcal{A}, \mathcal{C})$. If $F, G : \mathcal{A} \longrightarrow \mathcal{C}$ are functors, let $H : \mathcal{A} \longrightarrow \mathcal{C}$ by letting $H(A)$ be a sum for $F(A)$ and $G(A)$ and let $(j_F)_A = j_{F(A)} : F(A) \longrightarrow H(A)$ and $(j_G)_A = j_{G(A)} : G(A) \longrightarrow H(A)$ be the maps exhibiting $H(A)$ as the sum. Then j_F and j_G are natural transformations which exhibit H as a sum for F and G.

Let $\tilde{\partial} : \mathrm{Fun}(\mathcal{A}, \mathcal{C}) \longrightarrow \mathrm{Fun}(\mathcal{A}, \mathcal{C}) : F \longrightarrow \partial \circ F : \lambda \longrightarrow \partial(\lambda)$ and let $\tilde{i} : \tilde{\partial} \longrightarrow I$ be the natural transformation given by the map $\tilde{i}_F : \partial \circ F \longrightarrow F$ whose evaluation at any object A of $\mathcal{A}$ is $i_{F(A)} : \partial(F(A)) \longrightarrow F(A)$.

Then $(\mathrm{Fun}(\mathcal{A}, \mathcal{C}), \tilde{\partial}, \tilde{i})$ is a cobordism category.

<u>Remark</u>: Many standard examples fit this construction. Suppose $\mathcal{A}$ is the category with one object A whose maps are a finite group $G = \mathrm{Map}(A,A)$. A functor $F : \mathcal{A} \longrightarrow \mathcal{B}$ is given by selecting a manifold $X = F(A)$ and a homomorphism $G \longrightarrow \mathrm{Map}(X,X)$. Since G is finite, the induced map $G \times X \longrightarrow X$ is a differentiable action of G on X. Thus $\Omega(\mathrm{Fun}(\mathcal{A}, \mathcal{B}), \tilde{\partial}, \tilde{i})$ is the unoriented cobordism group of (unrestricted) G-actions as defined by P. E. Conner and E. E. Floyd: "Differentiable Periodic Maps", Springer, Berlin, 1964 (section 21).

<u>Relative Cobordism</u>

In order to study the relationship between two cobordism categories it is convenient to have available a 'relative cobordism' semigroup. In the geometric case this is made possible by joining together two manifolds with the same boundary to form a closed manifold. In the categorical situation, the idea is to replace a pair of objects having the same boundary by a pair of closed objects. For this one needs the idea of the Grothendieck group construction.

Recall that for any category with finite sums for which the isomorphism classes of objects form a set, $\mathcal{X}$, one defines $K(\mathcal{X})$, the Grothendieck group of $\mathcal{X}$, to be the set of equivalence classes of pairs (X,X') of objects of $\mathcal{X}$, where (X,X') is equivalent to (Y,Y') if there is an object A of $\mathcal{X}$ such that $X + Y' + A \cong X' + Y + A$. $K(\mathcal{X})$ is an abelian group under the operation induced by the sum in $\mathcal{X}$.

Let $(\mathcal{C},\partial,i)$ and $(\mathcal{C}',\partial',i')$ be two cobordism categories, $F : \mathcal{C} \longrightarrow \mathcal{C}'$ an additive functor, and $t : \partial' F \cong F\partial$ a natural equivalence of additive functors such that the diagram

$$\begin{array}{ccc} \partial' F(A) & \xrightarrow{t(A)} & F(\partial A) \\ & {\scriptstyle i'_{F(A)}} \searrow \quad \swarrow {\scriptstyle F(i_A)} & \\ & F(A) & \end{array}$$

commutes. Let $\mathcal{P}$ be the category whose objects are triples (X,Y,f) with $X \in \mathcal{C}'$, $Y \in \mathcal{C}$, Y closed, and $f : \partial' X \longrightarrow FY$ an isomorphism and with $\mathrm{Map}((X,Y,f), (X',Y',f'))$ the set of $(\phi,\psi) \in \mathrm{Map}(X,X') \times \mathrm{Map}(Y,Y')$ such that

$$\begin{array}{ccc} \partial' X & \xrightarrow{f} & FY \\ {\scriptstyle \partial'\phi} \downarrow & & \downarrow {\scriptstyle F\psi} \\ \partial' X' & \xrightarrow{f'} & FY' \end{array}$$

commutes. Then $\mathcal{P}$ has finite sums and a small subcategory $\mathcal{P}_0(X \in \mathcal{C}'_0, Y \in \mathcal{C}_0)$ such that each object of $\mathcal{P}$ is isomorphic to an object of $\mathcal{P}_0$.

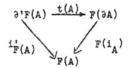

Let $\mathcal{S}$ be the collection of pairs $((X,Y,f), (X',Y',f'))$ of objects of $\mathcal{F}$ for which $Y \cong Y'$. Let $(x,x') \sim (y,y')$ if there are objects u and v of $\mathcal{F}$ such that $x + u \cong y + v$ and $x' + u \cong y' + v$. Then the set of equivalence classes $\mathcal{S}/\sim$ forms an abelian group under the operation induced by the sum.

One has a homomorphism $\beta : K(\mathcal{C}'_{cl}) \longrightarrow \mathcal{S}/\sim$, where $\mathcal{C}'_{cl}$ is the subcategory of closed objects of $\mathcal{C}'$ by sending (X,X') into $((X,\emptyset,j), (X',\emptyset,j'))$ where $\emptyset$ is an initial object of $\mathcal{C}$ and j,j' are the unique isomorphisms of initial objects.

If one has a homomorphism

$$\alpha : \mathcal{S}/\sim \longrightarrow K(\mathcal{C}'_{cl})/(\partial'_* K(\mathcal{C}') + F_* K(\mathcal{C}_{cl}))$$

such that the composition with β is the quotient homomorphism of $K(\mathcal{C}'_{cl})$, then one can define a relative cobordism semigroup as follows:

For objects (X,Y,f) and (X',Y',f') of $\mathcal{F}$, one writes $(X,Y,f) \equiv (X',Y',f')$ if there exist objects U and U' of $\mathcal{C}$ with $Y + \partial U \cong Y' + \partial U'$ and for which $\alpha((X+FU,Y+\partial U,f+tU),(X'+FU',Y'+\partial U',f'+tU')) = 0$. Using the fact that α is a homomorphism one easily sees that $\equiv$ is an equivalence relation. The relative cobordism semigroup $\Omega(F,t,\alpha)$ is the set of equivalence classes under $\equiv$ of elements of $\mathcal{F}$ with the sum induced by the sum in $\mathcal{F}$.

One has homomorphisms

$$\partial : \Omega(F,t,\alpha) \longrightarrow \Omega(\mathcal{C},\partial,i) : (X,Y,f) \longrightarrow Y,$$

$$F_* : \Omega(\mathcal{C},\partial,i) \longrightarrow \Omega(\mathcal{C}',\partial',i') : Y \longrightarrow FY, \quad \text{and}$$

$$i : \Omega(\mathcal{C}',\partial',i') \longrightarrow \Omega(F,t,\alpha) : X \longrightarrow (X,\emptyset,j)$$

and the triangle

$$\begin{array}{ccc} \Omega(\mathcal{C},\partial,i) & \xrightarrow{F_*} & \Omega(\mathcal{G}',\partial',i') \\ & {\scriptstyle\partial}\nwarrow \quad \swarrow{\scriptstyle i} & \\ & \Omega(F,t,\alpha) & \end{array}$$

is easily seen to have period 2 (i.e. $\partial i = iF_* = F_*\partial = 0$).

In order to clarify the relationship between the homomorphism α and the joining of two manifolds along their common boundary, consider elements (X,Y,f) of $\mathcal{P}$ as a manifold with boundary together with additional structure on its boundary. For $((X,Y,f), (X',Y',f')) \in \mathcal{S}$ choose an isomorphism $g : Y \xrightarrow{\cong} Y'$ and let $\alpha(x,x')$ be the class of $X \cup_k (-X')$, where $-X'$ is X' with its opposite structure (e.g. orientation), and the boundaries of X and X' are identified via $k = f'^{-1}F(g)f$. This class does not depend on the choice of g, for if g' is another isomorphism, one may attach $X' \times I$ to $(X \cup_k (-X')) \times I \cup [\pm(X \cup_{k'} (-X'))] \times I$ so that the difference of two representatives is cobordant to $X \cup_{k''} (-X)$, where $k'' = f^{-1}F(g^{-1}g')f$. Identifying $\partial X{\times}0$ with $\partial X{\times}1$ using k'' in $X{\times}I$ gives a cobordism of $X \cup_{k''} (-X)$ and $\partial X \times I$ with ends identified by k'' - but this is isomorphic via f to the image under F of $Y \times I$ with ends identified using $g^{-1}g'$. Thus α does not depend on the choice of g.

With this choice of α, suppose one has $(X,Y,f) \equiv (X',Y',f')$. One may then find a cobordism of Y and Y', say $\partial V = Y-Y'$ so that $X \cup (-V) \cup (-X')$ is cobordant to a closed manifold D with additional structure. Thus one may find a cobordism of Y and Y', $U = V+D$, $\partial U = Y-Y'$, so that $X \cup (-U) \cup (-X')$ bounds. This is the usual geometric description for cobordism of manifolds with boundary.

Remark: One may let $\mathcal{C}$ be the subcategory of $\mathcal{C}'$ consisting of initial objects, with F the inclusion. Then β is epic, uniquely determining α. The relative cobordism semigroup in this case is then identifiable with the cobordism semigroup of $\mathcal{C}'$.

Chapter II

Manifolds With Structure - The Pontrjagin-Thom Theorem

The standard cobordism theories are based on manifolds with additional structure on the tangent or normal bundle. The exposition given here is taken from the paper: R. K. Lashof: Poincaré duality and cobordism, Trans. Amer. Math. Soc., 109 (1963), 257-277.

Denote by $G_{r,n}$ the Grassmann manifold of unoriented r-planes in R^{r+n} and let γ_n^r be the r-plane bundle over $G_{r,n}$ consisting of pairs: an r-plane in R^{n+r} and a point in that r-plane. Then $BO_r = \lim_{n\to\infty} G_{r,n}$ with universal r-plane bundle $\gamma^r = \lim_{n\to\infty} \gamma_n^r$.

Definition: Let $f_n : B_n \longrightarrow BO_n$ be a fibration. If ξ is an n-dimensional vector bundle over the space X classified by the map $\xi : X \longrightarrow BO_n$, then a (B_n,f_n) structure on ξ is a homotopy class of liftings to B_n of the map ξ; i.e. an equivalence class of maps $\hat{\xi} : X \longrightarrow B_n$ with $f_n \circ \hat{\xi} = \xi$, where $\hat{\xi}$ and $\tilde{\xi}$ are equivalent if they are homotopic by a homotopy $H : X \times I \longrightarrow B_n$ such that $f_n \circ H(x,t) = \xi(x)$ for all $(x,t) \in X \times I$.

Note: A (B_n,f_n) structure depends on the specific map into BO_n. There is no way to make (B_n,f_n) structures correspond for equivalent bundles, since the correspondence is dependent upon the choice of the equivalence.

Let M^n be a compact differentiable (C^∞) manifold (with or without boundary) and let $i : M^n \longrightarrow R^{n+r}$ be an imbedding. The normal bundle of i is the quotient of the pullback of the tangent bundle of R^{n+r}, $i^*\tau(R^{n+r})$, by the subbundle $\tau(M)$. Giving $\tau(R^{n+r}) = R^{n+r} \times R^{n+r}$ the Riemannian metric obtained from the usual inner product in Euclidean space, the total space N of the normal bundle may be identified with the orthogonal complement of $\tau(M)$ in $i^*\tau(R^{n+r})$, or the fiber of N at m may be identified with the subspace of $R^{n+r} \times R^{n+r}$ consisting of vectors (m,x) with x orthogonal to $i_*\tau(M)_m$. The normal map of i is given by sending m into $N_m \in G_{r,n}$, covered by the bundle map $n : N \longrightarrow \gamma_n^r : (m,x) \longrightarrow (N_m,x)$. Composing with the inclusion into γ^r provides a map $\nu(i) : M \longrightarrow BO_r$ which classifies the normal bundle of the imbedding i.

<u>Lemma</u>: If r is sufficiently large (depending only on n), there is a one-to-one correspondence between the (B_r,f_r) structures for the normal bundles of any two imbeddings $i_1,i_2 : M^n \longrightarrow R^{n+r}$.

<u>Proof</u>: For r sufficiently large, any two imbeddings i_1, i_2 of M^n in R^{n+r} are regularly homotopic and any two such regular homotopies are homotopic through regular homotopies leaving endpoints fixed. (A regular homotopy is a homotopy $H : M \times I \longrightarrow R^{n+r}$ such that each $H(\ ,t)$ is an immersion and such that the differentials $H(\ ,t)_* : \tau(M) \longrightarrow \tau(R^{n+r})$ define a homotopy). See M. Hirsch: Immersions of manifolds, Trans. Amer. Math. Soc., 93 (1959), 242-276. Then a regular homotopy from i_1 to i_2 gives a homotopy from $\nu(i_1)$ to $\nu(i_2)$, and two homotopies defined in this way are themselves homotopic relative to endpoints. Thus one has a well-defined equivalence for the two bundles. Applying the homotopy lifting property for the map f_r then establishes the correspondence quite easily. **

<u>Definition</u>: Suppose one is given a sequence (B,f) of fibrations $f_r : B_r \longrightarrow BO_r$ and maps $g_r : B_r \longrightarrow B_{r+1}$ such that the diagram

$$\begin{array}{ccc} B_r & \xrightarrow{g_r} & B_{r+1} \\ {\scriptstyle f_r}\downarrow & & \downarrow{\scriptstyle f_{r+1}} \\ BO_r & \xrightarrow{j_r} & BO_{r+1} \end{array}$$

commutes, j_r being the usual inclusion. A (B_r,f_r) structure on the normal bundle of M^n in R^{n+r} defines a unique (B_{r+1},f_{r+1}) structure via the inclusion $R^{n+r} \subset R^{n+r+1}$. A (B,f) structure on M^n is an equivalence class of sequences of (B_r,f_r) structures $\xi = (\xi_r)$ on the normal bundle of M, two such sequences being equivalent if they agree for sufficiently large r. A (B,f) manifold is a pair consisting of a manifold M^n and a (B,f) structure on M^n.

If W^w is a manifold and M^m is a submanifold of W with trivialized normal bundle, one may imbed M in R^{m+r}, r large, and extend by means of the trivialization to an imbedding of a neighborhood of M in W into $R^{w+r} = R^{m+r} \times R^{w-m}$ so that the neighborhood meets R^{m+r} orthogonally along M. This may then be extended to an imbedding of W in R^{w+r}. The normal planes to M in R^{m+r} are then the restriction to M of the normal planes to W in R^{w+r}. If $\tilde{\nu} : W \longrightarrow B_r$ is a lifting of the normal map, then $\tilde{\nu}|_M$ is a lifting for the normal map of M. Thus a (B,f) structure on W induces a well-defined (B,f) structure on M.

Remarks: 1) The induced (B,f) structure depends only on the equivalence class of the trivialization, not on the specific choice of trivialization.

2) If $f : M \longrightarrow W$ is an isomorphism of manifolds, the normal bundle is trivialized, being zero dimensional. If $i : M \longrightarrow W$ is the inclusion

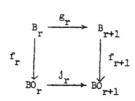

of the boundary, there are two choices of trivialization, via the choice of inner or outer normal. If $j : M \longrightarrow W$ is the inclusion of a direct summand, then the normal bundle is again zero dimensional, so trivialized.

Definition: The cobordism category of (B,f) manifolds is the category whose objects are compact differentiable manifolds with (B,f) structure and whose maps are the boundary preserving differentiable imbeddings with trivialized normal bundle such that the (B,f) structure induced by the map coincides with the (B,f) structure on the domain manifold. The functor ∂ applied to a (B,f) manifold W is the manifold ∂W with (B,f) structure induced by the inner normal trivialization, and ∂ on maps is restriction. The natural transformation i is the inclusion of the boundary with inner normal trivialization.

The cobordism semigroup of this category will be denoted $\Omega(B,f)$. The sub-semigroup of equivalence classes of n-dimensional closed manifolds will be denoted $\Omega_n(B,f)$. Clearly $\Omega(B,f)$ is the direct sum of the $\Omega_n(B,f)$.

Proposition: The cobordism semigroup $\Omega(B,f)$ is an abelian group.

Proof: Let M^n be a closed manifold imbedded in R^{n+r} for some large r with $\tilde{\nu} : M \longrightarrow B_r$ a lifting of the normal map. Extend to an imbedding of $M \times I$ in $R^{n+r} \times R = R^{n+r+1}$ by the usual inclusion of I in R. The normal map for $M \times I$ is the composition of the projection on M and the normal map of M. Thus the lifting for M defines a (B,f) structure on $M \times I$ which induces the given structure on $M \times 0$. The inner normal along $M \times 1$ gives rise to an induced structure on $M \times 1$, and with these structures one has $M + M \times 1 \cong \partial(M \times I)$ in the category. Thus the structure on $M \times 1$ is an inverse for the structure on M in $\Omega(B,f)$. **

Considering BO_r as the space of r-planes contained in some finite dimensional subspace R^s of R^∞ and taking the usual inner product on the subspace of R^∞ consisting of vectors with only finitely many non-zero components, one obtains a Riemannian metric on the universal bundle γ^r. If ξ is an r-plane bundle over a space X classified by the map $\xi : X \longrightarrow BO_r$, one has induced a Riemannian metric on ξ. (Note: For the normal bundle of a manifold this coincides with the metric obtained from the splitting). The <u>Thom space</u> of ξ, $T\xi$, is the space obtained from the total space of ξ by collapsing all vectors of length at least one to a point, denoted ∞. If ξ is the bundle induced from a bundle $\eta : Y \longrightarrow BO_r$, by a map $g : X \longrightarrow Y$, then the usual bundle map $\xi = g^*\eta \longrightarrow \eta$ induces a map $Tg : T\xi \longrightarrow T\eta$.

The map $j_r : BO_r \longrightarrow BO_{r+1}$ induces a vector bundle $j_r^*(\gamma^{r+1})$ over BO_r which may be identified as the Whitney sum of γ^r and a trivial line bundle. Then $Tj_r^*(\gamma^{r+1})$ may be identified as the suspension of $T\gamma^r$.

One then has a commutative diagram

$$\begin{array}{ccc}
\Sigma TB_r & \xrightarrow{Tg_r} & TB_{r+1} \\
{\scriptstyle \Sigma Tf_r}\downarrow & & \downarrow{\scriptstyle Tf_{r+1}} \\
\Sigma TBO_r & \xrightarrow{Tj_r} & TBO_{r+1}
\end{array}$$

and a homomorphism $Tg_r \circ \Sigma : \pi_{n+r}(TB_r,\infty) \longrightarrow \pi_{n+r+1}(TB_{r+1},\infty)$ of the homotopy groups, where Σ denotes suspension, and TBO_r, TB_r denote the Thom spaces $T\gamma^r$ and $Tf_r^*(\gamma^r)$.

The main theorem is the generalized Pontrjagin-Thom theorem:

<u>Theorem</u>: The cobordism group of n-dimensional (B,f) manifolds $\Omega_n(B,f)$ is isomorphic to $\lim\limits_{r\to\infty} \pi_{n+r}(TB_r,\infty)$.

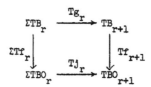

Proof:

A) Definition of the homomorphism $\Theta: \Omega_n(B,f) \longrightarrow \lim_{r\to\infty} \pi_{n+r}(TB_r,\infty)$.

Let $\sigma \in \Omega_n(B,f)$ be represented by a (B,f) manifold M^n. Let $i : M \longrightarrow R^{n+r}$ be an imbedding with a lifting $\tilde{\nu} : M \longrightarrow B_r$ which defines the given (B,f) structure on M. Let N denote the total space of the normal bundle of M, thought of as a subspace of $R^{n+r} \times R^{n+r}$. Under the evaluation map $e : R^{n+r} \times R^{n+r} \longrightarrow R^{n+r} : (a,b) \longrightarrow a+b$, the subspace N is mapped differentiably and on $M = M \times 0 \subset N$ this map restricts to the imbedding i. For some sufficiently small $\varepsilon > 0$, the subspace of N consisting of vectors of length less than or equal to ε, N_ε, is imbedded by this map $e|_N$.

To define a map $S^{n+r} \longrightarrow Tf_r^*(\gamma^r)$, begin by considering S^{n+r} as $R^{n+r} \cup \infty$ and let $c : S^{n+r} \longrightarrow N_\varepsilon/\partial N_\varepsilon$ by collapsing all points of S^{n+r} outside or on the boundary of N_ε to a point. Multiplication by $1/\varepsilon$ induces a map $N_\varepsilon/\partial N_\varepsilon \longrightarrow TN$, denoted by ε^{-1}. The map $\bar{n} \times (\tilde{\nu}\circ\pi) : N \to \gamma^r \times B_r$, where $\bar{n}$ is the composition of n with the inclusion of γ_n^r in γ^r and π is the projection of N on M, is a bundle map into $f_r^*(\gamma^r)$ and induces the map $T(\bar{n} \times (\tilde{\nu}\circ\pi)) : TN \longrightarrow TB_r$. The composition $\theta = T(\bar{n} \times (\tilde{\nu}\circ\pi)) \circ \varepsilon^{-1} \circ c$ is a map of pairs $(S^{n+r},\infty) \longrightarrow (TB_r,\infty)$.

Replacing ε by a smaller value does not change the homotopy class of θ since the maps $\varepsilon^{-1}\circ c$ will be homotopic. Replacing $\tilde{\nu}$ by an equivalent lifting simply gives a homotopy of $T(\bar{n} \times (\tilde{\nu}\circ\pi))$ and so does not change the homotopy class of θ. Clearly, the map $M \xrightarrow{i} R^{n+r} \subset R^{n+r+1}$ gives rise to $Tg_r \circ \Sigma\theta$ and thus one has defined an element of $\lim_{r\to\infty} \pi_{n+r}(TB_r,\infty)$ represented by the map θ.

To show that this element depends only on the cobordism class of the manifold M and not on the choice of the imbedding, let W be a (B,f) manifold and $j : M + \partial W \longrightarrow R^{n+r}$ an imbedding with a lift $\bar{\nu} : M + \partial W \longrightarrow B_r$ giving the same (B_r, f_r) structure on M (here r is assumed large). Let $H : M \times I \longrightarrow R^{n+r}$ be a regular homotopy of the imbeddings i and $j|_M$ chosen so that $H(x,t)$ is $i(x)$ if $t < \delta_1$ and is $j(x)$ if $t > 1-\delta_2$ and let $k : W \longrightarrow R^{n+r} \times (0,1]$ be a map agreeing with $j \times 1$ on ∂W and imbedding a tubular neighborhood of ∂W orthogonally along $j(\partial W) \times 1$. The map $(H \times \pi_2) + k : M \times I + W \longrightarrow R^{n+r} \times I$ is an imbedding on a closed neighborhood of the boundary and may be homotoped to an imbedding $F : M \times I + W \longrightarrow R^{n+r} \times I$ by a homotopy fixed on that neighborhood of the boundary. $F|_{M\times I}$ is a regular homotopy and corresponding to its normal map one may find a covering map $M \times I \longrightarrow B_r$ agreeing with $\tilde{\nu}$ on $M \times 0$. Since the normal map is constant near $M \times 1$, one may modify the lift to agree with $\bar{\nu}$ on $M \times 1$. Since the (B,f) structure on ∂W is induced from that of W one may find a lift of W agreeing with $\bar{\nu}$ on ∂W. Following the previous construction with the imbedding $F : M \times I + W \longrightarrow R^{n+r} \times I$, one has a collapse $S^{n+r} \times I \longrightarrow N_\varepsilon/\partial N_\varepsilon$ where N_ε is a neighborhood of (image F), a map $\varepsilon^{-1} : N_\varepsilon/\partial N_\varepsilon \longrightarrow TN$ and the map $T(\bar{n} \times (\tilde{\nu}\circ\pi)) : TN \longrightarrow TB_r$ which compose to give a map $S^{n+r} \times I \longrightarrow TB_r$. This provides a homotopy for the maps θ defined by i and j.

Taking W empty shows that the class of θ is independent of the imbedding of M. Further, if $M \equiv M'$ with $M + \partial W \cong M' + \partial W'$, then $\theta(M) \sim \theta(M+\partial W) = \theta(M'+\partial W') \sim \theta(M')$ so that the class of the map θ depends only on the cobordism class of M.

B) Θ is a homomorphism.

If M_1 and M_2 represent two classes in $\Omega_n(B,f)$, choose imbeddings $i_\alpha : M_\alpha \longrightarrow R^{n+r}$ for which the last coordinate is positive for i_1 and negative for i_2. If tubular neighborhoods are chosen small enough to lie in the same half spaces then $\Theta([M_1] + [M_2])$ is represented by $S^{n+r} \xrightarrow{d} S^{n+r} \vee S^{n+r} \xrightarrow{\theta_1 \vee \theta_2} TB_r$, where d collapses the equator of S^{n+r} and θ_α represents $\Theta([M_\alpha])$. Since this map represents the sum of the homotopy classes, Θ is a homomorphism.

C) Θ is epic.

Let $\theta : (S^{n+r},p) \longrightarrow (TB_r,\infty)$, r large, represent a class in $\lim_{r\to\infty} \pi_{n+r}(TB_r,\infty)$. Then $Tf_r \circ \theta : (S^{n+r},p) \longrightarrow (TBO_r,\infty)$, and since $TBO_r = T\gamma^r = \lim_{s\to\infty} T\gamma_s^r$ with S^{n+r} compact, $Tf_r \circ \theta(S^{n+r}) \subset T\gamma_s^r$ for some s. The map $Tf_r \circ \theta$ may be deformed to a map h_r so that

1) h_r is differentiable on the preimage of some open set of $T\gamma_s^r$ containing $G_{r,s}$ and is transverse regular on $G_{r,s}$. <u>Note</u>: $T\gamma_s^r - \infty$ is a differentiable manifold.

2) If $M^n = h_r^{-1}(G_{r,s})$, h_r is a bundle map on a normal tube of M^n in $R^{n+r} = S^{n+r} - p$.

3) The map $Tf_r \circ \theta$ agrees with h_r on the preimage V of a closed neighborhood of ∞.

Since $h_r|_M$ classifies the normal bundle of M, one may assume (by a further homotopy if necessary) that $h_r|_M$ is the normal map $\nu : M \longrightarrow G_{r,n} \subset G_{r,s}$ and that h_r is given by the usual translation of vectors to the origin on a normal tubular neighborhood of M.

Now $Tf_r : TB_r \longrightarrow TBO_r$ is a fibration except at the point ∞, and since $Tf_r \circ \theta(S^{n+r}$ - *interior* $V)$ *does not contain* ∞, *the covering homotopy* theorem applies so that the deformation of $Tf_r \circ \theta$ to h_r on S^{n+r} - interior V may be covered by a homotopy of θ on S^{n+r} - interior V which is pointwise fixed on the boundary of V. Taking the homotopy to be constant on V, one may cover the homotopy of $Tf_r \circ \theta$ to h_r by a homotopy of θ to a new map θ_1. The inverse image of B_r under θ_1 is the inverse image of BO_r under h_r, which is M. Further $\theta_1|_M$ is a lift of the normal map $h_r|_M$.

Thus one has a (B_r,f_r) structure on M in R^{n+r} and hence a (B,f) structure on M. Using the given imbedding of M in R^{n+r} with the lift $\theta_1|_M$, the resulting map θ_2 agrees with θ_1 on a neighborhood N_ε of M and since $TB_r - B_r$ deforms to ∞, one may homotope θ_1 to agree with θ_2 by pushing the complement of N_ε to ∞. Then $\Theta([M])$ is the class of θ.

D) Θ is monic.

Let M be a (B,f) manifold such that $\Theta([M]) = 0$. Thus for some large r the standard map $\theta_0 : S^{n+r} \longrightarrow TB_r$ defined by M is homotopic to the trivial map $\theta_1 : S^{n+r} \longrightarrow \infty$. One may choose the homotopy $L : S^{n+r} \times I \longrightarrow TB_r$ so that $L_t = \theta_0$ for $t \in [0,\eta]$. By compactness $Tf_r \circ L(S^{n+r} \times I) \subset T\gamma_s^r$ for some $s\ (\geq n)$. As above, one may homotope $Tf_r \circ L$ (relative to $N_\varepsilon(M) \times [0,\eta]$) in a neighborhood of $G_{r,s}$ to a map H_r which is differentiable near and transverse on $G_{r,s}$. $W = H_r^{-1}(G_{r,s})$ is a submanifold of $R^{n+r} \times I$ with $\partial W = M$ meeting $R^{n+r} \times 0$ orthogonally along M. One may also assume $H_r|_W$ is the normal map and H_r agrees with the usual translation of vectors map on a neighborhood of W. Applying covering homotopy, one may deform L to a map $\theta : S^{n+r} \times I \longrightarrow TB_r$ with $\theta_t = \theta_0$ for small t, $\theta_1 = \theta | S^{n+r} \times 1$ with $\theta|_W$ covering the normal map H_r of W. This defines a (B,f) structure on W which induces the original (B,f) structure on M. Thus $M + \partial\emptyset \tilde{=} \emptyset + \partial W$ and $[M]$ is the zero class of $\Omega_n(B,f)$. **

Tangential Structures

It is frequently desirable to define (B,f) structures on manifolds by means of structures on the stable tangent bundle. Let $B = \lim (B_r, g_r)$ and $BO = \lim (BO_r, j_r)$, with $f = \lim f_r : B \longrightarrow BO$. The map $I_{n,N} : G_{n,N} \longrightarrow G_{N,n}$ obtained by assigning to each n plane its orthogonal N plane induces a map $I : BO \longrightarrow BO$, with I^2 the identity. $f : B \longrightarrow BO$ is a fibration and one has the induced fibration $f^* : B^* = I^*B \longrightarrow BO$. Since $I^2 = 1$, I^*B^* is again B. The induced bundle maps give a diagram

$$\begin{array}{ccccc} B & \xrightarrow{I'} & B^* & \xrightarrow{I^*} & B \\ {\scriptstyle f}\downarrow & & {\scriptstyle f^*}\downarrow & & {\scriptstyle f}\downarrow \\ BO & \xrightarrow{I} & BO & \xrightarrow{I} & BO \end{array}$$

with I^*I' and $I'I^*$ both being identity maps.

If M^n is imbedded in R^{n+N}, N large, the maps $\nu_N : M^n \longrightarrow G_{N,n}$ and $\tau_N : M^n \longrightarrow G_{n,N}$ obtained by translation of normal and tangent vectors are related by $\tau_N = I_{N,n}\nu_N$. Following these by the inclusions one has maps $\tau : M \longrightarrow BO$ and $\nu : M \longrightarrow BO$ with $\tau = I\nu$.

A (B,f) structure on M as previously defined is precisely a homotopy class (through liftings) of liftings to B of the map $\nu : M \longrightarrow BO$. The maps I' and I^* define an obvious equivalence between these classes of liftings of ν and the homotopy classes (through liftings) of liftings to B^* of $\tau : M \longrightarrow BO$. Such a class of liftings of τ is a (B^*,f^*) structure on the stable tangent bundle of M.

Structures For Sequences Of Maps

If instead of fibrations one is given only spaces C_r and maps $f_r : C_r \longrightarrow BO_r$, $g_r : C_r \longrightarrow C_{r+1}$ such that $f_{r+1}g_r$ is homotopic to $j_r f_r$ one may replace the maps (C_r, f_r) by homotopy equivalent fibrations. The resulting maps g_r may be deformed inductively to give commutative diagrams by means of covering homotopy. A (C,f) structure is then a structure for some such fibration sequence (chosen). Since the cobordism group is given by homotopy of the Thom complex, which depends only on the homotopy type of the fibrations, the resulting cobordism group does not depend on the choice of equivalent fibration sequence.

Ring Structure

If one has an r plane in $r+s$ space and an r' plane in $r'+s'$ space, they span an $r+r'$ plane in $R^{r+r'+s+s'} = R^{r+s} \times R^{r'+s'}$. This defines a map $G_{r,s} \times G_{r',s'} \longrightarrow G_{r+r',s+s'}$ and induces a map $BO_r \times BO_{r'} \longrightarrow BO_{r+r'}$, corresponding to the Whitney sum of vector bundles. $G_{0,0}$ is a point and provides a base point in each $G_{r,s}$ (the usual $R^r \subset R^{r+s}$) so that $BO_r \vee BO_{r'}$ is mapped via standard inclusions.

The twisted map $BO_{r'} \times BO_r \longrightarrow BO_r \times BO'_r \longrightarrow BO_{r+r'}$ is homotopic to the usual map $BO_{r'} \times BO_r \longrightarrow BO_{r+r'}$ by a rotation of $R^{r+r'+s+s'}$ to interchange factors. This gives the usual homotopy commutative H-space structure on BO.

Having similar multiplications $B_r \times B_s \longrightarrow B_{r+s}$ so that the maps f_r preserve products up to homotopy, one may define a ring structure in (B,f) cobordism, for the multiplication defines a (B,f) structure on the product manifold $M^n \times M'^{n'} \subset R^{n+N} \times R^{n'+N'} = R^{n+n'+N+N'}$.

The map $B_r \times B_s \longrightarrow B_{r+s}$ induces a map $TB_r \wedge TB_s \longrightarrow TB_{r+s}$ giving a product in the stable homotopy, making it into a ring. This ring structure is the same as that of the cobordism groups.

Relative Groups

If one has commutative diagrams

$$\begin{array}{ccc} B_r & \xrightarrow{g_r} & B_{r+1} \\ \downarrow h_r & & \downarrow h_{r+1} \\ C_r & \xrightarrow{k_r} & C_{r+1} \\ \downarrow d_r & & \downarrow d_{r+1} \\ BO_r & \xrightarrow{j_r} & BO_{r+1} \end{array}$$

$$f_r : B_r \to BO_r, \qquad f_{r+1} : B_{r+1} \to BO_{r+1}$$

in which h_r and d_r are fibrations and g_r and k_r are fibre preserving maps, a (B,f) structure induces a (C,d) structure by means of the 'reduction' h. This gives a functor h from the cobordism category of (B,f) manifolds to that of (C,d) manifolds. An $(n+1)$ dimensional manifold W with boundary having a (B,f) structure on its boundary inducing via h the same (C,d) structure on ∂W as is induced by a (C,d)

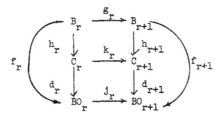

structure on W is a relative manifold. Using the standard 'piecing together' homomorphism one has defined a relative cobordism semigroup.

If one imbeds ∂W in R^{n+r} (r large) with lifting of the normal map to B_r, extends to an imbedding of W (orthogonally along the boundary using a tubular neighborhood) in $R^{n+r} \times [0,\infty) = H^{n+r+1}$, selecting a lifting of the normal map of W to C_r which agrees with the h-induced lifting on ∂W, then one may apply the tubular neighborhood map to this imbedding-lifting situation to construct a map

$$\theta : (D^{n+r+1}, S^{n+r}, \infty) = (H^{n+r+1} \cup \infty, R^{n+r} \cup \infty, \infty) \longrightarrow (TC_r, TB_r, \infty).$$

If W is cobordant to W' (relatively) one may find a (B,f) manifold U giving a cobordism of ∂W and $\partial W'$ so that with proper boundary identifications the closed (C,d) manifold $W \cup (-U) \cup (-W')$ bounds. One may imbed U in $R^{n+r} \times I$ to give the proper identifications at $\partial W \times 0$ and $\partial W' \times 1$, with a lifting of the normal map to B_r and fill in the manifold with (C,d) structure in $H^{n+r+1} \times I$ along its boundary $W \cup (-U) \cup (-W') \subset H^{n+r+1} \times 0 \cup R^{n+r} \times I \cup H^{n+r+1} \times 1$. Ignoring corners (which don't affect the homotopy situation, but rather involve the identification of $D^{n+r+1} \times I$) the normal maps and their liftings define a homotopy

$$L : (D^{n+r+1} \times I, S^{n+r} \times I, \infty \times I) \longrightarrow (TC_r, TB_r, \infty)$$

of the maps for W and W'.

Ignoring lots of details one sees easily that the $(n+1)$ dimensional relative cobordism group ($-W$ being constructed from $W \times I$) is isomorphic to the stable homotopy group $\lim_{r\to\infty} \pi_{n+r+1}(TC_r, TB_r, \infty)$. Further, the cobordism triangle is identifiable as the exact homotopy sequence of the 'pair' (TC, TB).

Chapter III

Characteristic Classes and Numbers

As mentioned in the introduction, the determination of invariants which distinguish manifolds in one of the principal aims of differential topology. In the framework of cobordism theory, the use of characteristic classes provides invariants called characteristic numbers which are cobordism invariants. In order to set up the machinery for these invariants, the ideas of generalized cohomology theory play a central role, and for this the basic reference is G. W. Whitehead: Generalized homology theories, Trans. Amer. Math. Soc., 102 (1962), 227-283.

Definition: A spectrum $\underset{\sim}{E}$ is a sequence $\{E_n | n \in Z\}$ of spaces with base point together with a sequence of maps $e_n : \Sigma E_n \longrightarrow E_{n+1}$, Σ being the suspension. If $\underset{\sim}{F} = \{F_n, f_n\}$ is another spectrum, a map h from $\underset{\sim}{E}$ to $\underset{\sim}{F}$ is a sequence of maps $h_n : E_n \longrightarrow F_n$ with $h_{n+1} \circ e_n = f_n \circ \Sigma h_n$.

Examples: 1) The sphere spectrum $\underset{\sim}{S} = \{S^n, \sigma_n\}$ where $\sigma_n : \Sigma S^n \longrightarrow S^{n+1}$ is the identity map.

2) If (B,f) is a sequence of fibrations $f_r : B_r \longrightarrow BO_r$ with maps $g_r : B_r \longrightarrow B_{r+1}$ as in Chapter II, then $\underset{\sim}{TB} = \{TB_r, Tg_r\}$ is a spectrum, known as the Thom spectrum of the family (B,f). In particular, the maps $Tf_r : TB_r \longrightarrow TBO_r$ define a map of spectra $Tf : \underset{\sim}{TB} \longrightarrow \underset{\sim}{TBO} = \{TBO_r, Tj_r\}$.

If one chooses base points $b_r \in B_r$ such that $g_r b_r = b_{r+1}$, then the bundle $f_r^*(\gamma^r)$ induces a trivial r-plane bundle δ^r over b^r, defining a map $Tb_r : T\delta^r \longrightarrow TB_r$. Since $T\delta^r = S^r$, this gives a map of spectra $Tb : \underset{\sim}{S} \longrightarrow \underset{\sim}{TB}$. Note: The identification of $T\delta^r$ with S^r requires a choice of framing of the fiber over b_r.

<u>Definition</u>: The homology and cohomology groups with coefficients in the spectrum $\underset{\sim}{E}$ are defined by

$$H_n(X,A;\underset{\sim}{E}) = \lim_{i\to\infty} \pi_{n+i}((X/A)\wedge E_i) \quad \text{and}$$

$$H^n(X,A;\underset{\sim}{E}) = \lim_{i\to\infty} [\Sigma^i(X/A), E_{n+i}]$$

where X/A is the space obtained from X by collapsing A to a point (the base point), $\wedge$ is the smash product $U\wedge V = U \times V/(U \times *) \cup (* \times V)$, and $[\ ,\ ]$ denotes homotopy classes of maps.

$H^*(\ ;\underset{\sim}{E})$ and $H_*(\ ;\underset{\sim}{E})$ are functors satisfying all the axioms of Eilenberg-Steenrod as cohomology and homology theories with the exception of the dimension axiom.

One defines $H_*(X;\underset{\sim}{E})$ to be $H_*(X,\emptyset;\underset{\sim}{E})$ where $\emptyset$ is the empty set and $X/\emptyset$ is the disjoint union of X and a point. If Y is a space with base point p, one writes $\tilde{H}_*(Y;\underset{\sim}{E})$ for $H_*(Y,p;\underset{\sim}{E})$.

<u>Definition</u>: A <u>ring spectrum</u> is a spectrum $\underset{\sim}{A} = \{A_p, a_p\}$ with a map $\alpha : \underset{\sim}{S} \longrightarrow \underset{\sim}{A}$ and a pairing $m : (\underset{\sim}{A},\underset{\sim}{A}) \longrightarrow \underset{\sim}{A}$; i.e. a collection of maps $m_{p,q} : A_p \wedge A_q \longrightarrow A_{p+q}$ such that the maps of the diagram

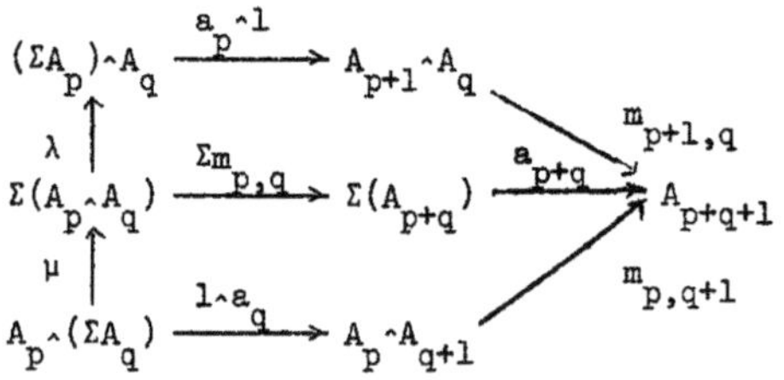

represent classes of the group $[\Sigma(A_p\wedge A_q), A_{p+q+1}]$ related by

$$[m_{p+1,q}\circ(a_p\wedge 1)\circ\lambda] = [a_{p+q}\circ\Sigma m_{p,q}] = (-1)^p[m_{p,q+1}\circ(1\wedge a_q)\circ\mu];$$

such that the diagram

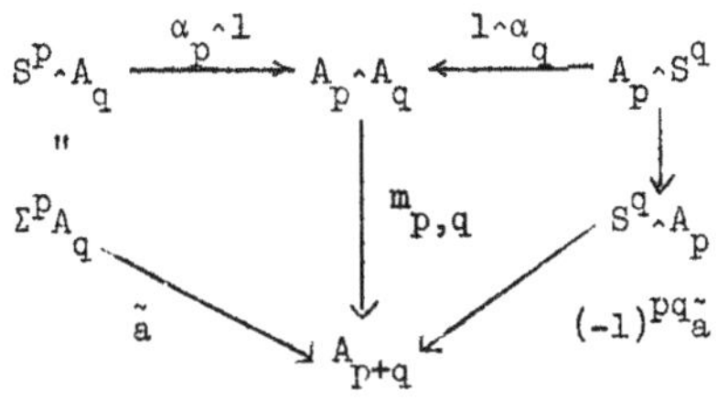

commutes, where $\tilde{a}$ is the multiple composition of the suspensions of maps a_j and $(-1)^{pq}\tilde{a}$ is a map whose class in the group $[\Sigma^q A_p, A_{p+q}]$ is $(-1)^{pq}$ times the class of $\tilde{a}$. [Note: If $q = 0$ this is not a group, but $(-1)^{pq} = 1$ so that one does not need the group structure to find a map.]

<u>Example</u>: Let R be a ring with unit, $K(R,n)$ an Eilenberg-MacLane space (the only non-zero homotopy group being R in dimension n) and let $\kappa_n : \Sigma K(R,n) \longrightarrow K(R,n+1)$ be a map corresponding to the identification $K(R,n) \longrightarrow \Omega K(R,n+1)$. The spectrum $\underset{\sim}{K}(R) = \{K(R,n),\kappa_n\}$ is a ring spectrum and $H^n(X;\underset{\sim}{K}(R))$ is the usual cohomology with coefficients in R.

With a ring spectrum one has the usual sorts of products, such as cup products in cohomology making $H^*(X;\underset{\sim}{A})$ into a commutative ring with unit.

<u>Definition</u>: Let (B,f) be a sequence of fibrations $f_r : B_r \longrightarrow BO_r$. A <u>Thom class</u> is a map of spectra $U : \underset{\sim}{TB} \longrightarrow \underset{\sim}{A}$, where $\underset{\sim}{A}$ is a ring spectrum.

Now let M^n be a (B,f) manifold, $i : M^n \longrightarrow H^{n+r}$ an imbedding with ∂M imbedded in R^{n+r-1} with the usual orthogonal framing along a tubular neighborhood of $\partial M = \partial M \times 0$ (denoting the neighborhood by $\partial M \times [0,1]$). Let N be the normal bundle of M and N' the normal bundle of ∂M. Let $\nu : M \longrightarrow B_r$ be a lifting defining the (B,f) structure on M with $T\nu : TN \longrightarrow TB_r$ the induced map on the Thom complex.

Consider the map

$$N \xrightarrow{\Delta} N \times N \xrightarrow{\pi \times p} M \times TN \longrightarrow (M/\partial M)\wedge TN$$

where Δ is the diagonal map, π the bundle projection, p is the collapse onto the Thom complex, and the last map is the obvious collapse. Under this map the vectors of norm at least one are sent to the base point, as are all vectors over ∂M, i.e. N'. Thus one induces a map $\emptyset : TN/TN' \longrightarrow (M/\partial M)\wedge TN$.

Letting $c : H^{n+r} \longrightarrow TN$ be the standard (scaled) collapse, the projection into TN/TN' sends R^{n+r-1} into the base point also and hence defines a collapse $c : S^{n+r} = (H^{n+r} \cup \infty)/(R^{n+r-1} \cup \infty) \longrightarrow TN/TN'$.

Letting $U = \{U_r\} : \underset{\sim}{TB} \longrightarrow \underset{\sim}{A}$ be a Thom class, one has a composite map

$$S^{n+r} \xrightarrow{c} TN/TN' \xrightarrow{\emptyset} (M/\partial M)\wedge TN \xrightarrow{1\wedge T\nu} (M/\partial M)\wedge TB_r \xrightarrow{1\wedge U_r} (M/\partial M)\wedge A_r$$

which represents an element of $\pi_{n+r}((M/\partial M)\wedge A_r)$. Letting r go to infinity defines a class $[M,\partial M] \in H_n(M,\partial M;\underset{\sim}{A})$. This element is easily seen to depend only on the (B,f) structure of M.

<u>Definition</u>: If M^n is a (B,f) manifold and $U : \underset{\sim}{TB} \longrightarrow \underset{\sim}{A}$ is a Thom class, then the <u>fundamental class</u> of $(M,\partial M)$ is the class $[M,\partial M] \in H_n(M,\partial M;\underset{\sim}{A})$. If ∂M is empty, this class will be denoted $[M] \in H_n(M;\underset{\sim}{A})$.

If one collapses the complement of the tubular neighborhood of the boundary by

$$d : M/\partial M \longrightarrow M/\{\partial M \cup (M-\partial M \times (0,1))\} = \Sigma(\partial M/\emptyset)$$

one has the map which defines the boundary homomorphism in homology

$$\partial : H_s(M,\partial M;\underset{\sim}{A}) \longrightarrow H_{s-1}(\partial M;\underset{\sim}{A}).$$

If one composes the map defining $[M,\partial M]$ with the map $d \wedge 1$, it is immediate that the resulting map is the suspension of the one which defines $[\partial M]$. Thus:

Proposition: Under the boundary homomorphism in $\underset{\sim}{A}$-homology, the fundamental class of $(M,\partial M)$ is sent into the fundamental class of ∂M.

Definition: A universal characteristic class with $\underset{\sim}{A}$ coefficients for (B,f) bundles is a class $x \in H^*(B;\underset{\sim}{A})$ where $B = \lim_{r\to\infty} \{B_r, g_r\}$. If ξ is an r-plane bundle over a space X with a (B_r,f_r) structure given by a lifting $\hat{\xi} : X \longrightarrow B_r$, then the x-characteristic class of the (B_r,f_r) bundle is the class $x(\hat{\xi}) = \hat{\xi}^*\bar{g}_r^*(x) \in H^*(X;\underset{\sim}{A})$, where $\bar{g}_r : B_r \longrightarrow B$ is the usual map into the limit space. If M^n is a (B,f) manifold, the x-normal characteristic class of M is the class $x(M) \in H^*(M;\underset{\sim}{A})$ defined by $x(M) = x(\nu)$ where $\nu : M \longrightarrow B_r$ is a lift of the normal map (of some imbedding) defining the (B,f) structure on M.

Definition: If M^n is a closed (B,f) manifold and $x \in H^p(B;\underset{\sim}{A})$, then the x-characteristic number of M is the class in $H^{p-n}(pt;\underset{\sim}{A})$ obtained by evaluating $x(M)$ on the fundamental class of M. Thus if $x(M) \in H^p(M;A)$ is represented by a map $\chi : \Sigma^i(M/\emptyset) \longrightarrow A_{p+i}$ and $[M] \in H_n(M;\underset{\sim}{A})$ is represented by a map $\mu : S^{n+r} \longrightarrow (M/\emptyset)\wedge A_r$, then $x[M] = \langle x(M),[M] \rangle \in H^{p-n}(pt;\underset{\sim}{A})$ is represented by the map

$$S^{n+r+i} \xrightarrow{\Sigma^i\mu} \Sigma^i(M/\emptyset)\wedge A_r \xrightarrow{\chi\wedge 1} A_{p+i}\wedge A_r \xrightarrow{m_{p+i,r}} A_{p+i+r}.$$

The usefulness of characteristic numbers in cobordism theory arises from the result of L. S. Pontrjagin: Characteristic cycles on differentiable manifolds, Math. Sbor. (N.S.), 21 (63) (1947), 233-284; Amer. Math. Soc. translation 32.

<u>Theorem</u>: If $x \in H^p(B;\underset{\sim}{A})$ and M^n is a closed (B,f) manifold, then the x-characteristic number of M depends only on the (B,f) cobordism class of M.

<u>Proof</u>: Since x-characteristic numbers are clearly additive, it suffices to show that for $M = \partial W$ one has $x[M] = 0$. Letting $i : M \longrightarrow W$ be the inclusion one has $x(M) = i^*x(W)$, so

$$< x(M),[M] > = < i^*x(W),\partial[W,\partial W] > = < \delta i^*x(W),[W,\partial W] >$$

where δ is the cohomology coboundary homomorphism induced by the collapse $d : W/\partial W \longrightarrow \Sigma(\partial W/\emptyset)$. Since the cohomology sequence

$$H^p(W;\underset{\sim}{A}) \xrightarrow{i^*} H^p(\partial W;\underset{\sim}{A}) \xrightarrow{\delta} H^{p+1}(W,\partial W;\underset{\sim}{A})$$

is exact, $\delta i^*x(W) = 0$; hence $x[M] = 0$. **

<u>Remark</u>: Being given fibration sequences $\overline{B} \xrightarrow{h} B \xrightarrow{f} BO$, one may think of $y \in H^*(B,\overline{B};\underset{\sim}{A})$ as a relative characteristic class. Being given a (B,f) manifold M with $(\overline{B},f\circ h)$ structure on ∂M, one has defined a relative characteristic number $y[M,\partial M] \in H^*(pt;\underset{\sim}{A})$ since the normal map gives $(M,\partial M) \longrightarrow (B,\overline{B})$. To see that such numbers are relative cobordism invariants, one may suppose by additivity that there is a (B,f) manifold W with $\partial W = M \cup (-U)$ joined along $\partial M \cong \partial U$, with U a $(\overline{B},f\circ h)$ manifold. One

then has

$$(W,\partial W) \xrightarrow{d} \Sigma(\partial W/\emptyset) \xrightarrow{\Sigma j} \Sigma(\partial W/U) \xleftarrow{\Sigma p} \Sigma(M/\partial M)$$

giving $p_*[M,\partial M] = j_*\partial[W,\partial W]$ by the orientation assumption in the decomposition of ∂W and $y(M,\partial M) = p^*q^*y(W,U)$ where

$$(M,\partial M) \xrightarrow{p} (\partial W,U) \xrightarrow{q} (W,U)$$

so $y[M,\partial M] = < q^*y, j_*\partial[W,\partial W] > = < \delta j^*q^*y, [W,\partial W] >$. However, from the exact sequence of the triple $(W,\partial W,U)$ the composition

$$H^*(W,U) \xrightarrow{q^*} H^*(\partial W,U) \xrightarrow{\delta j^*} H^*(W,\partial W)$$

is zero, so $y[M,\partial M] = 0$. Note: Taking $\overline{B}$ empty, this reduces precisely to the closed case.

In addition to the manifold theoretic treatment of characteristic numbers by using the Thom class to construct fundamental classes, one may also give homology and cohomology theoretic descriptions of characteristic numbers which are frequently useful. In particular, these will be needed later.

As in the construction of the map $\emptyset$, one has for any r-plane bundle γ over a space X the composition

$$\gamma \xrightarrow{\Delta} \gamma \times \gamma \xrightarrow{\pi\times p} X \times T\gamma \longrightarrow (X/\emptyset)\wedge T\gamma$$

giving a map $\emptyset : T\gamma \longrightarrow (X/\emptyset)\wedge T\gamma$.

Applying this to the bundle $f_r^*(\gamma^r)$ over B_r, and composing with the Thom class and inclusion of B_r in B gives

$$(1\wedge U_r)\circ(\overline{g}_r\wedge 1)\circ\emptyset : TB_r \longrightarrow (B/\emptyset)\wedge A_r$$

inducing on homotopy the homomorphism

$$\Omega_n(B,f) = \lim_{r\to\infty} \pi_{n+r}(TB_r,\infty) \longrightarrow \lim_{r\to\infty} \pi_{n+r}((B/\emptyset)\wedge A_r) = H_n(B;\underset{\sim}{A}).$$

If M^n is a closed (B,f) manifold, so that $M/\emptyset$ maps into $B_r/\emptyset$ under the normal map, one has the commutative diagram

$$\begin{array}{ccccccc}
S^{n+r} & \xrightarrow{c} & TN & \xrightarrow{\emptyset} & (M/\emptyset)\wedge TN & \xrightarrow{1\wedge U_r \circ T\nu} & (M/\emptyset)\wedge A_r \\
& \searrow & \downarrow{\scriptstyle T\nu} & & \downarrow{\scriptstyle \nu\wedge T\nu} & & \downarrow{\scriptstyle \bar{g}_r\circ\nu\wedge 1} \\
& & TB_r & \xrightarrow{\emptyset} & (B_r/\emptyset)\wedge TB_r & \xrightarrow{\bar{g}_r\wedge U_r} & (B/\emptyset)\wedge A_r
\end{array}$$

Thus the homotopy homomorphism is the homomorphism sending the cobordism class of a manifold M^n into the image under the normal map of the fundamental class of M^n. Thus the pairing of homology and cohomology of B into the cohomology of a point gives

$$\Omega_n(B,f) \otimes H^p(B;\underset{\sim}{A}) \longrightarrow H_n(B;\underset{\sim}{A}) \otimes H^p(B;\underset{\sim}{A}) \longrightarrow H^{p-n}(pt;\underset{\sim}{A})$$

which coincides with the evaluation of characteristic numbers.

In addition, the composition

$$TB_r\wedge A_s \xrightarrow{\emptyset\wedge 1} (B_r/\emptyset)\wedge TB_r\wedge A_s \xrightarrow{1\wedge U_r\wedge 1} (B_r/\emptyset)\wedge A_r\wedge A_s \xrightarrow{1\wedge m_{r,s}} (B_r/\emptyset)\wedge A_{r+s}$$

gives rise to the Thom homomorphism in homology

$$\begin{array}{ccc}
\Phi_U : \tilde{H}_j(TB_r;\underset{\sim}{A}) & \longrightarrow & H_{j-r}(B_r;\underset{\sim}{A}) \\
\Vert & & \Vert \\
\lim\limits_{s\to\infty} \pi_{j+s}(TB_r\wedge A_s) & & \lim\limits_{s\to\infty} \pi_{j+s}((B_r/\emptyset)\wedge A_{r+s})
\end{array}$$

determined by the Thom class U. (Note: This is given by $\pi_*\circ(\cap U)$ since the map $\emptyset$ is the composition of $\pi\wedge 1$ with the map used in defining cap products.) The homomorphism $\Omega_n(B,f) \longrightarrow H_n(B;\underset{\sim}{A})$ may then be interpreted as the composition of the Hurewicz homomorphism in $\underset{\sim}{A}$-homology $\pi_{n+r}(TB_r,\infty) \longrightarrow \tilde{H}_{n+r}(TB_r;\underset{\sim}{A})$ given by $TB_r = TB_r \wedge S^0 \xrightarrow{1 \wedge \alpha_0} TB_r \wedge A_0$ and the Thom homomorphism determined by U (at least after letting r go to infinity).

It is more common to consider the Thom homomorphism in cohomology theory. If one begins with the map

$$TB_r \xrightarrow{\emptyset} (B_r/\emptyset) \wedge TB_r \xrightarrow{1 \wedge U_r} (B_r/\emptyset) \wedge A_r$$

and chooses a map $\chi_r : \Sigma^i(B_r/\emptyset) \longrightarrow A_{p+i}$ representing a class $x_r \in H^p(B_r;\underset{\sim}{A})$ then the composition

$$\Sigma^i(TB_r) \longrightarrow \Sigma^i(B_r/\emptyset) \wedge A_r \xrightarrow{\chi_r \wedge 1} A_{p+i} \wedge A_r \xrightarrow{m_{p+i,r}} A_{p+i+r}$$

represents a class in $\tilde{H}^{p+r}(TB_r;\underset{\sim}{A})$. This defines the cohomology Thom homomorphism

$$\Phi^U : H^p(B_r;\underset{\sim}{A}) \longrightarrow \tilde{H}^{p+r}(TB_r;\underset{\sim}{A}).$$

The construction of $\emptyset$ shows that $\Phi^U(x_r) = \pi_r^*(x_r) \cup U_r$, where $\pi_r : f_r^*(\gamma^r) \longrightarrow B_r$ is the projection and $U_r : TB_r \longrightarrow A_r$ is interpreted as a class $U_r \in \tilde{H}^r(TB_r;\underset{\sim}{A})$.

In particular, for $x \in H^p(B;\underset{\sim}{A})$ one has the sequence of elements $x_r = \bar{g}_r^*(x) \in H^p(B_r;\underset{\sim}{A})$, lifting to elements $\pi_r^*(x_r) \cup U_r \in \tilde{H}^{p+r}(TB_r;\underset{\sim}{A})$. If M^n is a closed (B,f) manifold with cobordism class represented by the map $\phi_M : S^{n+r} \longrightarrow TB_r$, then $x[M] = \phi_M^*(\pi_r^*(x_r) \cup U_r) \in \tilde{H}^{p+r}(S^{n+r};\underset{\sim}{A}) = H^{p-n}(pt;\underset{\sim}{A})$. (<u>Note</u>: It is immediate that this agrees with the previous homology interpretation. All interpretations are really based on the naturality of the map $\emptyset$).

Definition: The Thom class $U : \underset{\sim}{TB} \longrightarrow \underset{\sim}{A}$ is said to be an $\underset{\sim}{A}$-orientation if for each point b_r of B_r there is a framing of δ^r so that the classes determined by $U_r \circ Tb_r$ and $\alpha_r : S^r \longrightarrow A_r$ are the same in $\tilde{H}^r(S^r;\underset{\sim}{A})$.

Remarks: 1) This is the assertion that the bundles $f_r^*(\gamma^r)$ are (uniformly) $\underset{\sim}{A}$-oriented in the sense of A. Dold: Relations between ordinary and extraordinary cohomology, Notes, Aarhus Colloquium on Algebraic Topology, Aarhus, 1962.

2) If the class of $\alpha : \underset{\sim}{S} \longrightarrow \underset{\sim}{A}$ as an element of $\tilde{H}^0(S^0;\underset{\sim}{A})$ does not have order 2, this gives a preferred orientation to the fiber δ^r. If this class has order 2, then all $\underset{\sim}{A}$ cohomology is of order 2 and orientation doesn't enter the situation.

Proposition: Let $U : \underset{\sim}{TB} \longrightarrow A$ be an A-orientation, ξ an r-plane bundle over a finite complex X with $\tilde{\xi} : X \longrightarrow B_r$ a (B_r, f_r) structure on ξ. Then the Thom homomorphism

$$\Phi^U : H^p(X;\underset{\sim}{A}) \longrightarrow \tilde{H}^{p+r}(T\xi;\underset{\sim}{A})$$

is an isomorphism.

Proof: The composition $U_r \circ T\tilde{\xi} : T\xi \longrightarrow A_r$ defines an $\underset{\sim}{A}$-orientation of ξ. Over any cell D^n of X the bundle ξ is trivial, and D^n being path connected and simply connected the class $U_r \circ T\tilde{\xi}$ orients the bundle ξ over D^n. Over D^n one then has the Thom space equivalent to $D^n{}_\wedge S^r$, with the Thom homomorphism being just the suspension isomorphism. Thus the Thom homomorphism defines a homomorphism of the $\underset{\sim}{A}$ spectral sequence of X into the reduced cohomology $\underset{\sim}{A}$ spectral sequence of $T\xi$ which is an isomorphism on E_2, hence also on E_∞. **

Corollary: If $U : \underset{\sim}{TB} \longrightarrow \underset{\sim}{A}$ is an $\underset{\sim}{A}$ orientation and M^n is a (B,f) manifold, then the Thom homomorphisms Φ^U

$$H^p(M;\underset{\sim}{A}) \longrightarrow \tilde{H}^{p+r}(TN;\underset{\sim}{A}),$$

$$H^p(\partial M;\underset{\sim}{A}) \longrightarrow \tilde{H}^{p+r}(TN';\underset{\sim}{A}), \text{ and}$$

$$H^p(M,\partial M;A) \longrightarrow \tilde{H}^{p+r}(TN,TN';\underset{\sim}{A})$$

are isomorphisms.

Let M^n be a manifold, and let M be imbedded in H^{n+r} with $\partial M \subset R^{n+r-1}$ in the usual way, and let ν be the normal bundle of M, ν' the normal bundle of ∂M.

Theorem: (Atiyah [14]) The pairs

a) $(M/\partial M)$ and $T\nu$, or

b) $(M/\emptyset)$ and $(T\nu/T\nu')$

are dual in S^{n+r+1}.

[B and C in S^k are dual if B and C are disjoint and each is a strong deformation retract of the complement of the other.]

Proof: Let N be a tubular neighborhood of M in H^{n+r}. Consider S^{n+r+1} as $H^{n+r} \times R$ with $(R^{n+r-1} \cup \infty) \times R \cup H^{n+r} \times \{-\infty,\infty\}$ collapsed to a base point ∞.

Consider $(T\nu/T\nu')$ as $(N \times 0 \cup \partial N \times [0,\infty) \cup \infty)$ and then one may collapse the complement which is $\{H^{n+r} \times (-\infty,0) \cup (-N) \times [0,\infty)\} \cup \text{interior}(N) \times (0,\infty)$ to $pt \cup (M-\partial M) \times 1$ by a strong deformation retract, and conversely, the complement of a point inside $H^{n+r} \times (-\infty,0) \cup (-N) \times [0,\infty)$ and $(M-\partial M) \times 1$ may be collapsed onto $(T\nu/T\nu')$ by a strong deformation retraction.

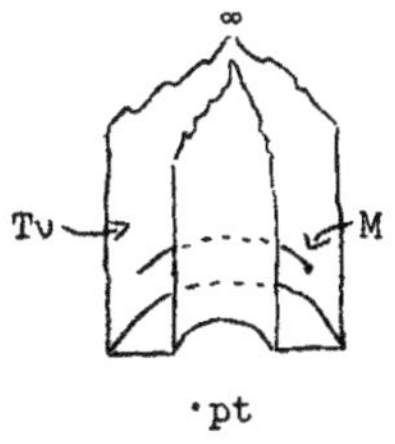

Considering $(M/\partial M)$ as $(M \times 1) \cup \infty$ one may collapse the complement by a strong deformation retract onto the subset $\bar{N} \times 0 \cup \partial\bar{N} \times [0,2] \cup c$, where c is the cone on $\partial\bar{N} \times 2$ with vertex some point with last coordinate larger than 2, where $\bar{N} = N - N \cap R^{n+r-1}$. This subset is clearly just $T\nu$. Similarly, removal of this set $T\nu$ gives a set collapsing onto $M/\partial M$.

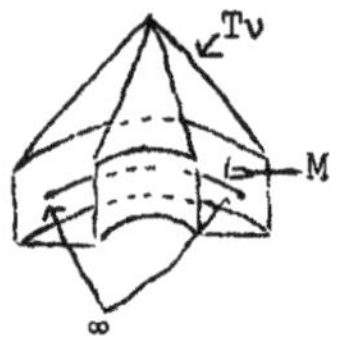

Note: All deformations are obtained by radial deformations toward points in question, and scalar multiplication expansions in the fibers of the normal bundles of M. **

If $B,C \subset S^k$ are disjoint sets as above, let $p_0 \in S^k - B \cup C$ and apply stereographic projection to map B, C into R^k as disjoint subsets. Let $f : B \times C \longrightarrow S^{k-1}$ by $f(b,c) = (b-c)/\|b-c\|$. Letting b, c be base points of B, C respectively, $f(B \times c) \cup f(b \times C)$ is a proper subset of S^{k-1} and f factors homotopically through a map $\bar{f} : B \wedge C \longrightarrow S^{k-1}$.

One then has defined a duality as follows: For $\alpha \in \tilde{H}_p(B;\underset{\sim}{A})$ choose a representative map $a : S^{p+i} \longrightarrow B \wedge A_i$ and then

$$S^{p+i} \wedge C \xrightarrow{a \wedge 1} B \wedge A_i \wedge C \xrightarrow{\bar{f} \wedge 1} S^{k-1} \wedge A_i \longrightarrow A_{i+k-1}$$

defines a class in $\tilde{H}^{k-1-p}(C;\underset{\sim}{A})$, denoted $D\alpha$.

Lemma: Let $b \in B \subset R^k$ with B an imbedded disc. Then $H^*(R^k-b, R^k-B;\underset{\sim}{A}) = 0$.

Proof: $H^*(R^k-b,R^k-B) \cong H^*(S^k-b,S^k-B)$ by excision. Since S^k-b and S^k-B are contractible, $\tilde{H}^*(S^k-b) = \tilde{H}^*(S^k-B) = 0$ and the exact sequence of the pair gives the result. **

<u>Corollary</u>: If $B \subset R^k$ is an imbedded disc then $D : H_p(R^k, R^k - B; \underset{\sim}{A}) \longrightarrow H^{k-p}(B; \underset{\sim}{A})$ is an isomorphism.

<u>Proof</u>: One has the diagram

$$\begin{array}{ccccccc}
H_p(R^k - b, R^k - B) & \longrightarrow & H_p(R^k, R^k - B) & \longrightarrow & H_p(R^k, R^k - b) & \longrightarrow & H_{p-1}(R^k - b, R^k - B) \\
\downarrow{\scriptstyle D'} & & \downarrow{\scriptstyle D} & & \downarrow{\scriptstyle D''} & & \downarrow{\scriptstyle D'} \\
H^{k-p}(B, b) & \longrightarrow & H^{k-p}(B) & \longrightarrow & H^{k-p}(b) & \longrightarrow & H^{k-p+1}(B, b)
\end{array}$$

in which the end maps are both isomorphisms since the groups are zero, and to complete the proof one need only check D'', which is clearly an isomorphism. **

<u>Theorem</u>: (Alexander Duality) For any polyhedral pair (B, B') in R^k, $D : H_p(R^k - B', R^k - B; \underset{\sim}{A}) \longrightarrow H^{k-p}(B, B'; \underset{\sim}{A})$ is an isomorphism.

<u>Proof</u>: By naturality it suffices to consider the case $B' = \emptyset$, and then one may apply a Mayer-Vietoris argument using induction on the number of cells of B, the corollary and the five lemma. **

<u>Theorem</u>: (Spanier-Whithead duality) For any pair $B, C \subset S^k$ as above $D : \tilde{H}_p(B; \underset{\sim}{A}) \longrightarrow \tilde{H}^{k-1-p}(C; \underset{\sim}{A})$ is an isomorphism.

<u>Proof</u>: D is given by the composition of isomorphisms

$$\tilde{H}_p(B; \underset{\sim}{A}) \cong \tilde{H}_p(S^k - C; \underset{\sim}{A}) \overset{\cong}{\leftarrow} H_{p+1}(S^k - c, S^k - C; \underset{\sim}{A}) \overset{\cong}{\leftarrow} H_{p+1}(R^k - c, R^k - C; \underset{\sim}{A}) \overset{D}{\rightarrow} \tilde{H}^{k-p-1}(C; \underset{\sim}{A}). \quad **$$

<u>Note</u>: See Spanier [110], pages 295 and 462.

<u>Theorem</u>: (Hsiang and Wall [58]) A class $\alpha \in \tilde{H}^r(T\nu; \underset{\sim}{A})$ is a Dold orientation if and only if the class $D^{-1}(\alpha) \in H_n(M, \partial M; \underset{\sim}{A})$ is an $\underset{\sim}{A}$ orientation in the sense of Whitehead; i.e. for each point $q \in M - \partial M$ the class in $H^n(M, \partial M; \underset{\sim}{S})$ obtained by collapsing $M/\partial M$ onto $D^n/\partial D^n$ where D^n is a disc

neighborhood of q and the class $D^{-1}(\alpha)$ have Kronecker product the class defined by the unit in $H^0(pt;\underset{\sim}{A})$.

Proof: The map $D^n \longrightarrow M - \partial M$ given by taking the disc neighborhood of q defines the collapse $M/\partial M \longrightarrow D^n/\partial D^n$ which is patently dual to the inclusion of TD^n in $T\nu$, but TD^n is homotopy equivalent to the Thom space of a fiber. **

Note: Similarly, the collapse $S^{n+r} \longrightarrow T\nu/T\nu'$ is dual to the map of M to a point.

Corollary: If the Thom class $U : \underset{\sim}{TB} \longrightarrow \underset{\sim}{A}$ is an $\underset{\sim}{A}$ orientation, then for any (B,f) manifold M^n, the fundamental class $[M,\partial M] \in H_n(M,\partial M;\underset{\sim}{A})$ is an $\underset{\sim}{A}$ orientation in the sense of Whitehead.

Theorem: (Poincaré-Lefschetz duality) If $U : \underset{\sim}{TB} \longrightarrow \underset{\sim}{A}$ is an $\underset{\sim}{A}$-orientation, then for any (B,f) manifold M^n one has isomorphisms

$$H^q(M;\underset{\sim}{A}) \longrightarrow H_{n-q}(M,\partial M;\underset{\sim}{A})$$

$$H^q(M,\partial M;\underset{\sim}{A}) \longrightarrow H_{n-q}(M;\underset{\sim}{A})$$

given by the cap-product with the fundamental class $[M,\partial M]$.

The cap-product relation is given by

$$u:S^{n+r} \xrightarrow{c} TN/TN' \xrightarrow{\emptyset} (M/\partial M)_\wedge TN \xrightarrow{1_\wedge U_r \circ T\nu} (M/\partial M)_\wedge A_r \xrightarrow{\Delta_\wedge 1} (M/\partial M)_\wedge (M/\emptyset)_\wedge A_r$$

where Δ is the map given from the diagonal. If $\Sigma^i(M/\emptyset) \xrightarrow{x} A_{q+i}$ represents a class in $H^q(M;\underset{\sim}{A})$, the cap product is represented by

$$S^{n+r+i} \xrightarrow{\Sigma^i u} (M/\partial M)_\wedge \Sigma^i(M/\emptyset)_\wedge A_r \xrightarrow{1_\wedge x_\wedge 1} (M/\partial M)_\wedge A_{q+i}{}_\wedge A_r \xrightarrow{1_\wedge m_{q+i,r}} (M/\partial M)_\wedge A_{r+q+i}.$$

A similar formula defines the other homomorphism.

Proof: These isomorphisms are just the composite of Spanier-Whitehead duality and the Thom isomorphisms. **

Chapter IV

The Interesting Examples - A Survey of the Literature

Since cobordism theory is a classificational tool, the interest really lies in the investigation of specific classification problems. Numerous examples have been considered and hence a vast literature exists, with few really central theoretical tools, largely due to the idiosyncrasies inherent in the examples. The purpose of this chapter is to list many of these examples and indicate briefly what is known and where to find it in the literature.

Example 1: Framed cobordism: Ω_*^{fr}.

Historically: First application of cobordism theory, intended to study the homotopy of spheres.

Objects: Framed manifolds, i.e. manifolds with an equivalence class of trivializations of the normal bundle.

Determination: (B,f) cobordism with each B_r contractible (classifying space for the identity subgroup 1_r of O_r), so $\Omega_n^{fr} \cong \lim_{r\to\infty} \pi_{n+r}(S^r)$ is the stable homotopy of spheres (Pontrjagin [101]).

Results: A vast literature exists but is largely unrelated to cobordism. Use of surgery (Milnor [83], Wallace [137]) to construct framed cobordisms shows that representatives frequently may be taken to be homotopy spheres (Kervaire-Milnor [61]). Recent work of Conner and Floyd [41] has placed the e-invariant of Adams [4] in a cobordism framework.

Example 2: Unoriented cobordism: $\mathfrak{N}_*$.

Historically: The turning point for cobordism theory.

Objects: All compact manifolds, i.e. the category $(\mathcal{O},\partial,i)$.

Determination: Equivalent to (B,f) cobordism with $B_r = BO_r$ and f_r the identity map. (Thom [127]).

Calculation: $\mathfrak{N}_*$ is the polynomial ring over Z_2 on classes x_i of dimension i for each integer i not of the form 2^s-1. Even dimensional generators may be taken to be the classes of real projective spaces. (Thom [127]). Odd dimensional generators were constructed by Dold [43].

Characteristic numbers: Z_2 cohomology characteristic numbers give complete invariants (Thom [127]). All relations among these numbers (expressed tangentially) are given by Wu's formulae (Wu [142]) relating to the action of the Steenrod algebra (Dold [44]).

Example 3: Complex cobordism: Ω_*^U.

Objects: Stably almost complex manifolds - manifolds with an equivalence class of complex vector bundle structures on the normal bundle.

Determination: (B,f) cobordism with $B_{2r} = B_{2r+1} = BU_r$ the classifying space for the unitary group U_r (limit of complex Grassmann manifolds).

Calculation: Ω_*^U is the integral polynomial ring on classes x_i of dimension $2i$ for each integer i, with x_i represented by a projective complex algebraic variety (Milnor [82], Novikov [92, 93]). In fact, every class is represented by such a variety (Milnor; see Hirzebruch [54] or Thom [129]).

Characteristic numbers: Cobordism is determined by integral cohomology characteristic numbers (Milnor [82]). All relations among these numbers are given by the Atiyah-Hirzebruch [17] form of Riemann-Roch theorem relating complex K-theory to rational cohomology (Stong [117], Hattori [52]).

Relation to $\mathfrak{N}_*$: Ω_*^U maps onto the squares of classes of $\mathfrak{N}_*$ (Milnor [87]).

Relation to Ω_*^{fr}: Every framed manifold bounds a complex manifold. The Todd class homomorphism $\Omega_*((BU,f),(B1,f)) \longrightarrow Q$ of the relative cobordism group induces the Adams e homomorphism (Conner and Floyd [41]).

Example 4: Oriented cobordism: Ω_*^{SO}.

Objects: Oriented manifolds.

Determination: (B,f) cobordism with $B_r = BSO_r$ the classifying space for the special orthogonal group SO_r (limit of Grassmannians of oriented planes) (Thom [127]).

Calculation: $\Omega_*^{SO} \otimes Q$ is the rational polynomial ring on classes x_{4i} of the complex projective spaces $CP(2i)$ (Thom [127]). Ω_*^{SO} has no odd torsion (Milnor [81], Averbuh [21]) and Ω_*^{SO}/Torsion is a polynomial ring over Z on $4i$ dimensional generators (Milnor [81], Novikov [92,93]). Ω_*^{SO} has only torsion of order two and the quotient $\Omega_*^{SO}/2\Omega_*^{SO}$ may be described as follows: Let $\mathcal{W}_*$ be a Z_2 polynomial ring on classes x_{2k}, x_{2k-1}, for k not a power of 2, and $x_{2^j}^2$. Let $\partial_1 : \mathcal{W}_* \longrightarrow \mathcal{W}_*$ be a derivation given by $\partial_1 x_{2k} = x_{2k-1}$, $\partial_1 x_{2k-1} = 0$, $\partial_1 x_{2^j}^2 = 0$. Then $\Omega_*^{SO}/2\Omega_*^{SO} \cong$ kernel ∂_1 and the image of the torsion will be the image of ∂_1 (Wall [130]).

Characteristic numbers: Cobordism is determined by Z and Z_2 cohomology, all relations among the Z_2 numbers being given by the relations of Wu together with the vanishing of the first Stiefel-Whitney class (Wall [130]). All relations among the Z numbers are given by the Riemann-Roch theorem (Stong [117]).

Relation to Ω_*^U: Ω_*^U maps onto Ω_*^{SO}/Torsion (Milnor [82]).

Relation to $\mathfrak{N}_*$: $\Omega_*^{SO}/2\Omega_*^{SO}$ is mapped isomorphically to the subring kernel ∂_1 described above, the x_i being (well-chosen) generators of $\mathfrak{N}_*$ (Wall [130]).

Example 5: w_1 spherical cobordism: $\mathcal{W}_*$

Historically: This cobordism theory arises in Wall's determination of oriented cobordism, and was completely determined by Wall [130].

Objects: Manifolds for which the first Stiefel-Whitney class w_1 is the reduction of an integral cohomology class; is induced by a map into the sphere S^1.

Determination: (B,f) cobordism with B_r the total space of the fibration over $BO_r \times S^1$ induced from the path fibration over $K(Z_2,1)$ by the map realizing the cohomology class $w_1 \otimes 1 + 1 \otimes \sigma$, $\sigma \in H^1(S^1;Z_2)$ being the generator.

Calculation: Given by the polynomial ring $\mathcal{W}_*$ described above.

Characteristic numbers: Z_2 cohomology determines cobordism, all relations being given by those of Wu together with the vanishing of $w_1{}^2$.

Relation of $\mathfrak{N}_*$ and Ω_*^{SO}: Maps monomorphically into $\mathfrak{N}_*$, with ∂_1 describing the image of Ω_*^{SO} as above.

Example 6: Bordism: $\Omega_*(B,f)[X,A]$.

Objects: Let $F : (B,f) \longrightarrow \mathfrak{X}$ be the forgetful functor from the cobordism category of (B,f) manifolds to the category of topological spaces which takes the underlying topological space. One then has the cobordism category of (B,f) manifolds 'over' a space X. If $A \subset X$ is a subspace one has a functor $J : (B,f)/A \longrightarrow (B,f)/X$ induced by the inclusion.

Determination: $(B,f)/X$ cobordism is just the cobordism theory based on the fibration $B_r \times X \xrightarrow{\pi} B_r \xrightarrow{f_r} BO_r$, π-being the projection. The relative bordism group of the pair (X,A) is $\Omega_n(B,f)[X,A] = \Omega_n(J,\alpha)$, where α is the piecing together previously described, and is given by

$$\lim_{r\to\infty} \pi_{n+r}(TB_r \wedge (X/A)) = H_n(X,A;\underset{\sim}{TB}).$$

Historically: These groups were originally defined by Atiyah [13], who called them the (B,f) bordism groups of the pair (X,A). He reserved the name cobordism for the dual cohomology theory with coefficients in the spectrum $\underset{\sim}{TB}$.

$\mathfrak{N}_*(X,A)$: The unoriented bordism of a pair (X,A) is essentially trivial, being isomorphic as $\mathfrak{N}_*$ module to $\mathfrak{N}_* \otimes H_*(X,A;Z_2)$. Cobordism is determined by Z_2 cohomology. (See Conner and Floyd [36]). Operations in this theory were determined by Landweber [63, 64]. Used by Brown and Peterson [29] to study relations among Stiefel-Whitney classes.

$\Omega_*^{SO}(X,A)$: Studied extensively by Conner and Floyd [36]. One has $\Omega_*^{SO}(X,A)$ isomorphic to $H_*(X,A;\Omega_*^{SO})$ modulo the Serre class of finite groups of odd order. If all torsion of $H_*(X;Z)$ is of order 2, then Z and Z_2 cohomology characteristic numbers determine cobordism class in $\Omega_*^{SO}(X)$. Künneth theorems for this homology theory are studied in Landweber [65]. An interesting application is the use of $\Omega_*^{SO}(BU)$ in the proof of the Atiyah-Singer index theorem (Atiyah-Singer [20], Palais [97]).

$\Omega_*^U(X,A)$: Studied by Conner and Floyd [35, 37]. If X has no torsion in its homology $\Omega_*^U(X) \cong \Omega_*^U \otimes H_*(X;Z)$ and cobordism is determined by integral cohomology characteristic numbers. The relation of $\Omega_*^U(X)$ to the complex K-theory of X is studied in Conner and Floyd [41]. Operations in Ω_*^U theory are studied in Novikov [96].

Example 7: Special unitary cobordism: Ω_*^{SU}.

Objects: Manifolds with an equivalence class of special unitary structures on the normal bundle.

Determination: (B,f) cobordism in which $B_{2r} = B_{2r+1} = BSU_r$ is the classifying space for the special unitary group SU_r.

Calculation: First partial results were by Novikov [93]. The main structure was determined by Conner and Floyd [39], who proved that Ω_n^{SU} is torsion free except for $n \equiv 1$ or 2 modulo 8, and $\Omega_{8k+1}^{SU} \cong \Omega_{8k+2}^{SU}$ is a Z_2 vector space whose dimension is the number of partitions of k. The multiplicative structure has been described by Wall [135].

Characteristic numbers: KO-theory characteristic numbers *determine* cobordism class (Anderson, Brown, and Peterson [6]). Ignoring torsion, integral cohomology suffices, and all relations among these follow from appropriate Riemann-Roch theorems (Stong [117]).

Relation to Ω_*^U: The kernel of $\Omega_*^{SU} \longrightarrow \Omega_*^U$ is the torsion subgroup. The image was described by Conner and Floyd.

Relation to Ω_*^{fr}: The image of Ω_*^{fr} in Ω_*^{SU} is zero except in dimensions $8k+1$ and $8k+2$ where it is Z_2 (Anderson, Brown, and Peterson [6]).

Relation to $\mathcal{N}_*$: The image of Ω_*^{SU} in $\mathcal{N}_*$ is the squares of classes containing an oriented manifold all of whose Pontrjagin numbers divisible by $\mathcal{P}_1$ are even. (Conner and Landweber [42]).

Example 8: c_1 spherical manifolds: $\mathcal{W}_*^U$.

Historically: The analog in the complex case of $\mathcal{W}_*$, used by Conner and Floyd in computation of Ω_*^{SU}.

Objects: Stably almost complex manifolds for which the first Chern class c_1 is induced by a map into a sphere.

Determination: (B,f) cobordism in which $B_{2r} = B_{2r+1}$ is the total space of the fibration over $S^2 \times BU_r$ induced from the path fibration over $K(Z,2)$ by the map realizing $\sigma \otimes 1 - 1 \otimes c_1$.

Calculation: Conner and Floyd determine $\mathcal{W}_*^U$ in [39].

Characteristic numbers: Integral cohomology characteristic numbers determine cobordism class.

Relation to Ω_*^U: $\mathcal{W}_*^U$ maps monomorphically into Ω_*^U with image all classes for which numbers divisible by c_1^2 are zero.

Relation to $\mathfrak{N}_*$: $\mathcal{W}_*^U$ has image in $\mathfrak{N}_*$ precisely the squares of classes of $\mathcal{W}_*$ (Stong [118], Conner and Landweber [42]).

Example 9: Spin cobordism: Ω_*^{Spin}.

Objects: Manifolds with an equivalence class of Spin structures on the normal bundle. ($Spin_n$ is the simply connected covering group of SO_n; see Atiyah, Bott, and Shapiro [16] and Milnor [85]).

Determination: (B,f) cobordism with B_r the classifying space of $Spin_r$; i.e. the two-connective covering space of BSO_r.

Calculation: Preliminary results were by Novikov [93]. The main calculation is due to Anderson, Brown, and Peterson [7, 8] who showed that all torsion is of order 2, being of two types: that arising by products with a framed S^1 (similar to the SU case) and that which maps monomorphically into unoriented cobordism. Ω_*^{Spin}/Torsion is the subring of an integral polynomial ring on classes x_{4i} (dimension $4i$) consisting of all classes of dimension a multiple of 8 and twice the classes whose dimension is not a multiple of 8.

Characteristic numbers: Cobordism is determined by Z_2 cohomology and KO-theory characteristic numbers. The relations in integral cohomology all follow from the Riemann-Roch theorem.

Relation to Ω_*^{fr}: The image of framed cobordism is the same as the image of framed in SU cobordism.

Relation to Ω_*^{SO}: The kernel of the map $\Omega_*^{Spin} \longrightarrow \Omega_*^{SO}$ is in dimensions $8k + 1$ and $8k + 2$ only and is the part generated by framed manifolds.

Relation to $\mathfrak{N}_*$: The image in $\mathfrak{N}_*$ is all classes for which the Z_2 characteristic numbers divisible by w_1 and w_2 are zero. Preliminary work in this direction was done by Milnor [87], P. G. Anderson [9], and Stong [118], showing that the class of the square of an oriented manifold is the class of a Spin manifold.

Example 10: Spin^c, Pin, and Pin^c cobordism: $\Omega_*^{Spin^c}$, Ω_*^{Pin}, and $\Omega_*^{Pin^c}$.

Objects: Manifolds with an equivalence class of Spin^c, Pin, or Pin^c structures (see Atiyah, Bott, and Shapiro [16]).

Determination: (B,f) cobordism with the obvious classifying spaces. (BSpin^c is BSO with w_2 made reduced integral, BPin^c is BO with w_2 made reduced integral, and BPin is BO with w_2 killed).

Calculation: Due to Anderson, Brown, and Peterson (announced in [8]), largely as a byproduct in the study of Spin. $\Omega_*^{Spin^c}$/Torsion is discussed in Stong [117].

Remarks: Results have not yet been published for Pin aand Pin^c. These groups are 2 primary, having elements of arbitrarily large order. Images in $\mathfrak{N}_*$ are those classes for which the appropriate Stiefel-Whitney numbers vanish.

Note: Pin gives the first example of a theory for which the tangential and normal structures are not of the same type. Specifically, if M has a Pin normal bundle, then the tangent bundle has $w_2 = w_1^2$ and hence the tangent bundle does not necessarily have a Pin structure.

Example 11: Complex-Spin cobordism: Ω_*^{c-s}.

Objects: Manifolds with both a stably almost complex and a Spin structure.

Determination: (B,f) cobordism with B the fibration over BU induced from the fibration of BSpin over BSO.

Calculation: Ω_n^{c-s} is the direct sum of Ω_n^{SU} and a free abelian group (Stong [121]).

Remark: Useful in trying to understand the relationship between SU and Spin cobordism.

Example 12: Symplectic cobordism: Ω_*^{Sp}.

Objects: manifolds with an equivalence class of quaternionic vector bundle structures on the normal bundle.

Determination: (B,f) cobordism with $B_{4r} = B_{4r+1} = \dots = B_{4r+3}$ the classifying space BSp_r of the symplectic group Sp_r of unitary quaternionic $r \times r$ matrices (limit of quaternionic Grassmann manifolds).

Calculation: Novikov [93] showed that $\Omega_*^{Sp} \otimes Z[1/2]$ is polynomial on 4i dimensional generators and calculated the low dimensional groups. Liulevicius [75] calculated more low groups using the Adams spectral sequence, and computations are still in progress using that method.

Relation with $\mathfrak{N}_*$: The image in unoriented cobordism is zero in dimensions less than 24 (Stong [123]).

Remark: The corresponding bordism theory is studied in Landweber [68].

Example 13: Quasi-symplectic cobordism.

Objects: Manifolds for which the normal bundle is a sum of tensor products of quaternionic vector bundles. (Note: The tensor product of quaternionic bundles in only a real bundle.)

Remark: Introduced by Landweber [67], this cobordism is the subgroup of $\mathfrak{N}_*$ consisting of the fourth powers of all classes. This was intended to fill the gap left by the fact that Ω_*^{Sp} maps into the fourth powers but not onto them (as one might at first guess). In particular, quaternionic projective spaces are quasi-symplectic but not symplectic (see Hirzebruch [54], Conner and Floyd [39], and Kraines [62]).

Example 14: Clifford algebra cobordism: $\Omega_*^{p,q}$.

Objects: Manifolds together with an equivalence class of actions of the Clifford algebra associated with the quadratic form $\sum_{i=1}^{p} x_i^2 - \sum_{i=p+1}^{p+q} x_i^2$ on the normal bundle (see Atiyah, Bott, and Shapiro [16]).

Determination: (B,f) cobordism for an appropriate classifying space. These may be decomposed into somewhat more standard objects.

a) $(p,q) = (0,0)$ is $\mathfrak{N}_*$.

b) $(p,q) = (0,1)$ is Ω_*^U.

c) $(p,q) = (0,2)$ is Ω_*^{Sp}.

d) $(p,q) = (0,3)$ is $\Omega_*^{Sp}(BSp)$.

e) $(p,q) = (1,0)$ is $\mathfrak{N}_*(BO)$.

f) $(p,q) = (2,0)$ is cobordism of manifolds for which the normal bundle is the complexification of a real bundle.

g) $(p,q) = (1,1)$ coincides with $(2,0)$.

Remarks: The odd primary structure is easily computable. One can give upper bounds for images in $\mathfrak{N}_*$, describable as at most 2^k-th powers of elements of $\mathfrak{N}_*$. An unstable version of the case (f) has been studied by R. Wells, the unstable form occuring in an exact sequence with immersion cobordism (see below).

Example 15: Self-conjugate cobordism: Ω_*^{SC}.

Objects: Manifolds having a stably almost complex structure, together with an isomorphism of that structure with its complex conjugate.

Determination: (B,f) cobordism in which B is the classifying space for self-conjugate K-theory, BSC, defined by Anderson [5] and Green [51].

Calculation: When tensored with $Z[1/2]$ this coincides with the symplectic bordism of Sp (studied in Landweber [68]) which is the symplectic self-conjugate cobordism analog. Except for low dimensions the 2-primary structure is unknown. (Smith and Stong [108]).

Remark: This provides a synthesis of symplectic and Clifford algebra (2,0)-type cobordism.

Example 16: Exotic theories associated with classical groups.

Objects: (B,f) manifolds with B formed as follows: Let G and H be topological groups, and $\theta : G \longrightarrow H$, $\rho : G \longrightarrow O$ representations (O being the orthogonal group). Let H/G denote the generalized homogeneous space which is the fiber of $H/G \xrightarrow{i} BG \xrightarrow{B\theta} BH$. Then $(B,f) = (H/G,\pi)$ where π is the composite $H/G \xrightarrow{i} BG \xrightarrow{B\rho} BO$.

Calculation: If ρ and θ are inclusions of classical groups, this reduces to the framed bordism of the space H/G. The case in which θ is given by complexification is studied in Smith and Stong [109]. The odd primary structure tends to be the framed bordism of H/G while the 2-primary structure is a direct summand of Ω_*^G.

Remarks: Many standard cases may be expressed in this form, as for example SO cobordism. When $G = H = U$ is the unitary group and θ is complexification H/G is the second loop space of BSC, so that one obtains theories related to self-conjugate cobordism.

Example 17: k-connective and k-parallelizeable cobordism.

Objects: k-connected and oriented manifolds.

Determination: When connectivity is large with respect to dimension these are groups of homotopy spheres (see Kervaire-Milnor [61]). In the other cases they coincide with k-parallelizeable cobordism, given by (B,f) cobordism in which B_r is the k-connective cover of BO_r. Relative groups and images of one group in the others are particularly intriguing.

Remarks: Next to nothing is know, and the problem is hard (the case $k = 2$ is Spin cobordism which isn't easy). Images in unoriented cobordism are zero in low dimensions (but higher than might be expected) (Stong [115, 116]). The complex analog is studied in Lashof [71].

Example 18: Wu class cobordism: $\Omega_*\langle v_k \rangle$.

Objects: Manifolds with a 'reduction' killing the Wu class v_k.

Determination: (B,f) cobordism with B_r the total space of the fibration over BO_r induced from the path fibration over $K(Z_2,k)$ by the map realizing the Wu class v_k.

Remarks: Defined and used by W. Browder [25] in work on the Arf-Kervaire invariant. Similar questions of killing classes are studied in Lashof [71] and Peterson [98].

All of the above examples are basically given by manifolds, and there has been no significant problem involved in determination of the theory. The next group of examples are not in this pattern.

Example 19: Cobordism of pairs: $\Omega_{n,k}(B,f;G_{n-k})$.

Objects: A (B,f) manifold M^n with a submanifold V^k of M whose normal bundle in M is reduced to the group G_{n-k}.

Determination: Studied by Wall [131], the problem requires only a (B,f) cobordism of M and a $(B \times BG_{n-k}, f \times \pi)$ cobordism of V (separately). This may be phrased as (B,f) bordism of the space TBG_{n-k}. If G_{n-k} is the identity group 1_{n-k}, this is the cobordism of the category $\mathrm{Fun}(\mathfrak{m},(B,f))$ where $\mathfrak{m}$ has two objects D and R and with $\mathrm{Map}(D,D) = \{1_D\}$, $\mathrm{Map}(R,R) = \{1_R\}$, $\mathrm{Map}(D,R) = \{x\}$, and $\mathrm{Map}(R,D) = \emptyset$; i.e. the category of maps in the category (B,f) (recall that maps are imbeddings with trivialized normal bundle).

Example 20: Cobordism of immersions: $\mathfrak{N}_*(k)$

Objects: Manifolds together with an immersion in Euclidean space having codimension k.

Determination: Studied by Wells [137]. Using Hirsch's work [53] on immersions this reduces to the stable homotopy of Thom spaces of finite classifying spaces; i.e. $\mathfrak{N}_n(k) \cong \pi^S_{n+k}(TBO_k)$. This is (B,f) cobordism in which $B_r = BO_k$ for all $r \geq k$.

Calculation: Results are available in low dimensions; i.e. near $n = k$. The case $k = 1$ is the stable homotopy of projective space, which has been studied by Liulevicius [74].

Example 21: Cobordism of maps: $\mathfrak{N}(m,n)$.

Objects: Maps of m-dimensional manifolds into n-dimensional manifolds.

Determination: Cobordism of $\mathrm{Fun}(\mathfrak{m},\mathfrak{R})$ with $\mathfrak{m}$ as in example 19. This reduces to the bordism group $\lim_{r\to\infty} \mathfrak{N}_n(\Omega^{r+m}TBO_{r+n})$, Ω being loop space. This is computable and cobordism is determined by Z_2 cohomology characteristic numbers easily obtained from the map itself. (Stong [119]).

Remarks: One may impose additional structure on the manifolds. Interesting variants are self-maps ($\mathcal{O}\!\mathcal{C}$ has one object X with $\mathrm{Map}(X,X) \cong Z^+$) and diffeomorphisms ($\mathcal{O}\!\mathcal{C}$ has one object X with $\mathrm{Map}(X,X) \cong Z$). The latter is cobordism of fibrations over S^1 (take the mapping torus), which has been studied by Conner and Floyd [38], Burdick [31], Browder and Levine [26], and Farrell [49]. (These all take a slightly different point of view; nothing is known about these cobordism problems).

Example 22: Cobordism with group action: $\Omega_*(\mathrm{Fun}(\mathcal{J}, \mathcal{B})), \Omega_*(G)$.

Objects: Manifolds on which one has a differentiable action of the group G (finite or compact Lie group).

Determination: As previously noted, if $\mathcal{J}$ has one object X with $\mathrm{Map}(X,X) \cong G$ a finite group, then $\mathrm{Fun}(\mathcal{J}, \mathcal{B})$ gives the cobordism category of unrestricted G actions. If G is assumed to act freely on a manifold M ($gx = x$ implies $g = 1$) one has a principal differentiable G-bundle $G \longrightarrow M \longrightarrow M/G$ which is classified by a map $M/G \longrightarrow BG$. Conversely, if $N \longrightarrow BG$ is any map of a manifold into the classifying space, one has induced a principal differentiable G-bundle over N, G acting freely on the total space. The cobordism of free G actions, $\Omega_*(G)$, is then identified with the bordism groups of BG. Variants may be found by restricting the isotropy groups ($G_x = \{g \in G | gx = x\}$) to lie in some family of subgroups.

Remarks: The groups $\Omega_*(G)$ may be handled by bordism methods. Less restrictive group actions are usually treated by means of exact sequences relating theories. The standard method involves the cobordism analysis of the fixed point sets and their normal bundles.

The primary workers in this area are Conner and Floyd [35, 36, 37, 40] (see also Conner [33]), who initiated this method of studying group actions. Other work may be found in Anderson [10], Boardman [22], Hoo [57], Landweber [69, 70], Stong [120, 122], and Su [124].

In all of the preceding examples the manifolds used have been differentiable. Many of the easy ideas carry over at once to non-differentiable manifolds, but there are technical problems to be overcome.

Example 23: Piecewise linear cobordism: Ω_*^{PL}, Ω_*^{SPL}.

Objects: Piecewise linear manifolds.

Remarks: Every differentiable manifold is triangulable (J. H. C. Whitehead [139], see also Munkres [91]), but a given PL manifold may have distinct differentiable structures (Milnor [77]) or no differentiable structure (Kervaire [60]). This leads one to consider cobordism of PL manifolds.

Determination: The notion of vector bundle is replaced by microbundles (Milnor [86]), giving a Pontrjagin-Thom construction analogous to the differentiable case (Williamson [140]). Thus unoriented and oriented PL cobordism groups are the stable homotopy groups of Thom spectra $\underset{\sim}{TBPL}$ and $\underset{\sim}{TBSPL}$.

Calculation: Explicit computation of cobordism groups have been made by Wall [133] (oriented and unoriented in dimensions ≤ 8) and Williamson [140] (oriented in dimensions ≤ 18, ignoring 2-primary difficulties above dimension 9). Browder, Liulevicius, and Peterson [27] have shown that $\Omega_*^{PL} \cong \mathfrak{N}_* \otimes C$, C being the dual of a Hopf algebra factor of the Z_2 cohomology of BPL. The groups $\Omega_*^{SPL} \otimes Q$ form a polynomial ring on $4i$ dimensional generators and Ω_*^{SPL}/Torsion is conjectured to be a polynomial ring over Z (true in dimensions ≤ 12).

<u>Characteristic numbers</u>: Ω_*^{PL} is detected by its Z_2 cohomology characteristic numbers (Browder, Liulevicius, and Peterson [27]). Stiefel-Whitney classes were defined in the combinatorial case by Wu and Thom [126], but do not give all Z_2 characteristic classes. Adams [2] showed that there are no new relations among Stiefel-Whitney numbers. Rational characteristic classes (called Pontrjagin classes) were defined in the combinatorial case by Thom [128] and Rohlin and Švarč [106]. Recent work has been done on the integral (non-torsion) cohomology of BSPL by Brumfiel and Sullivan (unpublished).

<u>Example 24</u>: <u>Topological cobordism</u>: Ω_*^{Top}, Ω_*^{STop}.

<u>Objects</u>: Topological manifolds.

<u>Remarks</u>: Of great interest, but practically nothing is known. This is primarily due to lack of transversality, hence of a Pontrjagin-Thom construction. What is known follows from the existence of classifying spaces BTop and BSTop for topological microbundles, giving a homomorphism $\Omega_*^{Top} \longrightarrow \pi_*^S(TBTop)$, hence characteristic numbers. The Z_2 characteristic classes are known to include Stiefel-Whitney classes (Thom [126]) with no new relations among their numbers (Adams [2]) so that $\mathfrak{N}_*$ is a direct summand of Ω_*^{Top}. Rational characteristic classes exist to map onto Pontrjagin classes (Novikov [95]), so that $\Omega_*^{SPL} \otimes Q \cong \Omega_*^{SO} \otimes Q$ is a direct summand of $\Omega_*^{STop} \otimes Q$.

<u>Example 25</u>: <u>Cobordism of Poincaré duality spaces</u>: Ω_*^F, Ω_*^{SF}.

<u>Objects</u>: Finite CW pairs satisfying Poincaré-Lefschetz duality.

<u>Remarks</u>: Initiated by Wall's question (see Novikov [94] page 152) at the Seattle conference, 1963. One has normal spherical fibrations (Spivak [111]) and hence maps into a classifying space BF, BSF (Stasheff [112]),

but the map from cobordism to the homotopy of the Thom spectrum is not an isomorphism (in the oriented case the index is an invariant of infinite order, but the homotopy groups are finite). The cohomology of BF has been studied by Milnor [89] and Gitler and Stasheff [50]. Examples are known of Poincaré duality spaces not having the homotopy type of manifolds (Gitler and Stasheff [50]).

Example 26: Cobordism of manifolds with singularity: Ω_*^C.

Objects: Manifolds with boundary and a decomposition of the boundary in the form $A \cup (B \times C)$, A and B being manifolds with boundary and C being a closed manifold $(\partial A \cong \partial(B \times C))$, the boundary of this object being A with boundary decomposition given by $\partial B \times C$.

Remarks: Introduced by Sullivan in studying the Hauptvermutung [125], and called 'introduction of a singularity of type C'. Successions of this operation may be performed (interpreting the term manifold above as 'object'). The main result is an exact sequence relating the theories before and after adding the singularity. Of particular interest is the case when C has n points $(C = Z_n)$ on oriented manifolds, when this becomes the usual cobordism with Z_n coefficients (as homology theory).

Finally, there is an example of a cobordism category involving no spaces, which is of interest in that one need not think of cobordism as a manifold theoretic phenomenon.

Example 27: Cobordism of algebras with duality: $\mathcal{M}_*(\text{alg})$.

Objects: Let $\mathcal{C}$ be the category whose objects are 7-tuples $(H,H',H'',i,j,\delta,\mu)$ in which: H and H' are finite dimensional (as Z_2 vector spaces) graded unstable left algebras (commutative with unit) over the Steenrod algebra $\mathcal{A}_2$, H'' is a graded unstable left $\mathcal{A}_2$ module (finite dimensional over Z_2) and an H' module such that $Sq^i(h'h'') = \sum_{j+k=i} Sq^j(h')Sq^k(h'')$ if $h' \in H'$, $h'' \in H''$; i, j, δ are $\mathcal{A}_2$ module homomorphisms of degree $0,0,1$ such that

$$\begin{array}{ccc} H' & \xrightarrow{\ i\ } & H \\ & {\scriptstyle j}\nwarrow \qquad \swarrow{\scriptstyle \delta} & \\ & H'' & \end{array}$$

is exact, i being an algebra homomorphism, j an H' module homomorphism, and $\delta(hi(h')) = \delta h \circ h'$, $h \in H$, $h' \in H'$; and $\mu : H''^k \longrightarrow Z_2$ (k is called the dimension) is a vector space homomorphism so that the pairings $H' \otimes H'' \longrightarrow H'' \xrightarrow{\mu} Z_2$ and $H \otimes H \longrightarrow H \xrightarrow{\mu\circ\delta} Z_2$ are non-singular.

A map $f : (H, H', H'') \longrightarrow (H_0, H_0', H_0'')$ (ignoring the maps) is a triple (f, f', f'') of homomorphisms $f : H_0 \longrightarrow H$, $f' : H_0' \longrightarrow H'$, $f'' : H_0'' \longrightarrow H''$ (with all algebraic structures preserved) such that the maps of the diagrams commute.

The boundary of the septtuple $(H, H', H'', i, j, \delta, \mu)$ is $(0, H, H, 0, 1, 0, \mu\circ\delta)$ and its inclusion is $(0, i, 0)$.

<u>Determination</u>: This cobordism category was studied by Brown and Peterson [28]. It is analogous to the cohomology of a pair consisting of a manifold and its boundary. In fact, the cohomology functor $H^* : (\mathcal{O}, \partial, i) \longrightarrow (\mathcal{C}, \partial, i)$ is a good cobordism functor (Note: This is covariant since maps in $\mathcal{C}$ are reversed from the usual direction). Following Adams [2] one has a classifying algebra for the characteristic classes defined by the Steenrod algebra (isomorphic to the Z_2 cohomology of BO). Brown and Peterson have shown that H^* induces isomorphisms of the cobordism groups.

Although not properly cobordism theories in the cobordism category sense, there are similar equivalence relations obtained by defining two manifolds to be equivalent if they bound (jointly) some manifold with additional structure. Two examples of this are:

Pseudo-example 1: h-cobordism

Two compact manifolds V and V' are h-cobordant if there is a compact manifold W with $\partial W = V \cup V'$ such that both V and V' are deformation retracts of W. See Milnor [88] for details of this theory.

Pseudo-example 2: Cobordism with vector fields

Two (oriented) closed manifolds V and V' are equivalent if there is a compact (oriented) manifold W with $\partial W = V \cup (-V')$ and a non-singular tangent vector field on W which is interior normal along V and exterior normal along V'. This was studied by Reinhart [102] who introduced this cobordism in order to make the Euler class a 'cobordism' invariant. In the unoriented case, two manifolds are 'cobordant' if and only if they have the same Stiefel-Whitney numbers and Euler characteristic. In the oriented case, two manifolds V^n and V'^n are 'cobordant' if and only if they have the same Stiefel-Whitney numbers and Pontrjagin numbers and

a) $(n \neq 4k+1)$ the same Euler characteristic

b) $(n = 4k+1)$ the manifold W with $\partial W = V \cup (-V')$ has even Euler characteristic. [This only depends on V and V', not on the choice of W.]

It should be noted that in both of these examples the additional structure on the manifolds with boundary is not inherited by the boundary.

Chapter V

Cohomology of Classifying Spaces

In order to study the interesting examples of cobordism theories it is essential to have a detailed knowledge of the cohomology of the classifying spaces for the classical Lie groups.

Vector Bundles

Let K be one of the fields $\mathbb{R}$ (real numbers), $\mathbb{C}$ (complex numbers), or $\mathbb{H}$ (quaternions). Let k be the dimension of K as vector space over the reals.

Definition: A K vector bundle ξ is a 5-tuple $(B,E,p,+,\cdot)$ in which B and E are topological spaces, $p : E \longrightarrow B$ is a continuous function and

$$+ : E \oplus E = \{(e,e') \in E \times E \mid pe = pe'\} \longrightarrow E$$

$$\cdot : K \times E \longrightarrow E$$

are continuous functions such that $p\circ+(e,e') = p(e) = p(e')$ and $p\circ\cdot(k,e) = p(e)$ such that for each $b \in B$, the operations induced by $+$ and $\cdot$ on $p^{-1}(b)$ make $p^{-1}(b)$ into a vector space over K.

B is called the base space of ξ, E the total space of ξ, and p the projection of ξ. For $b \in B$, $p^{-1}(b)$ is the fiber of ξ over b.

Definition: A section of the bundle ξ is a continuous map $s : B \longrightarrow E$ such that $ps = 1$.

Definition: A vector bundle ξ is locally trivial (of dimension n) if for each point $b \in B$ there is an open set U of B containing b and sections $s_1,\ldots,s_n$ of ξ such that the map $\Psi_U : K^n \times U \longrightarrow p^{-1}(U)$ given by $\Psi_U((k_1,\dots,k_n),x) = \sum_{i=1}^{n} k_i s_i(x)$ is a homeomorphism.

Proposition: Let ξ be a locally trivial vector bundle over a compact Hausdorff space B. There exists a finite dimensional vector space V over K and a surjective bundle map $e : V \times B \longrightarrow E$. There is also an injective bundle map $i : E \longrightarrow V \times B$ with $ei = 1$.

Proof: Let $\Gamma = \Gamma(E)$ be the vector space of sections of ξ and $e : \Gamma \times B \longrightarrow E : (s,b) \longrightarrow s(b)$. Since ξ is locally trivial this is surjective and for each point b of B there is an open neighborhood U_b of b and a finite dimensional subspace V_b of Γ (spanned by n sections) so that $e : V_b \times U_b \longrightarrow p^{-1}(U_b)$ is surjective. Since B is compact, a finite number of the U_b cover B, and let V be the span of the corresponding V_b. Then V is finite dimensional and $e : V \times B \longrightarrow E$ is surjective.

Let V be given an inner product (over K) and let $E^{\perp}$ be the orthogonal complement of the kernel of e. Then $e|_{E^{\perp}} : E^{\perp} \longrightarrow E$ is an isomorphism and one may let $i = (e|_{E^{\perp}})^{-1} : E \longrightarrow V \times B$. **

Corollary: A locally trivial vector bundle over a compact Hausdorff space admits a Riemannian metric.

Corollary: A locally trivial vector bundle over a compact Hausdorff space has an inverse.

Remark: To each point b of B one may assign the subspace $i(E_b) \subset V$ which defines a map $B \longrightarrow G_n(V)$ of B into the Grassmannian of n planes of V. The usual n-plane bundle over $G_n(V)$ then induces the bundle ξ over B, so this is a classifying map for ξ.

Let $\xi = (B,E,p,+,\cdot)$ be a locally trivial n-dimensional K vector bundle with B a finite CW complex and let

$P(\xi)$ = space of one dimensional subspaces of the fibers of E,

$\ell(\xi)$ = space consisting of pairs (σ, e) where σ is a one dimensional subspace of a fiber and e is a point in that subspace.

The map $\pi : P(\xi) \longrightarrow B$ sending the one dimensional subspaces of $p^{-1}(b)$ into b is the projection of a locally trivial bundle with fiber $P(n-1)$, the '(n-1)-dimensional' projective space over K. The map $\lambda : \ell(\xi) \longrightarrow P(\xi)$ sending (σ, e) into σ is a one dimensional K vector bundle over $P(\xi)$, which restricts to precisely the canonical line bundle over each fiber $P(n-1)$ of ξ.

One then has commutative diagrams

$$\begin{array}{ccccc} E_b & \hookrightarrow & E & \longrightarrow & V \times B \\ \downarrow & & \downarrow{\scriptstyle p} & & \downarrow{\scriptstyle \pi_B} \\ b & \hookrightarrow & B & \xrightarrow{1} & B \end{array}$$

and

$$\begin{array}{ccccccc} P(n-1) = P(E_b) & \longrightarrow & \overbrace{P(E) & \longrightarrow & P(V \times B) \cong P(V) \times B & \xrightarrow{\pi_{P(V)}} & P(V)}^{j} \\ \downarrow & & \downarrow{\scriptstyle \pi} & & \downarrow{\scriptstyle \pi_B} & & \\ b & \longrightarrow & B & \xrightarrow{1} & B & & \end{array}$$

where $b \in B$. The bundle $\ell(\xi)$ is the restriction to $P(E)$ of the line bundle over $P(V \times B)$, which is induced from the canonical bundle over $P(V)$, and hence $j : P(E) \longrightarrow P(V)$ classifies $l(\xi)$. The induced map $P(E_b) \longrightarrow P(V)$ is just the inclusion induced by the inclusion $i : E_b \longrightarrow V$.

Definition of Characteristic Classes

Suppose: $\underset{\sim}{A} = \{A_i, a_i\}$ is a ring spectrum such that for each finite dimensional vector space V over K there is a class $\alpha_V \in H^k(P(V); \underset{\sim}{A})$ so that $H^*(P(V); \underset{\sim}{A})$ is the free $H^*(pt; \underset{\sim}{A})$ module on $1, \alpha_V, \ldots, \alpha_V^{\dim V - 1}$

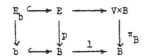

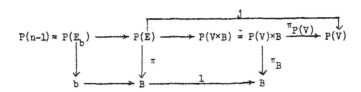

$(\alpha_V^{\dim V} = 0)$ and such that:

1) If $i : T \longrightarrow V$ is the inclusion of a subspace, then $i^*(\alpha_V) = \alpha_T$, and

2) If $q : P(K^{n+1}) \longrightarrow P(K^{n+1})/P(K^n) = S^{kn}$ is the collapse, then the image under q^* of $(-1)^n i \in \tilde{H}^{kn}(S^{kn};\underset{\sim}{A})$ is α^n.

<u>Note</u>: This choice of sign is made to try to get the 'usual' sign conventions. Unfortunately, signs vary wildly in the literature.

<u>Note</u>: If $K = \mathbb{R}$, one has a cofibration $S^1 \longrightarrow P(\mathbb{R}^2) \longrightarrow S^2$ and in the cohomology exact sequence the induced map on $H^*(pt;\underset{\sim}{A}) = \tilde{H}^*(S^n;\underset{\sim}{A})$ is multiplication by 2. In order that such classes α exist it is necessary that the unit class in $H^*(pt;\underset{\sim}{A})$ have order 2; hence $H^*(X;\underset{\sim}{A})$ has every element of order 2 for all X. (See Araki and Toda [11]).

<u>Examples</u>: If $K = \mathbb{R}$, $\underset{\sim}{A} = \underset{\sim}{K}(Z_2)$; if $K = \mathbb{C}$, $\underset{\sim}{A} = \underset{\sim}{K}(Z)$ or $\{BU,\Omega BU\}$-the spectrum for complex K-theory; if $K = \mathbb{H}$, $\underset{\sim}{A} = \underset{\sim}{K}(Z)$ or the spectrum for real K-theory.

<u>Theorem</u>: Let $c \in H^k(P(E);\underset{\sim}{A})$ be the class $j^*(\alpha_V)$. Then $H^*(P(E);\underset{\sim}{A})$ is the free $H^*(B;\underset{\sim}{A})$ module (via π^*) on the classes $1,c,\ldots,c^{n-1}$ and there exist unique classes $\sigma_i(\xi) \in H^{ki}(B;\underset{\sim}{A})$, $\sigma_0(\xi) = 1$, such that

$$c^n - c^{n-1}\pi^*(\sigma_1(\xi)) +\ldots+ (-1)^{n-1}c\pi^*(\sigma_{n-1}(\xi)) + (-1)^n\pi^*(\sigma_n(\xi)) = 0.$$

<u>Proof</u>: One has two spectral sequences E_r and E'_r with $E_2 = H^*(B;H^*(pt;\underset{\sim}{A}))$ converging to $H^*(B;\underset{\sim}{A})$ and $E'_2 = H^*(B,H^*(P(E_b);\underset{\sim}{A}))$ converging to $H^*(P(E);\underset{\sim}{A})$. The direct sum of n copies of E_* is mapped into E'_* by sending $(x_i) \longrightarrow \sum_{i=0}^{n-1} c^i\pi^*(x_i)$. Since $1,c,\ldots,c^{n-1}$ restrict to a base for $H^*(P(E_b);\underset{\sim}{A})$ as $H^*(b;\underset{\sim}{A})$ module, this is an isomorphism at the E_2 level, so that $\bigoplus_{i=0}^{n-1} H^*(B;\underset{\sim}{A}) \longrightarrow H^*(P(E);\underset{\sim}{A})$ is an isomorphism.

The relation exists since c^n is uniquely expressible in terms of the base $1, c, \ldots, c^{n-1}$. **

<u>Remarks</u>: 1) The structure of $H^*(P(E);\underset{\sim}{A})$ as $H^*(B;\underset{\sim}{A})$ algebra is completely determined by the relation for c^n.

2) The class $\sigma(\xi) = 1 + \sigma_1(\xi) + \ldots + \sigma_n(\xi) \in H^*(B;\underset{\sim}{A})$ is the total 'characteristic' class of ξ. These classes are given the following names:

a) $K = \mathbb{R}$: Stiefel-Whitney class $w(\xi)$,

b) $K = \mathbb{C}$: Chern class $c(\xi)$, and

c) $K = \mathbb{H}$: (Symplectic) Pontrjagin class $\mathfrak{P}^s(\xi)$.

<u>Theorem</u>: The total characteristic class $\sigma(\xi) \in H^*(B;\underset{\sim}{A})$ has the following properties:

1) $\sigma_i(\xi) = 0$ if $i > \dim\xi$; $\sigma_0(\xi) = 1$;

2) $\sigma(\xi)$ is natural; i.e. if $f : B' \longrightarrow B$ is a map, then $\sigma(f^*\xi) = f^*\sigma(\xi)$;

3) (Whitney sum formula) If ξ and η are two vector bundles over B, then $\sigma(\xi \oplus \eta) = \sigma(\xi) \cup \sigma(\eta)$; and

4) If ℓ is the canonical line bundle over $P(V)$, then $\sigma_1(\ell) = \alpha_V$.

<u>Proof</u>: 1) is immediate from the definition. For 2), one has the commutative diagram

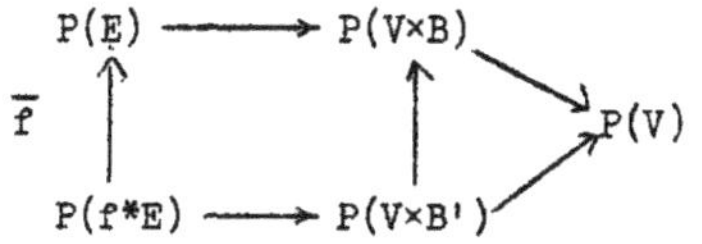

so that $c' = \bar{f}^*(c)$ and hence $0 = \sum_{i=0}^{n} (-1)^i c'^{n-i} \bar{f}^*\pi^*(\sigma_i(\xi)) = \sum_{i=0}^{n} (-1)^i c'^{n-i} \pi'^*(f^*\sigma_i(\xi))$ which by uniqueness gives $\sigma_i(f^*\xi) = f^*\sigma_i(\xi)$.

To prove 3), recall that $E(\xi\oplus\eta)$ is $\{(x,y) \in E(\xi) \times E(\eta) | p_\xi(x) = p_\eta(y)\}$ so that one has subspaces $E(\xi) \times 0$ and $0 \times E(\eta)$ of $E(\xi\oplus\eta)$ and hence may consider $P(\xi)$ and $P(\eta)$ as subspaces of $P(\xi\oplus\eta)$ with $\ell(\xi\oplus\eta)$ restricting to $\ell(\xi)$ on $P(\xi)$ and to $1(\eta)$ on $P(\eta)$. Let $U = P(\xi\oplus\eta) - P(\xi)$, $V = P(\xi\oplus\eta) - P(\eta)$. Then $P(\eta)$ is a deformation retract of U and $P(\xi)$ is a deformation retract of V, while $P(\xi\oplus\eta) = U \cup V$. Consider the class in $H^*(P(\xi\oplus\eta);\underset{\sim}{A})$ given by:

$$\theta = \sum_{j=0}^{m+n} (-1)^j c^{m+n-j} \sum_{r+s=j} \pi^*(\sigma_r(\xi) \cup \sigma_s(\eta)) =$$

$$\{\sum_{j=0}^{n} (-1)^j c^{n-j}\pi^*(\sigma_j(\xi))\} \{\sum_{k=0}^{m} (-1)^k c^{m-k}\pi^*(\sigma_k(\eta)\}.$$

[Note: If $K = R$ signs don't matter, while for $K \neq R$, k is even so that c^p and $\pi^*(\sigma_q(\xi))$ commute]. The factor $\theta_1 = \sum_{j=0}^{n} (-1)^j c^{n-j}\pi^*(\sigma_j(\xi))$ maps to zero in $H^*(P(\xi);\underset{\sim}{A})$, hence also in $H^*(V;\underset{\sim}{A})$ so θ_1 comes from $H^*(P(\xi\oplus\eta),V;\underset{\sim}{A})$. Similarly, $\theta_2 = \sum_{k=0}^{m} (-1)^k c^{m-k}\pi^*(\sigma_k(\eta))$ comes from $H^*(P(\xi\oplus\eta),U;\underset{\sim}{A})$ and hence $\theta = \theta_1\theta_2$ comes from $H^*(P(\xi\oplus\eta),U \cup V;\underset{\sim}{A}) = H^*(P(\xi\oplus\eta),P(\xi\oplus\eta);\underset{\sim}{A}) = 0$. Since $\theta = 0$, it is the relation for $H^*(P(\xi\oplus\eta);\underset{\sim}{A})$ and $\sigma_j(\xi\oplus\eta) = \sum_{r+s=j} \sigma_r(\xi) \cup \sigma_s(\eta)$. To prove 4), let $\xi = (B,E,p,+,\cdot)$ be any line bundle. Then since E_b is one dimensional there is only one one-dimensional subspace in each fiber and $\pi : P(E) \longrightarrow B$ is a homeomorphism. Further $1(E)$ is identified with E. The relation of $H^*(P(E);\underset{\sim}{A})$ is then $c - \pi^*(\sigma_1(\xi)) = 0$, and with π interpreted as identification, $\sigma_1(\xi) = j^*(\alpha_V)$, where $j : B \longrightarrow P(V)$ classifies E. **

<u>Remark</u>: The non-standard part of this proof is part 3), this proof being taken from Conner and Floyd [34], page 437. (See also Conner and Floyd [41], page 47).

Proposition: If ξ is a trivial bundle then $\sigma_i(\xi) = 0$ for $i > 0$. Consequently, if ξ and η are stably equivalent then $\sigma(\xi) = \sigma(\eta)$.

Proof: Since every trivial bundle is induced by a map into a point, it suffices by naturality to prove $\sigma_i(\xi) = 0$, $i > 0$, for the trivial bundle $p : K^n \longrightarrow pt$. For this one has $\alpha^n_{(K^n)} = 0$ in $H^*(P(K^n);\underset{\sim}{A})$ which is the usual relation and hence $\sigma_i(\xi) = 0$ if $i > 0$.

If ξ and η are stably equivalent, $\xi \oplus o^p \cong \eta \oplus o^q$ for some trivial bundles o^p and o^q, so $\sigma(\xi) = \sigma(\xi)\cdot 1 = \sigma(\xi \oplus o^p) = \sigma(\eta + o^q) = \sigma(\eta)\cdot 1 = \sigma(\eta)$. **

Proposition: (Splitting Lemma) Let $\xi = (B,E,p,+,\cdot)$ be an n-dimensional vector bundle. There is a space B' and a map $f : B' \longrightarrow B$ such that

1) $f^*\xi$ splits as a Whitney sum of line bundles, and

2) $f^* : H^*(B;\underset{\sim}{A}) \longrightarrow H^*(B';\underset{\sim}{A})$ is a monomorphism.

Proof: Let $B_0 = B$, $E_0 = E$ and suppose for $i \leq k$ one has defined spaces B_i, maps $f_i : B_i \longrightarrow B_{i-1}$ and bundles E_i over B_i so that $f_i^*(E_{i-1}) \cong E_i \oplus \ell_i$, ℓ_i a line bundle and $f_i^* : H^*(B_{i-1};\underset{\sim}{A}) \longrightarrow H^*(B_i;\underset{\sim}{A})$ is monic. Let $B_{k+1} = P(E_k)$, f_{k+1} the projection π. Then $f^*_{k+1}(E_k)$ has the Riemannian metric induced from E and $\ell(E_k)$ is a subbundle of $f^*_{k+1}(E_k)$, so that one may take E_{k+1} to be the orthogonal complement of $\ell(E_k)$. Since π^* is monic in cohomology this completes the induction. Then let $B' = B_{n-1}$, $f = f_1 \circ \ldots \circ f_{n-1}$, so that f^* is trivially monic on cohomology, while $f^*(E)$ splits as the sum of the line bundles E_{n-1}, $\ell(E_{n-2}) = \ell_{n-1}$, and $(f_i \circ \ldots \circ f_{n-1})^*(\ell_i)$ for $1 \leq i \leq n-2$. **

Remarks: 1) Taking the properties of $\sigma(\xi)$ from the principal theorem as axioms, the splitting lemma permits one to see that these characterize the characteristic classes σ_i. By knowledge of σ_1 for the canonical bundle, one knows $\sigma(\xi)$ for ξ any line bundle by naturality, but the Whitney sum formula and the splitting lemma show that $\sigma(\xi)$ is known for all ξ once it is known for line bundles.

2) If $\xi = \ell_1 \oplus \ldots \oplus \ell_n$ is a sum of line bundles, $\sigma(\xi) = \prod_{i=1}^{n} (1+\sigma_1(\ell_i))$ so $\sigma_j(\xi)$ is the j-th elementary symmetric function of the k-dimensional cohomology classes $\sigma_1(\ell_i)$. The splitting principle permits one to consider the class $\sigma_j(\xi)$ as the j-th elementary symmetric function of (formal) classes of dimension k when ξ is an arbitrary vector bundle.

Thom Spaces

Let $\xi = (B,E,p,+,\cdot)$ be given the Riemannian metric from V and denote by $D(\xi)$ the disc bundle $\{e \in E \mid \|e\| \leq 1\}$ and $S(\xi)$ the sphere bundle $\{e \in E \mid \|e\| = 1\}$. Let $\phi : D(\xi) \longrightarrow P(\xi \oplus 1)$ by sending e_x into the one dimensional subspace of $(E \times K)_x$ generated by $e_x - [1-\|e_x\|^2]^{1/2} 1_x$, where as before one thinks of ξ and 1 as subspaces of the total space of $\xi\oplus 1$. Then ϕ is a homeomorphism of $D(\xi) - S(\xi)$ with $P(\xi\oplus 1) - P(\xi)$ and maps $S(\xi)$ onto $P(\xi)$ by the usual identification map. Thus one may consider $P(\xi\oplus 1)/P(\xi)$ as the Thom space of ξ by means of ϕ.

Theorem: In cohomology with $\underset{\sim}{A}$ coefficients there is an exact sequence

$$0 \longleftarrow H^*(P(\xi);\underset{\sim}{A}) \xleftarrow{\alpha} H^*(P(\xi\oplus 1);\underset{\sim}{A}) \xleftarrow{\beta} \tilde{H}^*(T\xi;\underset{\sim}{A}) \longleftarrow 0$$

and the image of β is the ideal generated by the class

$$\tilde{U} = \sum_{j=0}^{n} (-1)^{n-j} c^{n-j} \pi^*(\sigma_j(\xi)) \in H^{kn}(P(\xi\oplus 1);\underset{\sim}{A}).$$

Proof: One has the exact sequence

$$\begin{array}{ccc} H^*(P(\xi);\underset{\sim}{A}) & \xleftarrow{\ \alpha\ } & H^*(P(\xi\oplus 1);\underset{\sim}{A}) \\ & \searrow \qquad \nearrow & \\ & H^*(P(\xi\oplus 1),P(\xi);\underset{\sim}{A}) & \\ & \cong & \\ & \tilde{H}^*(T\xi;\underset{\sim}{A}) & \end{array}$$

and since $\ell(\xi\oplus 1)$ restricts to $\ell(\xi)$ on $P(\xi)$, α is an epimorphism, giving the desired short exact sequence. The class $\tilde{U}$ clearly lies in the kernel of α (since it maps into the defining relation: Note: $c\tilde{U} = 0$ is the defining relation in $H^*(P(\xi\oplus 1);\underset{\sim}{A})$.), and clearly generates the kernel as $H^*(B;\underset{\sim}{A})$ module. **

Denote by U the class $\beta^{-1}(\tilde{U}) \in \tilde{H}^{kn}(T\xi;\underset{\sim}{A})$. The class U is clearly natural and its restriction to a fiber of ξ is the generator ι of $\tilde{H}^*(S^{kn};\underset{\sim}{A})$ as a module over $H^*(pt;\underset{\sim}{A})$. (Note: This restriction gives $\tilde{U} = (-1)^n\alpha^n$ in $\tilde{H}^{kn}(P(K^{n+1});\underset{\sim}{A})$ which pulls back to $\iota \in \tilde{H}^{kn}(S^{kn};\underset{\sim}{A})$.)

It should be noted that the class U is multiplicative, in the sense that $U(\xi\oplus\eta) = U(\xi) \cup U(\eta)$ where $T(\xi\oplus\eta)$ is identified with $T(\xi)\wedge T(\eta)$. This is immediate from the Whitney sum formula.

For computations involving the Thom spaces, it is convenient to have:

Proposition: a) $U \cup U = \pi^*(\sigma_n(\xi)) \cup U = \Phi^U(\sigma_n(\xi))$, where Φ^U is the Thom isomorphism defined by U.

b) If $i : B \longrightarrow T(\xi)$ is the map given by the zero section of ξ, then $i^*(U) = \sigma_n(\xi)$.

Proof: For a), one has $\tilde{U} \cup \tilde{U} = \Sigma\, (-1)^{n-j}c^{n-j}\pi^*(\sigma_j(\xi))\cdot\tilde{U}$, but $c\tilde{U} = 0$ so $\tilde{U} \cup \tilde{U} = \pi^*(\sigma_n(\xi)) \cup \tilde{U}$. For b) one has the commutative diagram

$$\begin{array}{ccccc} P(\xi\oplus 1) & \longrightarrow & P(V\times K)\times B & \longrightarrow & P(V\times K) \\ s\uparrow\downarrow\pi & & s\uparrow\downarrow\pi & & s\uparrow\downarrow\pi \\ B & \longrightarrow & B & \longrightarrow & pt \end{array}$$

from which $s^*j^*\alpha_{V\times K} = s^*c = 0$ for the section defined for pt may be thought of as the inclusion of $P(K)$ and $\alpha_K = \alpha_K^1 = 0$. Thus $i^*U = s^*\tilde{U} = s^*\pi^*(\sigma_n(\xi)) = \sigma_n(\xi)$. **

<u>Proposition</u>: Let V be a vector space over K and $\ell \longrightarrow P(V)$ the canonical line bundle. Then the Thom space of ℓ may be identified with $P(V\times K)$ so that the zero section inclusion is the standard map given by the inclusion of V in $V \times K$. Further, the Thom class U is identified with α.

<u>Proof</u>: Let $\pi : P(\ell\oplus 1) \longrightarrow P(V)$ be the projection. Then $\ell(\ell\oplus 1)$ is a subbundle of $\pi^*(\ell\oplus 1) = \pi^*(\ell) \oplus 1$. Denote the orthogonal complement of $\ell(\ell\oplus 1)$ by σ. If $p \in P(\ell\oplus 1)$ then p is a line in $V \times K$ which lies in the space $\pi(p) \times K$ and the orthogonal complement $p^\perp$ of p in $\pi(p) \times K$ is a line in $V \times K$. The correspondence $p \longrightarrow p^\perp$ defines a map $f : P(\ell\oplus 1) \longrightarrow P(V\times K)$ with σ induced from the canonical bundle. If p is in the image of $P(\ell)$ then $p^\perp$ is the line K and so $P(\ell)$ is mapped into the point $P(K)$. Thus f induces a map $\hat{f} : T(\ell) \longrightarrow P(V\times K)$. If $\mu \in P(V\times K) - P(K)$, let q be the point of $P(\ell\oplus 1)$ given by the line orthogonal to μ in $\mu + K$, and then $f(q) = \mu$. Thus $\hat{f}$ is a homeomorphism. If $x \in P(V)$ then $\phi(0_x)$ is the line generated by K in $x \times K$ so $\phi(0_x)^\perp = x$, so the zero section map is just the inclusion of $P(V)$ in $P(V\times K)$. Finally, $f^*(\alpha) = \sigma_1(\sigma)$ and $\sigma_1(\sigma) = \sigma_1(\pi^*(\ell)\oplus 1 - \ell(\ell\oplus 1)) = \pi^*(\alpha)-c = \tilde{U}$ so $\alpha = U$. **

Cohomology of Grassmann Manifolds

Let $G_{n,r}$ denote the Grassmann manifold of n planes in K^{n+r}. One has bundles γ_r^n (n-plane, point in it) and $\bar{\gamma}_r^n$ (n-plane, point of the orthogonal r-plane) over $G_{n,r}$ with $\gamma_r^n \oplus \bar{\gamma}_r^n$ trivial. One then has the cohomology classes $\sigma_i = \sigma_i(\gamma_r^n) \in H^{ki}(G_{n,r};\underset{\sim}{A})$ and $\bar{\sigma}_i = \sigma_i(\bar{\gamma}_r^n) \in H^{ki}(G_{n,r};\underset{\sim}{A})$, related by the equation $\sigma \cup \bar{\sigma} = 1$.

Proposition: $H^*(G_{n,r};\underset{\sim}{A})$ is the quotient of the polynomial algebra over $H^*(pt;\underset{\sim}{A})$ on σ_i, $i \leq n$, by the relations imposed by $\bar{\sigma}_j = 0$ for $j > r$. (Note: $\bar{\sigma}_j$ is the polynomial of degree j in the σ_i given by formal inversion of σ.)

Proof: The asserted polynomial algebra is certainly mapped into $H^*(G_{n,r};\underset{\sim}{A})$, and to prove an isomorphism one may induct on n. For $n = 1$, $G_{1,r} = P(K^{r+1})$ so that $H^*(G_{1,r};\underset{\sim}{A})$ is generated by $\alpha = \sigma_1$ with relation $\alpha^{r+1} = 0$, but $\bar{\sigma} = 1 - \alpha + \alpha^2 - \ldots + (-1)^r\alpha^r + (-1)^{r+1}\alpha^{r+1} + \ldots$ and all relations are given by $\bar{\sigma}_j = 0$ if $j > r$.

If the result holds for all $G_{n,r}$ with $n < s$, then consider $G_{s,t}$. A point in $P(\gamma_t^s)$ is a line a in an s-plane μ. The orthogonal complement of a in μ is an $s-1$ plane $a^{\perp}$, hence a point of $G_{s-1,t+1}$. The points of $P(\gamma_t^s)$ mapping into $\nu \in G_{s-1,t+1}$ are precisely those lines orthogonal to ν. Thus $P(\gamma_t^s)$ is exactly $P(\bar{\gamma}_{t+1}^{s-1})$ giving the diagram

$$\begin{array}{ccc} P(\gamma_t^s) = P(\bar{\gamma}_{t+1}^{s-1}) & \xrightarrow{\bar{\pi}} & G_{s-1,t+1} \\ \pi \downarrow & & \\ G_{s,t} & & \end{array}$$

Letting $\ell = \ell(\gamma_t^s) = \ell(\gamma_{t+1}^{s-1})$ one has $\pi^*(\gamma_t^s) = \xi \oplus \ell$, $\bar{\pi}^*(\bar{\gamma}_{t+1}^{s-1}) = \ell \oplus \eta$ with $\xi \oplus \ell \oplus \eta$ trivial. Now noting that $c = \sigma_1(\ell)$, the relation $\Sigma\, (-1)^i c^{s-i} \pi^*(\sigma_i(\gamma_t^s)) = 0$ is precisely $\sigma_s(\xi) = 0$. Looking at $P(\gamma_t^s)$ as a bundle over $G_{s-1,t+1}$ one has that $H^*(P(\gamma_t^s);\underset{\sim}{A})$ is generated by the characteristic classes of ξ, ℓ, and η subject only to the relations imposed by the dimensions of ξ, ℓ, and η and that their sum is trivial. Then looking at $P(\gamma_t^s)$ as a bundle over $G_{s,t}$ one sees that $H^*(G_{s,t};\underset{\sim}{A})$ must be generated by the characteristic classes of γ_t^s subject only to the relation imposed by the dimension of $\bar{\gamma}_t^s$. This completes the induction. **

Using $B_n = \lim_{r\to\infty} G_{n,r}$, $B = \lim_{n\to\infty} B_n$ and inverse limit cohomology (this being all that affects characteristic numbers) one has

<u>Proposition</u>: a) $H^*(B_n;\underset{\sim}{A})$ is the formal power series algebra over $H^*(pt;\underset{\sim}{A})$ generated by the universal characteristic classes σ_i, $1 \le i \le n$.

b) $H^*(B;\underset{\sim}{A})$ is the formal power series algebra over $H^*(pt;\underset{\sim}{A})$ generated by the universal characteristic classes σ_i, $1 \le i$.

The Whitney sum of vector bundles induces a map $B_n \times B_m \longrightarrow B_{n+m}$ or $B \times B \longrightarrow B$. Applying the Whitney sum formula for characteristic classes gives

<u>Proposition</u>: $H^*(B;\underset{\sim}{A})$ is a Hopf algebra over $H^*(pt;\underset{\sim}{A})$, which as algebra is formal power series on classes σ_i, $i \ge 1$, and has diagonal map given by $\Delta(\sigma) = \sigma \otimes \sigma$; i.e. $\Delta(\sigma_i) = \sum_{j+k=i} \sigma_j \otimes \sigma_k$.

<u>Note</u>: If one wishes to be thorough, one notes that the Künneth theorem $H^*(X\times Y;\underset{\sim}{A}) = H^*(X;A) \otimes_{H^*(pt;\underset{\sim}{A})} H^*(Y;\underset{\sim}{A})$ holds if $H^*(Y;\underset{\sim}{A})$ is a free $H^*(pt;\underset{\sim}{A})$ module. For a proof see Conner and Floyd [36], page 131. In particular, one should note that $H^*(B\times B;\underset{\sim}{A})$ is the <u>completed</u> tensor product $H^*(B;\underset{\sim}{A})\hat{\otimes} H^*(B;\underset{\sim}{A})$.

It is frequently convenient to use other characteristic classes, formed from the σ_i. For any set $\omega = (i_1,\ldots,i_r)$ of positive integers, called a partition of $n(\omega) = \sum_{\beta=1}^{r} i_\beta$, one defines the s_ω symmetric function of variables t_j, $1 \leq j \leq s$ to be the smallest symmetric function in the t_j which contains the monomial $t_1^{i_1}\ldots t_r^{i_r}$ $(s_\emptyset = 1)$. Then $s_\omega(t)$ is expressible uniquely as a polynomial with integral coefficients in the elementary symmetric functions $\theta_i = s_{\underbrace{(1,\ldots,1)}_{i}}(t)$ of the t's. If $s \geq n(\omega)$, this polynomial is independent of s, and write $s_\omega(t) = P_\omega(\theta_1,\ldots,\theta_{n(\omega)})$.

One defines classes $s_\omega(\sigma) \in H^{kn(\omega)}(B;\underset{\sim}{A})$ by $s_\omega(\sigma) = P_\omega(\sigma_1,\ldots,\sigma_{n(\omega)})$. Since by the splitting principle one may consider σ_i as the i-th elementary symmetric function of k dimensional classes $\sigma_1(\ell_j)$ (ℓ_j being line bundles), $s_\omega(\sigma)$ is represented as the s_ω symmetric function of these classes.

The usefulness of these classes follows from

<u>Proposition</u>: *In* $H^*(B;\underset{\sim}{A})$ *one has* $\Delta s_\omega(\sigma) = \sum_{\omega' \cup \omega''=\omega} s_{\omega'}(\sigma) \cup s_{\omega''}(\sigma)$, the sum being over all pairs of partitions ω' and ω'' for which $\omega = \omega' \cup_D \omega''$. In particular, for each integer i, $s_{(i)}(\sigma)$ is primitive. The dual $\mathrm{Hom}^{\mathrm{finite}}_{H^*(pt;\underset{\sim}{A})}(H^*(B;\underset{\sim}{A}),\ H^*(pt;\underset{\sim}{A}))$ is the polynomial algebra over $H^*(pt;\underset{\sim}{A})$ on classes x_i, $i \geq 1$, of degree $(-ki)$, where x_i is dual to $s_i(\sigma)$ with respect to the base consisting of the $s_\omega(\sigma)$ (i.e. $s_\omega(\sigma)[x_i] = 0$ if $\omega \neq (i)$ and $s_{(i)}(\sigma)[x_i] = 1 \in H^0(pt;\underset{\sim}{A})$.) $\mathrm{Hom}^{\mathrm{finite}}_{H^*(pt;\underset{\sim}{A})}(H^*(B;\underset{\sim}{A}),H^*(pt;\underset{\sim}{A}))$denotes homomorphisms vanishing on all but a finite number of monomials $\sigma_{i_1}\ldots\sigma_{i_r}$.)

<u>Proof</u>: If $\{z_j\}$ is the union of two collections of classes $\{u_i\}$ and $\{v_k\}$ then $\sum z_1^{i_1}\ldots z_r^{i_r}$ splits into symmetric functions in u's and v's, and this is given by $s_\omega(z) = \sum_{\omega' \cup \omega''=\omega} s_{\omega'}(u)\cdot s_{\omega''}(v)$. If bundles ξ and η split as sums of line bundles ℓ_i and m_k, then $\xi \oplus \eta$ splits into the

union of the two collections, or $s_\omega(\sigma(\xi\oplus\eta)) = \sum_{\omega'\cup\omega''=\omega} s_{\omega'}(\sigma(\xi))s_{\omega''}(\sigma(\eta))$ giving the diagonal formula. If as above x_i is dual to $s_i(\sigma)$, then by the diagonal formula one has $s_\omega(\sigma)[x_{i_1}\dots x_{i_r}] = \delta_{\omega,(i_1,\dots,i_r)}$, where $\delta_{\omega,\omega'} = 0$ if $\omega \neq \omega'$, 1 if $\omega = \omega'$. Thus the products of the x_i form a base over $H^*(pt;\underset{\sim}{A})$ for the dual of $H^*(B;\underset{\sim}{A})$. **

<u>Note</u>: $\mathrm{Hom}^{\text{finite}}_{H^*(pt;\underset{\sim}{A})}(H^*(B;\underset{\sim}{A}), H^*(pt;\underset{\sim}{A}))$ is clearly identifiable with the direct limit of the groups $\mathrm{Hom}_{H^*(pt;\underset{\sim}{A})}(H^*(G_{r,s};\underset{\sim}{A}); H^*(pt;\underset{\sim}{A}))$. It is clear that the characteristic number homomorphism defined by a manifold belongs to this set of homomorphisms.

<u>Remark</u>: There is another construction frequently used in determining $H^*(G_{r,s};\underset{\sim}{A})$ provided $\underset{\sim}{A}$ is a "good" theory (i.e. a theory for which one can compute the cohomology of sphere bundles). One considers the K^r bundle $E(\gamma^r_s) \longrightarrow G_{r,s}$ with $E_0(\gamma^r_s)$ the unit sphere bundle. To each point $x \in E_0(\gamma^r_s)$ one may associate the $r-1$ plane orthogonal to x in $\pi(x)$. This defines a projection $E_0(\gamma^r_s) \longrightarrow G_{r-1,s+1}$ and one may identify $E_0(\gamma^r_s)$ with $E_0(\gamma^{r-1}_{s+1})$. The bundle $\pi^*\gamma^r_s$ splits off a line bundle by means of the section $x \longrightarrow (x,x)$ over E_0 with the orthogonal complement of this section being identifiable to $\pi'^*(\gamma^{r-1}_{s+1})$.

Letting s become arbitrarily large one has

$$\begin{array}{ccccc} E_0(\gamma^r) & \longrightarrow & E(\gamma^r) & \longrightarrow & T(\gamma^r) . \\ {\scriptstyle \pi'}\big\downarrow & & {\scriptstyle \pi}\big\downarrow & & \\ B_{r-1} & & B_r & & \end{array}$$

The map π is a homotopy equivalence, with the zero section as inverse, while π' is a weak homotopy equivalence, the "infinite" sphere being contractible.

One then has an exact sequence

$$H^*(B_{r-1};\underset{\sim}{A}) \xleftarrow{\alpha} H^*(B_r;\underset{\sim}{A}) \xleftarrow{\beta} H^*(T\gamma^r;\underset{\sim}{A})$$

and since α corresponds to pulling γ^r back to $\gamma^{r-1} \oplus 1$, α is epic. Further β being identified with the zero section, $\beta U = \sigma_r(\gamma^r)$ identifying $H^*(T\gamma^r;\underset{\sim}{A})$ with the ideal generated by σ_r in $H^*(B_r;\underset{\sim}{A})$.

Relationship Between Fields

Let K and K' be two of the fields $\mathbb{R}, \mathbb{C}$, and $\mathbb{H}$, with $K \subset K'$ and let $\underset{\sim}{A}$ be a ring spectrum for which projective spaces over K have proper cohomology. Let r be the dimension of K' over $K(r=k'/k)$ and choose a base $1, x_1, \ldots, x_{r-1}$ for K' over K with $x_i^2 = -1$ (from among the standard $1,i,j,k$) so that $\phi_i : K \longrightarrow K$ defined by $x_i \cdot t = \phi_i(t) \cdot x_i$ is an automorphism of K $(\phi_i^2 = 1)$.

Let V be an n-dimensional vector space over K' - hence also a vector space over K of dimension rn. The assignment to a one dimensional K subspace p of V of the one dimensional K' subspace $K'p$ containing it defines a map $\pi : KP(V) \longrightarrow K'P(V)$. If q is a K' line of V, $\pi^{-1}(q)$ consists of all K lines in q, hence of all K-lines in the fiber of the canonical K' line bundle λ' of $K'P(V)$. Thus $KP(V)$ is identified with $KP(\lambda')$. In addition, the K-line bundle $\ell(\lambda')$ is trivially the canonical bundle λ of $KP(V)$.

Thus $H^*(KP(V);\underset{\sim}{A})$ is the free $H^*(K'P(V);\underset{\sim}{A})$ module (via π^*) on the classes $1, \alpha_V, \ldots, \alpha_V^{r-1}$ and has relation $\sum_{i=0}^{r} (-1)^i \alpha_V^{r-i} \pi^*(\sigma_i(\lambda')) = 0$.

The map $\theta_i : V \longrightarrow V : v \longrightarrow x_i v$ is $\mathbb{R}$ linear and semi-linear over K by means of the automorphism ϕ_i of K. In particular θ_i sends one dimensional K subspaces into one dimensional K subspaces and so defines a map $\theta_i : KP(V) \longrightarrow KP(V)$ $(\theta_i^2 = 1)$ which pulls the canonical bundle λ back to a bundle $\theta_i^*\lambda$. Pulling the bundle λ' back over KP(V) by π^* one has $\pi^*\lambda' = \sum_{i=0}^{r-1} \theta_i^*\lambda$; i.e. λ is a subbundle of $\pi^*\lambda'$ and the subsets $x_i\cdot\lambda$ decompose $\pi^*\lambda'$ into a Whitney sum over K.

Thus $\pi^*(\sigma_i(\lambda')) = \sigma_i(\pi^*\lambda')$ is the i-th elementary symmetric function of the variables $\sigma_1(\theta_j^*\lambda) = \theta_j^*\sigma_1(\lambda)$.

Case I: $K = \mathbb{R}$. Then K is central in K' so all ϕ_i are the identity and $\pi^*\lambda' = r\lambda$. In particular $\pi^*\sigma_r(\lambda') = \alpha_V^r$, all lower classes being zero since r = 2 or 4 and all elements of $\underset{\sim}{A}$ cohomology have order 2. Since $(\alpha_V^r)^n = \alpha_V^{rn} = 0$, $H^*(K'P(V);\underset{\sim}{A})$ contains the free $H^*(pt;\underset{\sim}{A})$ module on $1, \sigma_r(\lambda'),\ldots,\sigma_r(\lambda')^{n-1}$ with $\sigma_r(\lambda')^n = 0$, and since $H^*(KP(V);\underset{\sim}{A})$ is the free module on $1,\ldots,\alpha_V^{r-1}$ over this one has that $H^*(K'P(V);\underset{\sim}{A})$ is the free module on the powers of $\sigma_r(\lambda')$.

Under the map $KP(K^{r(n+1)}) \longrightarrow K'P(K'^{n+1})$ one has $KP(K^{rn+1})/KP(K^{rn}) = S^{rnk}$ sent onto $K'P(K'^{n+1})/K'P(K'^n) = S^{k'n}$ by a map of degree 1 and so $(-1)^{rn}\iota \in \tilde{H}^{nk'}(S^{nk'};\underset{\sim}{A})$ pulls back to $\sigma_r(\lambda')$ (since it pulls back to α_V^{rn}). Since r is even, ι maps to $\sigma_r(\lambda')$ - but $\iota = -\iota$ since every element has order 2.

Thus $H^*(K'P(V);\underset{\sim}{A})$ has proper cohomology, and with this theory the i-th K' characteristic class $\sigma_i^{K'}$ reduces to the ir-th characteristic class σ_{ri}^K.

Case II: $K = \mathbb{C}$, $K' = \mathbb{H}$. In this case one must consider the effect of action by $x_1 = j$, the automorphism ϕ_1 being complex conjugation. Under the map $\theta_1 : KP(V) \longrightarrow KP(V)$ the bundle λ pulls back to its complex conjugate bundle $\bar{\lambda}$, the problem being to compute $\theta_1^*(\alpha_V)$.

Since θ_1 is natural for inclusion of vector spaces, there is a power series $h(x) = \sum_{i=0}^{\infty} a_i x^i$, $a_i \in H^{2-2i}(pt;\underset{\sim}{A})$, such that $\theta_1^*(\alpha_V) = h(\alpha_V)$. (This is a finite sum in $H^*(KP(V);\underset{\sim}{A})$.)

Since $\alpha_V = 0$ if $\dim V = 1$, one must have $a_0 = 0$. If $\dim V = 2$, $CP(V)$ is just the two sphere S^2 and the map θ_1 is of degree -1, so $a_1 = -1$. In higher dimensions one can say nothing except under restrictions on $\underset{\sim}{A}$, since in particular one can make different choices for α_V. (It will be seen later that for complex K-theory the behaviour is not good).

If $a_i = 0$ for $i > 1$, then $\theta_1^*(\alpha_V) = -\alpha_V = \sigma_1(\bar{\lambda})$. Thus $\pi^*(\sigma_1(\lambda')) = 0$, $\pi^*(\sigma_2(\lambda')) = -\alpha_V^2$. Hence $H^*(\mathbb{H}P(V);\underset{\sim}{A})$ is the free $H^*(pt;\underset{\sim}{A})$ module on $1, \sigma_2(\lambda'),\ldots,\sigma_2(\lambda')^{n-1}$ with $\sigma_2(\lambda')^n = 0$. Further $(-1)^{2n}\iota \in \tilde{H}^{4n}(S^{4n};\underset{\sim}{A})$ maps to α^{2n} in $H^{4n}(\mathbb{C}P(\mathbb{C}^{2n+1});A)$, $(-1)^n\iota$ maps to $(-\alpha^2)^n$ and hence $(-1)^n\iota$ maps to $\sigma_2(\lambda')^n$ in $H^{4n}(\mathbb{H}P(\mathbb{H}^{n+1});\underset{\sim}{A})$. Thus $H^*(\mathbb{H}P(V);\underset{\sim}{A})$ has proper cohomology and with this theory, the i-th characteristic class $\sigma_i^{\mathbb{H}}$ reduces to $\sigma_{2i}^{\mathbb{C}}$.

There are two interesting cases for which $a_i = 0$ if $i > 1$. Trivially this holds when $H^j(pt;\underset{\sim}{A}) = 0$ for all $j < 0$. Another case in which the result is valid is when $1/2 \in H^0(pt;\underset{\sim}{A})$; for in this case one may take $\alpha_V' = 1/2(\alpha_V - h(\alpha_V))$, which is another acceptable generator, with $\theta_1^*(\alpha_V') = -\alpha_V'$.

Characteristic Numbers of Manifolds

Proposition: For $K = \mathbb{R}$ or $\mathbb{C}$, the tangent bundle of $KP(n) = KP(K^{n+1})$ is a K vector bundle satisfying $\tau \oplus 1 = (n+1)\xi$ where ξ is the canonical bundle if $K = \mathbb{R}$ and the complex conjugate of the canonical bundle if $K = \mathbb{C}$.

<u>Proof</u>: Let $< , >$ denote the usual K inner product on K^{n+1}, with $\mathrm{Re} < , >$ (its real part) being the usual R inner product. One may consider $KP(n)$ as the quotient of the sphere $S^{kn+k-1} = \{u \in K^{n+1} \mid |u| = 1\}$ under the action of $S^{k-1} = \{t \in K \mid |t| = 1\}$. The tangent bundle to S^{kn+k-1} has total space identifiable with $\{(u,v) \in K^{n+1} \times K^{n+1} \mid |u| = 1, \mathrm{Re} <u,v> = 0\}$ and the pullback of the tangent bundle of $KP(n)$ may be identified with those tangent vectors (u,v) orthogonal to the orbits of the action of S^{k-1}; hence those (u,v) with $< u,v > = 0$. This is a K-vector bundle by $s(u,v) = (u,sv)$.

Under the action of S^{k-1} the total space of the pull-back collapses to τ, and thus $E(\tau)$ may be represented as pairs $(u,v) \in K^{n+1} \times K^{n+1}$ with $|u| = 1$, $< u,v > = 0$, where (tu,tv) is identified with (u,v) if $t \in S^{k-1}$. This is compatible with the K vector bundle structure since K is commutative.

Let $c : P(V) \longrightarrow P(V)$ be the map induced by conjugation. Then the total space of $c^*(\ell) = \xi$ may be identified with the pairs $(x,s) \in K^{n+1} \times K$ with $|x| = 1$ and (x,s) identified with (tx,st) for $t \in S^{k-1}$ (<u>Note</u>: (x,s) represents the point of $c^*(\ell)$ given by the line through x and the point $s\bar{x}$ in the image of that line.). The total space of $(n+1)\xi$ may be thought of as pairs $(u,v) \in K^{n+1} \times K^{n+1}$ with $|u| = 1$ with (u,v) identified with (tu,tv) for $t \in S^{k-1}$ and with scalar multiplication given by $s(u,v) = (u,sv)$. Thus τ is a subbundle of $(n+1)\xi$, and may be considered as the fiberwise orthogonal complement of the set of all pairs (u,su), which is a trivial line bundle. Thus $(n+1)\xi = \tau \oplus 1$. **

Thus the normal bundle of $KP(n)$ admits a stable K-vector bundle structure, given by the 'negative' of the tangent bundle, giving a (B,f) structure to $KP(n)$. Since the $\underset{\sim}{A}$ cohomology of B is known, and since an orientation class U has been constructed, it should be possible to compute the characteristic numbers of $KP(n)$.

Let $KP(n) \subset S^{r+kn}$ be an imbedding with normal bundle ν having a K vector bundle structure, and let $[KP(n)] \in H_{kn}(KP(n);\underset{\sim}{A})$ be the fundamental homology class of $KP(n)$ defined by the orientation U of $T\nu$.

<u>Lemma</u>: $\alpha^n[KP(n)] = (-1)^n \in H^0(pt;\underset{\sim}{A})$.

<u>Proof</u>: One has the diagram

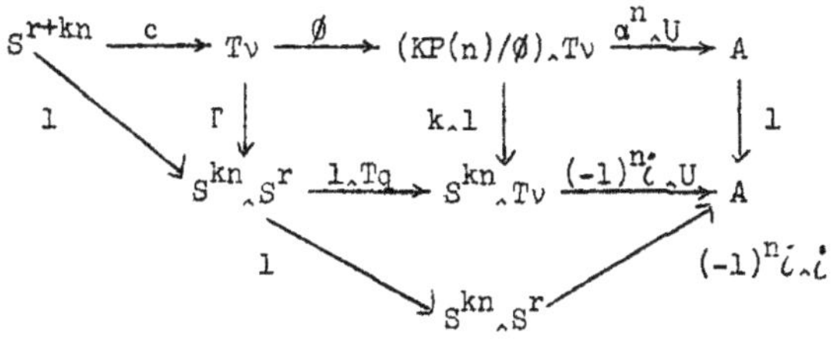

in which the maps given by cohomology class names properly exist after suspension, and after suspension the diagram commutes up to homotopy. Thus the map representing $\alpha^n[KP(n)]$ (the top line) gives the same class as $(-1)^n$ (the bottom line). **

<u>Proposition</u>: The tangential characteristic numbers of $KP(n)$ are given by

$$\sigma_\omega(\tau)[KP(n)] = \binom{n+1}{i_1}\cdots\binom{n+1}{i_r} \in H^0(pt;\underset{\sim}{A})$$

if $\omega = (i_1,\ldots,i_r)$ is a partition of n, where $\sigma_\omega = \sigma_{i_1}\cdots\sigma_{i_r}$. In addition

$$s_{(n)}(\sigma(\tau))[KP(n)] = n+1.$$

<u>Proof</u>: Letting $c : KP(n) \longrightarrow KP(n)$ be given by conjugation, one has $c^*(\alpha) = -\alpha + \sum_{i\geq 2} a_i\alpha^i$ and $\sigma(\tau) = (1+c^*\alpha)^{n+1}$. Thus $\sigma_\omega(\tau) = \binom{n+1}{i_1}\cdots\binom{n+1}{i_r}c^*(\alpha)^n$ and $s_{(n)}(\sigma(\tau)) = (n+1)c^*(\alpha)^n$. Since $\alpha^{n+1} = 0$, $c^*(\alpha)^n = (-\alpha)^n = (-1)^n\alpha^n$ and $\{c^*(\alpha)\}^n[KP(n)] = 1$. **

If $n(\omega) > n$, $\sigma_\omega(\tau)[KP(n)] = 0$ since $\{c^*(\alpha)\}^{n+1} = 0$. For $n(\omega) < n$, the difficulty lies in evaluation of α^k on $KP(n)$. Since α^k does not come from the map into the sphere, this evaluation is not feasible without additional assumptions.

Using projective spaces one may construct other manifolds for which the characteristic numbers are computable and which will be needed later.

Let $K = \mathbf{R}$ or $\mathbf{C}$ and let B be the classifying space for K vector bundles. Suppose M^n is a closed (B,f) manifold and ρ is a K line bundle over M with characteristic class $\sigma(\rho) = 1 + \theta$, $\theta \in H^k(M^n;\underset{\sim}{A})$.

Let $f : M \longrightarrow KP(N)$ for some N be a map for which $f^*(\xi) = \rho$. By deforming f if necessary one may assume f is transverse regular on $KP(N-1)$. Then $L = f^{-1}(KP(N-1)) \subset M$ is a closed submanifold of codimension k; the normal bundle of L in M being induced from that of $KP(N-1)$ in $KP(N)$ (i.e. $\xi|_{KP(N-1)}$), the normal bundle is $\rho|_L$.

The stable normal bundle of L, ν_L, admits a (B,f) structure identifying it with $i^*(\rho\oplus\nu_M)$, i being the inclusion of L in M. Thus $\sigma(\nu_L) = (1+i^*\theta)i^*(\sigma(\nu_M))$ or $\sigma(\tau_L) = i^*\sigma(\tau_M)/(1+i^*\theta)$.

Let $x \in H^*(L;\underset{\sim}{A})$ be given by $x = i^*y$, $y \in H^*(M;\underset{\sim}{A})$ (for example a characteristic class) and consider the number $x[L] \in H^*(pt;\underset{\sim}{A})$. Imbedding M and with K normal bundle in R^{n+r} imbeds L also, with $T\nu_M$ collapsing to $T\nu_L$ to give a commutative diagram

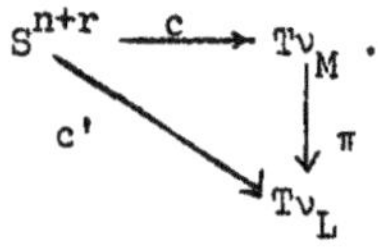

The number $x[L]$ is represented by the composition

$$S^{n+r} \xrightarrow{c'} T\nu_L \xrightarrow{\emptyset} (L/\emptyset)_\wedge T\nu_L \xrightarrow{\subset} (L/\emptyset)_\wedge T\rho|_L{}_\wedge T\nu_M|_L \xrightarrow{i_\wedge Tf_\wedge Ti} (M/\emptyset)_\wedge T\xi|_{KP(N-1)}{}_\wedge T\nu_M$$

$$\xrightarrow{y_\wedge U_\wedge U} A_\wedge A_\wedge A \xrightarrow{m} A$$

but the collapse diagram

$$\begin{array}{ccc} M & \xrightarrow{f} & KP(N) \\ {\scriptstyle \pi_1}\downarrow & & \downarrow{\scriptstyle \pi_2} \\ T\rho|_L & \xrightarrow{Tf} & T\xi|_{KP(N-1)} \end{array}$$

commutes, with $\pi_2^*(U) = \sigma_1(\xi)$. Thus $x[L]$ is represented by

$$S^{n+r} \xrightarrow{c} T\nu_M \xrightarrow{\emptyset} (M/\emptyset)_\wedge T\nu_M \xrightarrow{\Delta_\wedge 1} (M/\emptyset)_\wedge (M/\emptyset)_\wedge T\nu_M \xrightarrow{1_\wedge f_\wedge 1} (M/\emptyset)_\wedge (KP(N)/\emptyset)_\wedge T\nu_M$$

$$\xrightarrow{y_\wedge \sigma_1(\xi)_\wedge U} A_\wedge A_\wedge A \xrightarrow{m} A$$

which is $(y \cup \theta)[M]$, or $x[L] = (y \cup \theta)[M]$. (Note: This is commonly referred to as 'naturality of Poincaré duality').

The above construction is called 'dualization of the cohomology class θ', but is more properly 'dualization of the line bundle ρ'.

As an example one has:

Proposition: Let $\pi_i : KP(n_1) \times KP(n_2) \longrightarrow KP(n_i)$, $i = 1,2$, with $n_i > 1$ and let ℓ be the line bundle $\pi_1^*(\xi)\otimes\pi_2^*(\xi)$, with $H_{n_1,n_2} \subset KP(n_1) \times KP(n_2)$ the submanifold dual to $\sigma_1(\ell)$. Then

$$s_{n_1+n_2-1}(\sigma(\tau))[H_{n_1,n_2}] = -\binom{n_1+n_2}{n_1} .$$

Proof: Since the cohomology of $KP(n_i)$ is free over that of a point, one has $H^*(KP(n_1) \times KP(n_2);\underset{\sim}{A}) \cong H^*(KP(n_1);\underset{\sim}{A}) \otimes_{H^*(pt;\underset{\sim}{A})} H^*(KP(n_2);\underset{\sim}{A})$. The inclusion $KP(n_1) \xrightarrow{\times p} KP(n_1) \times KP(n_2)$ pulls ℓ back to ξ so $\sigma_1(\ell) = \pi_1^*\bar{\alpha} + \pi_2^*\bar{\alpha} + \sum_{i,j\geq 1} b_{ij}(\pi_1^*\bar{\alpha})^i(\pi_2^*\bar{\alpha})^j$, where $\bar{\alpha} = c^*(\alpha)$ of the previous discussion of projective spaces, $b_{ij} \in H^{k-ki-kj}(pt;A)$. Then $\sigma(\tau_H)$ is the restriction to H of

$$\frac{(1+\pi_1^*\bar{\alpha})^{n_1+1}(1+\pi_2^*\bar{\alpha})^{n_2+1}}{1+\sigma_1(\ell)}$$

or $s_{n_1+n_2-1}(\sigma(\tau_H))$ is the restriction of

$$(n_1+1)(\pi_1^*\bar{\alpha})^{n_1+n_2-1} + (n_2+1)(\pi_2^*\bar{\alpha})^{n_1+n_2-1} - (\sigma_1(\ell))^{n_1+n_2-1}$$
$$= -(\sigma_1(\ell))^{n_1+n_2-1}$$

(for since $n_i > 1$, $\bar{\alpha}^{n_1+n_2-1} = 0$). Hence

$$s_{n_1+n_2-1}(\sigma(\tau_H))[H] = -(\sigma_1(\ell))^{n_1+n_2}[KP(n_1) \times KP(n_2)],$$
$$= -(\pi_1^*\bar{\alpha} + \pi_2^*\bar{\alpha})^{n_1+n_2}[KP(n_1) \times KP(n_2)]$$

since all other terms are of degree greater than n_1+n_2 in the $\pi_i^*\bar{\alpha}$ and so are zero. But this is

$$-\binom{n_1+n_2}{n_1}(\pi_1^*\bar{\alpha})^{n_1}(\pi_2^*\bar{\alpha})^{n_2}[KP(n_1) \times KP(n_2)] = -\binom{n_1+n_2}{n_1}\bar{\alpha}^{n_1}[KP(n_1)]\bar{\alpha}^{n_2}[KP(n_2)],$$
$$= -\binom{n_1+n_2}{n_1}. \quad **$$

<u>Remarks</u>: The manifold $H_{m,n}$ is the non-singular hypersurface of degree $(1,1)$ in $KP(m) \times KP(n)$. If one uses homogeneous coordinates $(w_0,\ldots,w_m)$, $(z_0,\ldots,z_n)$ this can be defined as the locus of points satisfying $w_0z_0 + \ldots + w_rz_r = 0$, where $r = \inf(m,n)$. If one considers $KP((m+1)(n+1)-1)$

with homogeneous coordinates u_{ij}, $0 \leq i \leq m$, $0 \leq j \leq n$ and imbeds $KP(m) \times KP(n)$ by $u_{ij} = w_i \cdot z_j$, then H is the hyperplane section given by $\sum_{i=0}^{r} u_{ii} = 0$ and for this imbedding ξ pulls back to $\pi_1^*\xi \otimes \pi_2^*\xi$. The use of these manifolds was the idea of Milnor (see Milnor [87] or Hirzebruch [54]).

Ordinary Cohomology of BO and BSO

For $K = \mathbb{R}$, the only cohomology theories to which the previous methods are applicable are the 2-primary theories. Unfortunately, this does not give sufficient information to describe the spaces BO_n and $T\gamma^n$. The object of this section is to conquer this technical difficulty by computing the cohomology of BO with other coefficients.

Proposition: Let ξ be a real vector bundle over B. Then $\xi \oplus \xi$ admits a complex vector bundle structure given by

$$i(x,y) = (-y,x), \qquad (x,y) \in E \oplus E .$$

This complex vector bundle is the complexification of ξ, denoted $\xi \otimes \mathbb{C}$, and $\xi \otimes \mathbb{C}$ is isomorphic to its complex conjugate.

If ξ is itself a complex bundle, then $\xi \otimes \mathbb{C}$ is isomorphic as complex vector bundle with $\xi \oplus \bar{\xi}$, $\bar{\xi}$ the conjugate of ξ. In fact, $\xi \otimes \mathbb{C}$ admits a quaternionic vector bundle structure given by

$$i(x,y) = (-y,x) \qquad \text{and} \qquad j(x,y) = (ix,-iy).$$

Proof: $i^2(x,y) = i(-y,x) = (-x,-y)$ so $i^2 = -1$, giving a complex structure. Let $\phi : \xi \oplus \xi \longrightarrow \xi \oplus \xi : (x,y) \longrightarrow (-x,y)$. Then $\phi i = -i\phi$, so ϕ is an isomorphism of $\xi \otimes \mathbb{C}$ with its complex conjugate. If ξ is a complex bundle, let $f : \xi \longrightarrow \xi \otimes \mathbb{C}$ by $f(x) = (x,-ix)$ and $g : \xi \longrightarrow \xi \otimes \mathbb{C}$ by $g(x) = (x,ix)$. Then $fi = if$ and $gi = -ig$, so that f and g are maps of ξ and its complex conjugate into $\xi \otimes \mathbb{C}$. This is a direct sum decomposition since $(x,y) = 1/2\, f(x+iy) + 1/2\, g(x-iy)$. Finally with the given maps $i^2 = j^2 = -1$ and $ij = -ji$, so that i and j give a quaternionic structure on $\xi \otimes \mathbb{C}$. **

Now suppose $\underset{\sim}{A}$ is a ring spectrum for which $CP(V)$ has proper cohomology and such that $c^*(\alpha) = -\alpha$. For ξ a real vector bundle over B, one then has the Chern class $c(\xi \otimes \mathbb{C}) \in H^*(B;\underset{\sim}{A})$. Since $\xi \otimes \mathbb{C}$ is isomorphic to its conjugate

$$c_i(\xi \otimes \mathbb{C}) = c_i(\overline{\xi \otimes \mathbb{C}}) = (-1)^i c_i(\xi \otimes \mathbb{C})$$

so $2c_{2j+1}(\xi \otimes \mathbb{C}) = 0$. Since the odd Chern classes have order 2, and hence are amenable to 2-primary structure theorems, one ignores these and considers:

<u>Definition</u>: If $\underset{\sim}{A}$ is a ring spectrum for which $CP(V)$ has proper cohomology and such that $c^*(\alpha) = -\alpha$, and ξ is a real vector bundle over B, then <u>the i-th Pontrjagin class</u> $\mathcal{P}_i(\xi)$ is defined by

$$\mathcal{P}_i(\xi) = (-1)^i c_{2i}(\xi \otimes \mathbb{C}) \in H^{4i}(B;\underset{\sim}{A}).$$

The <u>total Pontrjagin class</u> is the 'formal' sum

$$\mathcal{P}(\xi) = 1 + \sum_{i=1}^{\infty} \mathcal{P}_i(\xi) \in H^*(B;\underset{\sim}{A}).$$

<u>Lemma</u>: If ξ is a complex vector bundle with $c(\xi) = \prod_{i=1}^{n} (1+x_i)$, $\dim x_i = 2$, then $\mathcal{P}(\xi) = \prod_{i=1}^{n} (1+x_i^2)$.

<u>Proof</u>: $c(\bar{\xi}) = \prod_{i=1}^{n} (1-x_i)$ so $c(\xi \otimes \mathbb{C}) = \prod_{i=1}^{n} (1-x_i^2)$. **

Lemma: If $\underset{\sim}{A}$ is a ring spectrum as above and if $1/2 \in H^0(pt;\underset{\sim}{A})$ then

$$\mathcal{P}(\xi \oplus \eta) = \mathcal{P}(\xi) \cup \mathcal{P}(\eta).$$

Proof: Since $1/2 \in H^0(pt;\underset{\sim}{A})$, $H^*(B;\underset{\sim}{A})$ has no 2-primary torsion, so $c_{2j+1}(\xi \otimes \mathbb{C}) = 0$. From $c(\xi \otimes \mathbb{C} \oplus \eta \otimes \mathbb{C}) = c(\xi \otimes \mathbb{C}) \cup c(\eta \otimes \mathbb{C})$ one has this formula after making the proper sign modifications. **

Proposition: If $\underset{\sim}{A}$ is a ring spectrum for which $CP(V)$ has proper cohomology and $1/2 \in H^0(pt;\underset{\sim}{A})$ with $c^*(\alpha) = -\alpha$, then $H^*(BO_n;\underset{\sim}{A})$ contains a subring isomorphic to the polynomial algebra over $H^*(pt;\underset{\sim}{A})$ on the universal Pontrjagin classes $\mathcal{P}_i$ for $1 \leq i \leq [n/2]$, ([ ] denotes 'integral part of').

Proof: The classes $\mathcal{P}_i$ are defined for $2i \leq n$, and it suffices to show this ring maps monomorphically into $H^*(BO_n;\underset{\sim}{A})$. For this one has the map $f : BU_{[n/2]} \longrightarrow BO_n$ obtained by considering a complex $[n/2]$-plane bundle as a real bundle, with $f^*(\mathcal{P}) = c \cup \bar{c}$. Thus

$$f^*(\mathcal{P}_i) = 2c_{2i} - 2c_{2i-1}c_1 + 2c_{2i-2}c_2 - \cdots \pm 2c_{i+1}c_{i-1} \pm c_i^2$$

and these generate a polynomial subalgebra of $H^*(BU_{[n/2]};\underset{\sim}{A})$. **

In order to prove equality, one needs further restrictions on $\underset{\sim}{A}$. Thus, it will be assumed throughout the remainder of this section that $\underset{\sim}{A} = \underset{\sim}{K}(\mathbb{S})$ where $\mathbb{S}$ is a commutative ring containing $1/2$. The proof in this case is fairly involved and depends upon a study of oriented vector bundles.

Definition: If V is an n-dimensional vector space over R with inner product $\langle \, , \, \rangle$, an orientation of V is a unit vector $\mathcal{O}$ in the n-th exterior power of V, $\Lambda^n(V)$. If $e_1,\ldots,e_n$ is an orthonormal base of V

such that $e_1 \wedge \cdots \wedge e_n = \mathcal{O}$; then $\{e_1,\ldots,e_n\}$ as an ordered base may be thought of as defining the orientation of V.

Lemma: If W is a complex vector space with inner product then W has a canonical orientation given by $f_1 \wedge if_1 \wedge \cdots \wedge f_n \wedge if_n$ where $f_1,\ldots,f_n$ is any orthonormal base for W over $\mathbb{C}$.

Definition: An oriented vector bundle ξ is a vector bundle together with a consistent choice of orientations in the fibers; i.e. a bundle ξ together with a cross section of the unit sphere bundle of the determinant bundle $\det\xi = \Lambda^n(\xi)$, $n = \dim\xi$.

Proposition: Let $\xi = (B,E,p,+,\cdot)$ be an oriented n-plane bundle. There is a unique class $U \in \tilde{H}^n(T\xi;Z)$ such that for each point $b \in B$ and orientation preserving isomorphism $f : R^n \longrightarrow E_b$, the induced class $(Tf)^*(U) \in \tilde{H}^n(S^n;Z)$ is the standard generator.

Proof: Let S be the unit sphere bundle of $\xi \oplus 1$. Then S is an oriented sphere bundle; i.e. if $\phi : [0,1] \longrightarrow B$, $\phi(0) = \phi(1) = b$ and ϕ^*S is trivialized then the maps of the fibers over 0 and 1 into the fiber of S over b are homotopic (rel base points).

Thus $\pi_1(B)$ acts trivially on the cohomology of the fiber of S. In the Serre spectral sequence one has $E_2^{p,q} \cong H^p(B;H^q(S^n)) \cong H^p(B;Z) \otimes H^q(S^n)$. Let $s : B \longrightarrow S : b \longrightarrow (0_b,1) \in E_b \times R$. Then s is a section, and the spectral sequence collapses. If $V \in H^n(S)$ represents $1 \otimes i_n \in E_2^{0,n} = E_\infty^{0,n}$ then V restricts to a generator of the cohomology of each fiber. Since V is determined only up to image π^*, $\pi : S \longrightarrow B$ the projection, one may assume $s^*V = 0$ (which characterizes V completely). Then $H^*(S)$ is the free $H^*(B)$ module on 1 and V.

Using the section $s : B \longrightarrow S$, one may form S/B which is identifiable with $T\xi$ and one has the exact sequence of the pair (S,B) giving

$$0 \longleftarrow H^*(B) \xleftarrow{s^*} H^*(S) \xleftarrow{\chi} \tilde{H}^*(T\xi) \longleftarrow 0$$

identifying $\tilde{H}^*(T\xi)$ with kernel s^*, the free $H^*(B)$ module on V. Let $U = \chi^{-1}(V)$. Since $H^0(pt;Z) \cong Z$, U is unique except for sign, i.e. choice of orientation over the components of B. **

Definition: If $\xi = (B,E,p,+,\cdot)$ is an oriented n-plane bundle, with $t : B \longrightarrow T\xi$ the map induced from the zero section of ξ, then the Euler class $e(\xi)$ is $t^*(U) \in H^n(B;Z)$.

Note: If $\tilde{s} : B \longrightarrow S$ is the section $b \longrightarrow (0_b,-1)$, the map into $S/B = T\xi$ is t. Thus $e(\xi) = \tilde{s}^*V$. Unless ξ admits a nonzero cross section it is not clear that s and $\tilde{s}$ should be homotopic, so $\tilde{s}^*V$ may be non-zero.

Lemma: $U \cup U = p^*(e(\xi))\cdot U$.

Proof: Let D and D_0 denote the disc and sphere bundles of ξ, $i : D \longrightarrow D/D_0$ the collapse and $z : B \longrightarrow D$ the zero section. Identifying $T\xi$ with D/D_0, $p^*(e(\xi))U = (p^*z^*i^*U)U$. Since B is a deformation retract of D, zp is homotopic to the identity and $p^*z^* = 1$. Thus $p^*(e(\xi))U=(i^*U)U$ but the external product $(i^*U)U$ coincides with the internal product $U \cup U$. **

Corollary: If n is odd, $e(\xi)$ is of order 2, hence is zero in $H^*(B;\mathbb{S})$ if $1/2 \in \mathbb{S}$.

Proof: $U \cup U = (-1)^{\dim U} U \cup U$. **

Lemma: If ξ is the underlying oriented bundle of a complex n-plane bundle ω, then $e(\xi) = c_n(\omega)$.

Proof: This assumes the fact (which will be verified later) that $CP(V)$ has proper integral cohomology. Then the orientation classes constructed for the oriented bundle and the complex bundle coincide by uniqueness, and hence the images under the zero section map must coincide. **

For a Whitney sum of oriented bundles, the product orientation class is again an orientation class, and by the same uniqueness argument:

Lemma: The Euler class of a Whitney sum of oriented bundles is the product of the Euler classes.

Corollary: If ξ is an oriented $2n$ plane bundle, then $e(\xi)^2 = \mathcal{P}_n(\xi)$.

Proof: If $e_1,\ldots,e_{2n}$ is an ordered base of the fiber over b defining the orientation, the fiber of $\xi \oplus \xi$ has orientation given by the base $(e_1,0),\ldots,(e_{2n},0),(0,e_1),\ldots,(0,e_{2n})$ while $\xi \otimes \mathbb{C}$ has orientation given by $(e_1,0),(0,e_1),\ldots,(e_{2n},0),(0,e_{2n})$. Thus $\xi \otimes \mathbb{C} = (-1)^n(\xi \oplus \xi)$ as oriented bundles [$n(2n-1)$ sign changes]. Thus

$\mathcal{P}_n(\xi) = (-1)^n c_{2n}(\xi \otimes \mathbb{C}) = (-1)^n e(\xi \otimes \mathbb{C}) = e(\xi \oplus \xi) = e(\xi)^2$. **

Definition: $\tilde{G}_{n,r}$ is the Grassmannian of oriented n planes in R^{n+r}; i.e. pairs consisting of an n-plane and an orientation of that plane. (Equivalently $\tilde{G}_{n,r}$ is the sphere bundle of the determinant bundle of γ_r^n). The limit space is $BSO_n = \lim_{r\to\infty} \tilde{G}_{n,r}$.

BSO_n is the classifying space for oriented n-plane bundles, which coincides with the sphere bundle of the determinant bundle of γ^n over BO_n. The pullback of γ^n to BSO_n is an oriented bundle $\tilde{\gamma}^n$.

<u>Proposition</u>: If $\mathbb{S}$ is a commutative ring containing 1/2, then $H^*(BSO_n; \mathbb{S})$ is the polynomial ring over $\mathbb{S}$ on

$$\mathcal{P}_1, \ldots, \mathcal{P}_k \qquad \text{for } n = 2k+1,$$
$$\mathcal{P}_1, \ldots, \mathcal{P}_{k-1}, e(\tilde{\gamma}^n) \qquad \text{for } n = 2k$$

and in the latter case $\mathcal{P}_k = e(\tilde{\gamma}^n)^2$, where $\mathcal{P}_i$ is the Pontrjagin class $\mathcal{P}_i(\tilde{\gamma}^n)$.

<u>Proof</u>: Let $h : BU_{[n/2]} \longrightarrow BSO_n$ be the map obtained by ignoring complex structure. The classes listed map to algebraically independent elements of $H^*(BU_{[n/2]}; \mathbb{S})$ and hence $H^*(BSO_n; \mathbb{S})$ contains this polynomial subring. Equality is proved by induction.

For $n = 1$, BSO_1 is the double cover of $RP(\infty)$ which is the infinite sphere, hence contractible.

For $n = 2$, an oriented n-plane bundle is just a complex line bundle. In fact, if V is an oriented 2 plane with orientation $\mathcal{O}$, and $v \in V$, there is a unique vector w with w orthogonal to v, with $|w| = |v|$ and such that $v \wedge w = k\mathcal{O}$ with $k \geq 0$ ($k = |v|^2$). This w is $i \cdot v$. Since $w \wedge v = -v \wedge w$, $i^2 = -1$. The orientation given by a complex structure gives that complex structure back. Thus $BSO_2 = BU_1$ and $H^*(BSO_2; \mathbb{S}) = \mathbb{S}[e(\tilde{\gamma}^2)] = \mathbb{S}[c_1]$.

To make the induction step one has

$$S(\tilde{\gamma}^n) \longrightarrow D(\tilde{\gamma}^n) \longrightarrow T\tilde{\gamma}^n.$$

At a point of $S(\tilde{\gamma}^n)$ one has a non-zero vector in an oriented n-plane and the complementary n-1 plane is then oriented. This defines a fibration $S(\tilde{\gamma}^n) \longrightarrow BSO_{n-1}$ with fiber an infinite sphere. The pullback to $S(\tilde{\gamma}^n)$ of

$\tilde{\gamma}^n$ splits off a section to give $\tilde{\gamma}^{n-1} \oplus 1$. The projection and zero section of $D(\tilde{\gamma}^n)$ give a homotopy equivalence of BSO_n and $D(\tilde{\gamma}^n)$. Thus the exact sequence of the pair $(D(\tilde{\gamma}^n), S(\tilde{\gamma}^n))$ gives an exact sequence

$$\begin{array}{ccccc} H^*(S(\tilde{\gamma}^n)) & \longleftarrow & H^*(D(\tilde{\gamma}^n)) & \longleftarrow & \tilde{H}^*(T\tilde{\gamma}^n) \\ \wr\| & & \wr\| & & \uparrow \Phi \\ H^*(BSO_{n-1}) & \xleftarrow{\ b\ } & H^*(BSO_n) & \xleftarrow{\ a\ } & H^*(BSO_n) \end{array}$$

(with connecting homomorphisms $H^*(S(\tilde{\gamma}^n)) \to \tilde{H}^*(T\tilde{\gamma}^n)$ and $H^*(BSO_{n-1}) \to H^*(BSO_n)$)

where Φ is the Thom isomorphism. The homomorphism a has degree n, and since BSO_n is identified with $D(\tilde{\gamma}^n)$ by the zero section, a is multiplication by the Euler class. The homomorphism b sends $\mathcal{P}_i(\tilde{\gamma}^n)$ to $\mathcal{P}_i(\tilde{\gamma}^{n-1})$ since under the identifications $\tilde{\gamma}^n$ pulls back to $\tilde{\gamma}^{n-1} \oplus 1$. The computation for the induction is then straightforward, giving the exact sequences

$(n = 2k)$

$$0 \longleftarrow \mathbb{S}[\mathcal{P}_1, \ldots, \mathcal{P}_{k-1}] \longleftarrow \mathbb{S}[\mathcal{P}_1, \ldots, \mathcal{P}_{k-1}, e] \xleftarrow{\times e} \mathbb{S}[\mathcal{P}_1, \ldots, \mathcal{P}_{k-1}, e] \longleftarrow 0$$

in which the equality for $H^*(BSO_n)$ is proved inductively using dimension, and

$(n = 2k+1)$

$$0 \longleftarrow \mathbb{S}[\mathcal{P}_1, \ldots, \mathcal{P}_k] \longleftarrow \mathbb{S}[\mathcal{P}_1, \ldots, \mathcal{P}_{k-1}, e] \longleftarrow \mathbb{S}[\mathcal{P}_1, \ldots, \mathcal{P}_k] \longleftarrow 0$$

since $e(\tilde{\gamma}^n) = 0$ implies $a = 0$, so that $H^*(BSO_n)$ is a subring of $H^*(BSO_{n-1})$ containing $\mathbb{S}[\mathcal{P}_1, \ldots, \mathcal{P}_k]$, but equality must hold by rank over $\mathbb{S}$, $\mathbb{S}[\mathcal{P}_1, \ldots, \mathcal{P}_{k-1}, e]$ having a base over $\mathbb{S}[\mathcal{P}_1, \ldots, \mathcal{P}_k]$ given by 1 and e. **

Proposition: If $\mathbb{S}$ is a commutative ring containing 1/2, then $H^*(BO_n; \mathbb{S})$ is the polynomial ring over $\mathbb{S}$ on the Pontrjagin classes $\mathcal{P}_i$, $1 \leq i \leq [n/2]$.

Proof: BSO_n is a double cover of BO_n, so $H^*(BO_n; \mathbb{S})$ is the subring of $H^*(BSO_n; \mathbb{S})$ consisting of the classes fixed under the homomorphism induced by the map $x : BSO_n \longrightarrow BSO_n$ which interchanges the sheets of the cover. Since $x^*(\mathcal{P}_i) = \mathcal{P}_i$ ($\mathcal{P}_i$ comes from BO_n) while $x^*(e(\tilde{\gamma}^n)) = e(x^*(\tilde{\gamma}^n)) = -e(\tilde{\gamma}^n)$ since $x^*(\tilde{\gamma}^n)$ is just $\tilde{\gamma}^n$ with reversed orinetation, the result is clear. **

Note: In the fibration $\pi : BO_{n-1} \longrightarrow BO_n$ with fiber S^{n-1}, BO_{n-1} being $S(\gamma^n)$, the homotopy group $\pi_1(BO_n) \cong Z_2$ acts non-trivially on the cohomology of the fiber.

Chapter VI

Unoriented Cobordism

In many respects the most interesting cobordism theory is unoriented cobordism; i.e. the cobordism problem associated to the category $(\mathcal{D},\partial,i)$ of all compact differentiable manifolds. It has additional interest in that its solution by Thom [127] illustrates all of the basic techniques for dealing with cobordism problems, without encountering excessive technicality.

First note that $\Omega(\mathcal{D},\partial,i)$ decomposes as a direct sum of semigroups $\Omega_n(\mathcal{D},\partial,i)$, n being the dimension of the manifold. This semigroup is usually denoted $\mathfrak{N}_n$, with $\mathfrak{N}_*$ denoting the direct sum. The first structure theorem is:

Proposition: $\mathfrak{N}_n$ is an abelian group in which every element has order 2. $\mathfrak{N}_*$ is a graded commutative ring, multiplication being induced by the product of manifolds, with unit, given by the cobordism class of a point.

Proof: For any closed M, $M + M + \partial\emptyset \tilde{=} \emptyset + \partial(M \times I)$ where $I = [0,1]$, so the class of M is its own inverse. If M, N_1 and N_2 are closed with $N_1 \equiv N_2$, say $N_1 + \partial U_1 \tilde{=} N_2 + \partial U_2$, then $M \times N_1 + \partial(M \times U_1) \tilde{=} M \times N_2 + \partial(M\times U_2)$ so $M \times N_1 \equiv M \times N_2$. Since $M \times (N_1 + N_2) \tilde{=} M \times N_1 + M \times N_2$ and $M \times N \tilde{=} N \times M$ this gives $\mathfrak{N}_*$ the structure of a graded commutative ring. If p is a point, $M \times p \tilde{=} p \times M \tilde{=} M$, so the class of p is a unit. **

The next standard step is to replace the cobordism problem by a homotopy problem. This is accomplished for unoriented cobordism by noting that every manifold has a unique $(BO,1)$ structure (1 being the sequence of identity maps $1_r : BO_r \longrightarrow BO_r$). The forgetful functor from the category of $(BO,1)$ manifolds to $(\mathcal{D},\partial,i)$ which ignores the $(BO,1)$ structure for objects and the normal trivialization for maps induces an isomorphism of cobordism semigroups. (Note: Isomorphisms, inclusions of boundaries, and summands are preserved). If $M_1^{n_1} \subset R^{n_1+r_1}$ and $M_2^{n_2} \subset R^{n_2+r_2}$ are imbeddings, then the normal bundle of the product imbedding $M_1 \times M_2 \subset R^{n_1+r_1+n_2+r_2}$ is the Whitney sum of the normal bundles of the factors. Thus one has the determination theorem:

Theorem: The cobordism group $\mathfrak{N}_n$ is isomorphic to $\lim_{r\to\infty} \pi_{n+r}(TBO_r, \infty)$. The ring structure in $\mathfrak{N}_*$ is induced by the maps $TBO_r \wedge TBO_s \longrightarrow TBO_{r+s}$ obtained from the Whitney sum operation on vector bundles.

The next step is clearly to try to solve the homotopy problem. It is here that the most ingenuity is required since the various cobordism theories differ widely at this point. The guidance one obtains from Thom's work is: Make use of the cohomology theories for which the manifolds in question are orientable.

For unoriented cobordism one makes use of ordinary cohomology with Z_2 coefficients; i.e. the cohomology theory for the spectrum $\underset{\sim}{K}(Z_2)$. One needs a knowledge of the operations in this theory, which may be summarized:

The mod 2 Steenrod algebra $\mathcal{A}_2$ is the graded algebra defined by

$$(\mathcal{A}_2)^i = H^{n+i}(K(Z_2,n);Z_2) \qquad i < n.$$

Then:

a) $\mathcal{A}_2$ is the associative graded algebra over Z_2 generated by symbols Sq^i of dimension i, and all relations are given by the Adem relations

$$Sq^a Sq^b = \sum_{i=0}^{[a/2]} \binom{b-i-1}{a-2i} Sq^{a+b-i} Sq^i$$

if $a < 2b$ ($Sq^0 = 1$).

b) For any pair (X,A) there is a natural pairing $\mathcal{A}_2 \otimes H^*(X,A;Z_2) \longrightarrow H^*(X,A;Z_2)$ such that

1) $Sq^i : H^n(X,A;Z_2) \longrightarrow H^{n+i}(X,A;Z_2)$ is additive;

2) $Sq^0\, u = u$ for all u,

$Sq^d\, u = u^2$ if dimension $u = d$, and

$Sq^d\, u = 0$ if dimension $u < d$; and

3) (Cartan formula) $Sq^d(a \cup b) = \sum_{e+f=d} (Sq^e\, a) \cup (Sq^f\, b)$.

(See Steenrod and Epstein [114]).

Following Milnor [79] one defines a diagonal map $\Delta : \mathcal{A}_2 \longrightarrow \mathcal{A}_2 \otimes \mathcal{A}_2$ by $\Delta(Sq^i) = \sum_{j+k=i} Sq^j \otimes Sq^k$, which makes $\mathcal{A}_2$ into a connected Hopf algebra over Z_2. (Connected means that the unit defines an isomorphism of $(\mathcal{A}_2)^0$ with the ground field Z_2).

It is well known that the Z_2 cohomology of real projective space $P(R^n)$ is the truncated polynomial algebra over Z_2 on the unique non-zero class α of dimension one, $\alpha^n = 0$, and that α^{n-1} is the image of the non-zero class $\iota \in \tilde{H}^{n-1}(S^{n-1};Z_2)$. From Chapter V, one then knows the full cohomology structure of BO_r and TBO_r using Z_2 coefficients. The following line of proof is due to Browder, Liulevicius, and Peterson [27]. (See also Liulevicius [73]).

Denote by $\tilde{H}^*(\underset{\sim}{TBO};Z_2)$ the direct sum of the groups $\tilde{H}^n(\underset{\sim}{TBO};Z_2) = \lim_{r\to\infty} \tilde{H}^{n+r}(TBO_r;Z_2)$.

Lemma: The maps $TBO_r \wedge TBO_s \longrightarrow TBO_{r+s}$ obtained from the Whitney sum of vector bundles induce a diagonal map

$$\psi : \tilde{H}^*(\underset{\sim}{TBO};Z_2) \longrightarrow \tilde{H}^*(\underset{\sim}{TBO};Z_2) \otimes \tilde{H}^*(\underset{\sim}{TBO};Z_2)$$

making $\tilde{H}^*(\underset{\sim}{TBO};Z_2)$ into a connected coalgebra over Z_2 with counit

$U \in \tilde{H}^0(\underset{\sim}{TBO};Z_2)$. $\tilde{H}^*(\underset{\sim}{TBO};Z_2)$ is a left module over the Hopf algebra $\mathcal{A}_2$ with ψ a homomorphism of $\mathcal{A}_2$ modules such that the homomorphism

$$\nu : \mathcal{A}_2 \longrightarrow \tilde{H}^*(\underset{\sim}{TBO};Z_2) : a \longrightarrow a(U)$$

is a monomorphism.

Proof: This is all obvious with the exception of the assertion that ν is monic. For this one may apply the splitting principle and the calculation of the Thom class of a line bundle to express U formally as the product $V = x_1 x_2 x_3 \ldots$ of one dimensional classes x_i, $i \geq 1$. By the Adem formulae $\mathcal{A}_2$ has a base consisting of operations $Sq^I = Sq^{i_1} Sq^{i_2} \ldots Sq^{i_k}$ with $i_\alpha \geq 2i_{\alpha+1}$ and evaluation of such an operation on V gives a sum of monomials $x_1^{r_1} x_2^{r_2} \ldots$ with only a finite number of $r_i \neq 1$. Order such monomials by $x_1^{r_1} x_2^{r_2} \ldots > x_1^{s_1} x_2^{s_2} \ldots$ if for some $j, r_i = s_i$ for all $i<j$ and $r_j > s_j$. Since $Sq^i x^{2^s} = x^{2^s}, x^{2^{s+1}}$, or 0 as $i = 0$, 2^s or any other if $\dim x = 1$, one has that the largest monomial of $Sq^I V$ is $x_1^{r_1} x_2^{r_2} \ldots$ where $r_1 \geq r_2 \geq \ldots$ and the sequence of r_i has $i_\alpha - 2i_{\alpha+1}$ copies of the integer 2^α, $\alpha = 1, \ldots, k$. In addition, each sequence of r_i occuring consists of powers of 2. Thus $Sq^I U = s_\omega U + \Sigma\, s_{\omega'} U$ where $\omega = \omega(I)$ contains $i_\alpha - 2i_{\alpha+1}$ copies of $2^\alpha - 1$ and ω' runs through a set of partitions into integers of the form $2^s - 1$ with $\omega' < \omega$ in lexicographic order (if $\omega = (j_1, \ldots, j_t)$ with $j_1 \geq j_2 \geq \ldots$, say $\omega > \omega'$ if for some γ $j_\beta = j'_\beta$ for all $\beta < \gamma$ and $j_\gamma > j'_\gamma$). Since the partitions $\omega(I)$ are distinct ($\omega(I)$ determines I) ν is monic. (Note: This is the standard argument needed to determine the dual of $\mathcal{A}_2$. See Steenrod and Epstein [114], Chapter I, 3.3). **

One then has the result of Milnor and Moore [90], Theorem 4.4:

<u>Lemma</u>: Let A be a connected Hopf algebra over a field F. Let M be a connected coalgebra over F with counit $1 \in M_0$ and a left module over A such that the diagonal map is a map of A modules. Suppose $\nu : A \longrightarrow M : a \longrightarrow a\cdot 1$ is a monomorphism. Then M is a free left A module.

<u>Proof</u>: Let $\bar{A} \subset A$ be the elements of positive degree and let $\pi : M \longrightarrow N = M/\bar{A}M$ be the projection. Let $f : N \longrightarrow M$ be any vector space splitting $(\pi f = 1_N)$ and define $\phi : A \otimes N \longrightarrow M : a \otimes n \longrightarrow af(n)$. ϕ is a homomorphism of A modules.

1) <u>ϕ is epic</u>: $\phi : A_0 \otimes N_0 \longrightarrow M_0$ is the identity map of F for $(\bar{A}M) \cap M_0 = 0$. Suppose $\phi : (A \otimes N)_i \longrightarrow M_i$ is epic if $i < k$. If $c \in M_k$, $c - \phi(1 \otimes \pi c)$ maps to zero under π so $c - \phi(1 \otimes \pi c) = \sum_i \bar{a}_i(c_i)$ with $\bar{a}_i \in \bar{A}$, $c_i \in M$. Since $\dim c_i < k$, $c_i = \phi(x_i)$ for some x_i and $c = \phi(1 \otimes \pi c) + \Sigma\, \bar{a}_i\phi(x_i) = \phi(1 \otimes \pi c + \Sigma\, \bar{a}_i x_i)$. Thus $\phi : (A \otimes N)_k \longrightarrow M_k$ is epic.

2) <u>ϕ is monic</u>: Consider

$$A \otimes N \xrightarrow{1 \otimes f} A \otimes M \xrightarrow{\text{mult}} M \xrightarrow{\Delta} M \otimes M \xrightarrow{1 \otimes \pi} M \otimes N.$$
$$\underbrace{\qquad\qquad\qquad\qquad}_{\phi} \qquad \underbrace{\qquad\qquad\qquad\qquad}_{\tilde{\Delta}}$$

Clearly ϕ, Δ, and $1 \otimes \pi$ are A module homomorphisms (Δ by assumption; $1 \otimes \pi$ is an A module homomorphism since $a(m \otimes n) = am \otimes n$, while if $\pi m' = n$ then $(1 \otimes \pi)a(m \otimes m') = (1 \otimes \pi)(\Sigma\, a'm \otimes a''m') = (1 \otimes \pi)(am \otimes 1\cdot m') = am \otimes n$ since $\pi(a''m') = 0$ if $\deg a'' > 0$; here $\Delta a = \Sigma\, a' \otimes a''$).

Then $1 \otimes n \longrightarrow 1 \otimes f(n) \longrightarrow f(n) \longrightarrow f(n) \otimes 1 + 1 \otimes f(n) +$ other $\longrightarrow f(n) \otimes 1 + 1 \otimes n +$ other, or $a \otimes n \longrightarrow a\cdot 1 \otimes n + e$ where $e \in \bigcup_{p<\dim n} M \otimes N_p$. Projection of $M \otimes N$ on $M \otimes N_{\dim n}$ gives $A \otimes N_{\dim n} \longrightarrow M \otimes N_{\dim n} : a \otimes n \longrightarrow a(1) \otimes n$ which is monic since ν is monic. Thus $\tilde{\Delta}\circ\phi$ is monic and so ϕ is monic.

3) Thus $\phi : A \otimes N \cong M$ is an isomorphism of A modules, so M is a free A module. **

Combining these lemmas one gets

<u>Theorem</u>: In dimensions less than or equal to $2r$, $\tilde{H}^*(TBO_r;Z_2)$ is a free module over the Steenrod algebra $\mathcal{A}_2$ and, in fact, in dimensions less than $2r$ TBO_r has the homotopy type of a product of Eilenberg-MacLane spaces $K(Z_2,n)$.

Thus $\mathcal{N}_n$ is a Z_2 vector space whose rank is the number of non-dyadic partitions of n ($\omega = (i_1,\ldots,i_r)$ is non-dyadic if none of the i_β are of the form $2^s - 1$) and two manifolds are unoriented cobordant if and only if they have the same Stiefel-Whitney numbers.

<u>Proof</u>: The free module structure follows from the stability $\tilde{H}^{r+i}(TBO_r;Z_2) \cong \tilde{H}^{r+i+1}(TBO_{r+1};Z_2)$ for $i \leq r$. There is then a map of TBO_r into a product of spaces $K(Z_2,n)$, $n \geq r$, inducing an isomorphism of Z_2 cohomology in dimensions less than or equal to $2r$, and by the generalized Whitehead theorem (Spanier [110] page 512) the homomorphism on homotopy is an isomorphism modulo odd torsion in dimensions less than $2r$. For p an odd prime, one has the exact sequence $0 \leftarrow H^*(BO_{r-1};Z_p) \leftarrow H^*(BO_r;Z_p) \leftarrow \tilde{H}^*(TBO_r;Z_p) \leftarrow 0$ arising from the pair $(D\gamma^r, S\gamma^r)$ from which $\tilde{H}^*(TBO_r;Z_p) = 0$ in dimensions less than $2r$. Thus the map into the product of $K(Z_2,n)$'s is a homotopy equivalence in dimensions less than $2r$. Since the rank of $\tilde{H}^n(\underset{\sim}{TBO};Z_2)$ is the number of partitions of n, while the rank of $\mathcal{A}_2^i$ is the number of dyadic partitions of i, the rank of $\pi_{n+r}(TBO_r,\infty)$ is the number of non-dyadic partitions of n if $r > n$. Since this homotopy group is isomorphic to $\mathcal{N}_n$ for r large, the rank of $\mathcal{N}_n$ is as asserted.

Since the Hurewicz homomorphism is monic for a product of $K(Z_2,n)$'s it is monic for TBO_r below dimension $2r$, and hence cobordism class is determined by Z_2 cohomology characteristic numbers. **

The complete structure theorem is then:

<u>Theorem</u>: $\mathfrak{N}_*$ is a polynomial algebra over Z_2 (with unit) on classes x_i of dimension i, i not of the form $2^s - 1$; the class x_i may be chosen to be the cobordism class of any closed manifold M^i for which the s-number $s_{(i)}(w(\nu))[M] = s_{(i)}(w(\tau))[M] \neq 0$.

<u>Note</u>: With ν the normal bundle and τ the tangent bundle, $\nu \oplus \tau$ is trivial, and since $s_{(i)}$ is primitive, $s_{(i)}(\nu) + s_{(i)}(\tau) = 0 = s_{(i)}(\text{trivial})$ and $s_{(i)}(\nu) = -s_{(i)}(\tau) = s_{(i)}(\tau)$ (mod 2).

<u>Note</u>: One may prove this as did Thom by showing that the $s_\omega U$ for ω non-dyadic form a base of $\tilde{H}^*(\underset{\sim}{TBO};Z_2)$ over $\mathcal{A}_2$ (see also Wall [130] pages 301-302). Since one wishes to have explicit constructions of generators an indirect proof will be given here, in a sequence of lemmas. This will in fact show that these $s_\omega U$ form a base over $\mathcal{A}_2$.

<u>Lemma</u>: Suppose there are manifolds M^i of dimension i $(i \neq 2^s-1)$ such that $s_{(i)}(w(\nu))[M^i] \neq 0$. Then $\mathfrak{N}_*$ is the polynomial algebra on the classes of the M^i. If manifolds N^i $(i \neq 2^s-1)$ also give a system of generators, then $s_{(i)}(w(\nu))[N^i] \neq 0$.

<u>Proof</u>: Totally order the non-dyadic partitions of n by an order $\prec$ compatible with the partial ordering $\omega \leq \omega'$ if ω' refines ω (i.e. if $\omega = (i_1,\ldots,i_r)$ then $\omega' = \omega_1 \cup \ldots \cup \omega_r$ where ω_β is a partition of i_β). If $\omega = (i_1,\ldots,i_r)$ let $M_\omega = M^{i_1} \times \ldots \times M^{i_r}$. Then $s_{\omega'}(w(\nu))[M_\omega]$ is zero if ω' does not refine ω and is nonzero if $\omega = \omega'$. (For this

$s_{\omega'}(\nu)[M_\omega] = \sum_{\omega_1 \cup \dots \cup \omega_r = \omega'} s_{\omega_1}(\nu)[M^{i_1}] \dots s_{\omega_r}(\nu)[M^{i_r}]$ and since $s_{\tilde{\omega}}(\nu)[M^i] = 0$ if $n(\tilde{\omega}) \neq i$, this must be zero unless there is some expression $\omega' = \omega_1 \cup \dots \cup \omega_r$ with $n(\omega_\beta) = i_\beta$. If $\omega' = \omega$, this gives $s_\omega(\nu)[M_\omega] = \Pi\, s_{(i_\beta)}(\nu)[M^{i_\beta}] \neq 0$.) Considering the matrix $\|s_{\omega'}(w(\nu))[M_\omega]\|$ for ω, ω' non-dyadic partitions of n, one has $s_{\omega'}(w(\nu))[M_\omega] = 0$ if $\omega' \prec \omega$ since ω' cannot then refine ω, so the matrix is triangular, and since $s_\omega(w(\nu))[M_\omega] \neq 0$ the diagonal entries are all one (in Z_2). Thus the manifolds M_ω are linearly independent over Z_2 and having the proper number they are a base for $\mathcal{N}_n$, which gives the polynomial structure.

If $\{N^i\}$ is another family of generators then N^i cannot be decomposable so one must have $N^i = M^i + \Sigma\, a_\omega M_\omega$, $a_\omega \in Z_2$, $\omega = (i_1, \dots, i_r)$ with $r > 1$. Since $s_{(i)}(w(\nu))$ vanishes on decomposable elements, this gives $s_{(i)}(w(\nu))[N^i] = s_{(i)}(w(\nu))[M^i] \neq 0$. **

<u>Lemma</u>: If $i = 2k$, then $s_{(i)}(w(\tau))[RP(i)] \neq 0$.

<u>Proof</u>: By the computations of Chapter V, $s_{(i)}(w(\tau))[RP(i)] = i + 1 \neq 0$ mod 2. **

<u>Lemma</u>: If i is odd and not of the form $2^s - 1$, write $i = 2^p(2q+1)-1$ with $p \geq 1$, $q \geq 1$ and let $H_{2^{p+1}q, 2^p} \subset RP(2^{p+1}q) \times RP(2^p)$ be the hypersurface of degree $(1,1)$. Then

$$s_{(i)}(w(\tau))[H_{2^{p+1}q, 2^p}] \neq 0.$$

<u>Proof</u>: From Chapter V the value of this s-class is $-\binom{2^p(2q+1)}{2^p}$. This is

$$-\frac{(2^{p+1}q+2^p)(2^{p+1}q+2^p-1)\dots(2^{p+1}q+1)}{(2^p)\qquad(2^p-1)\qquad\dots\qquad(1)} \quad \text{and} \quad \frac{2^{p+1}q+2^a(2b+1)}{2^a(2b+1)} = \frac{2^{p+1-a}q+2b+1}{2b+1}$$

so that factors of 2 divide out, making this an odd integer. **

Note: This choice of generators is due to Milnor [87]. The first construction of odd dimensional generators was by Dold [43].

To complete the determination of $\mathfrak{N}_*$ it is desirable to know the complete set of relations among the Stiefel-Whitney numbers. This problem was solved by Dold [44] who showed that all relations follow from formulae of Wu [142] relating the characteristic classes with the action of the Steenrod algebra. A thorough study of this situation appears in Atiyah and Hirzebruch [19].

Let M^n be a closed n-dimensional manifold and let $[M] \in H_n(M;Z_2)$ be the orientation class. Since Z_2 is a field, the universal coefficient theorem (Spanier [110] page 243) shows that $\mathrm{Hom}(H_k(M,Z_2);Z_2) \cong H^k(M;Z_2)$ while Poincaré duality gives that the cap product with $[M]$ is an isomorphism of $H^{n-k}(M;Z_2)$ with $H_k(M;Z_2)$. Thus the pairing

$$H^k(M;Z_2) \otimes H^{n-k}(M;Z_2) \longrightarrow Z_2 : a \otimes b \longrightarrow (a \cup b)[M]$$

is a dual pairing. (Note: This characterizes $[M]$ for on a component M_0 of M there will be a unique nonzero class in $H^n(M_0;Z_2)$ or $H_n(M_0;Z_2)$ and $[M]$ must restrict to this class. Thus $[M]$ is a homotopy type invariant.)

The operation $H^{n-k}(M;Z_2) \longrightarrow Z_2 : a \longrightarrow (Sq^k a)[M]$ is a homomorphism and so by the dual pairing there is a unique class $v_k \in H^k(M;Z_2)$ such that for all $a \in H^{n-k}(M;Z_2)$ one has $(Sq^k a)[M] = (v_k \cup a)[M]$. Since $Sq^k a = 0$ if $k > \dim a$, one has $v_k = 0$ if $k > n - k$ or $2k > n$. The class $v(M) = 1 + v_1 + \dots + v_{[n/2]}$ in $H^*(M;Z_2)$ is called the total Wu class of M.

It is useful to form the total Steenrod operation $Sq = 1 + Sq^1 + Sq^2 + \dots$ so that for $x \in H^*(M;Z_2)$ one has $(Sq\ x)[M] = (v \cup x)[M]$.

Remarks: For computations to be performed it is desirable to have a few properties of the total Steenrod operation. By linearity of Sq^i, one has $Sq(x + y) = Sqx + Sqy$. From the Cartan formula, $Sq(x \cup y) = (Sqx) \cup (Sqy)$. Considering Sq as a formal power series beginning with 1, one may invert Sq to define an operation Sq^{-1}. By dimension considerations $Sq^i x = 0$ if $i > 0$ or $Sq^i x = x$ if $i = 0$ for $x \in H^*(S^n;Z_2)$ so for $x \in H^*(S^n;Z_2)$, $Sqx = x$. (In the terminology of Atiyah and Hirzebruch, Sq is a cohomology automorphism.)

In order to relate the Wu class to characteristic classes one needs the result of Thom [126]:

Theorem: Let $U \in \tilde{H}^r(TBO_r;Z_2)$ be the Thom class. Then

$$Sq\ U = (\pi^* w) \cup U,$$

i.e. $Sq^i U = (\pi^* w_i) \cup U$.

Proof: By the splitting principle one may write U as a product $x_1 \dots x_r$ of one-dimensional classes. Then $Sq^i(x_1 \dots x_r)$ is the sum of all monomials $x_1 \dots x_{j_1}^2 \dots x_{j_i}^2 \dots x_r$ for $1 \leq j_1 < \dots < j_i \leq r$. This is the i-th elementary symmetric function of the x_β multiplied by $x_1 \dots x_r$, and hence $Sq^i U = (\pi^* w_i) \cup U$. **

One then has the result of Wu [142]:

Theorem: If M^n is a closed differentiable manifold with Wu class v and tangential Stiefel-Whitney class $w(\tau)$, then

$$w(\tau) = Sq\ v \qquad \text{or} \qquad v = Sq^{-1} w(\tau).$$

In particular, the Stiefel-Whitney classes are homotopy type invariants and the Wu class is a characteristic class. If one expresses $w(\tau) = \prod_{i=1}^{n} (1 + x_i)$, then $v = \prod_{i=1}^{n} (1 + x_i + x_i^2 + x_i^4 + \ldots)$

Proof: Let $\tilde{w} = Sq\, v$, $w(\nu)$ the normal Stiefel-Whitney class of M, $U \in \tilde{H}^*(T\nu; Z_2)$ the Thom class, $c : S^{n+r} \longrightarrow T\nu$ the collapse given by an imbedding, and y any class in $H^*(M;Z_2)$. Let $x \in H^*(M;Z_2)$ be $Sq^{-1}y$ so $y = Sq\, x$. Then

$$
\begin{aligned}
y\tilde{w}w(\nu)[M] &= c^*(\pi^*(y\tilde{w}w(\nu))U)[S^{n+r}],\\
&= c^*(Sqx \cdot Sqv \cdot SqU)[S^{n+r}], \quad \text{(dropping } \pi^*\text{'s from notation)}\\
&= c^*(Sq(xvU))[S^{n+r}],\\
&= Sqc^*(xvU)[S^{n+r}],\\
&= c^*(xvU)[S^{n+r}],\\
&= (x \cup v)[M],\\
&= (Sq\, x)[M],\\
&= y[M].
\end{aligned}
$$

Since this holds for all $y \in H^*(M;Z_2)$, one has from the dual pairing that $\tilde{w}w(\nu) = 1$. Since $w(\tau)w(\nu) = 1$, this gives $\tilde{w} = w(\tau)$. Since $[M]$ and Sq are homotopy invariant, so is $w(\tau) = Sq\, v$. Finally, $v = Sq^{-1}w$ is given by a universal class in $H^*(BO;Z_2)$, while the formula in one dimensional classes follows from $Sq(\sum_{i=0}^{\infty} x^{2^i}) = \sum_{i=0}^{\infty} (x^{2^i} + x^{2^{i+1}}) = x$ (dim $x = 1$). **

Corollary: Homotopy equivalent manifolds are unoriented cobordant.

With this machinery one can prove the Dold theorem:

Theorem: All relations among the Stiefel-Whitney numbers of closed n dimensional differentiable manifolds are given by the Wu relations; i.e.

if $\phi : H^n(BO;Z_2) \longrightarrow Z_2$ is a homomorphism, there is an n dimensional closed manifold M^n with $\phi(a) = (\tau^*(a))[M]$ for all a ($\tau : M \longrightarrow BO$ classifying the tangent bundle of M) if and only if $\phi(Sq\, b + vb) = 0$ for all $b \in H^*(BO;Z_2)$ where $v = Sq^{-1}w$.

Proof: By Wu's theorem $\tau^*(Sqb + vb)[M] = 0$ so the condition $\phi(Sqb + vb) = 0$ is necessary. To prove sufficiency, let $\chi : BO \longrightarrow BO$ be the map classifying the negative of the universal bundle (for any M, $\chi \circ \tau$ classifies the normal bundle) and let $\rho = \phi \circ \chi^* \circ \Phi^{-1} : \tilde{H}^n(\underset{\sim}{TBO};Z_2) \to Z_2$ where Φ is the Thom homomorphism. From the calculation of $\mathcal{N}_n$, ϕ is the tangential characteristic number homomorphism of some manifold if and only if $\rho(\bar{\mathcal{A}}_2\tilde{H}^*(\underset{\sim}{TBO};Z_2)) = 0$, or for all $y \in H^*(BO;Z_2)$, $\rho[Sq(yU) + yU] = 0$. But one has

$$\begin{aligned}
\rho[Sq(yU) + yU] &= \rho[SqySqU + yU], \\
&= \rho[Sqy \cdot wU + yU], \\
&= \rho[Sq(y \cdot Sq^{-1}w)U + yU], \\
&= \phi\chi^*[Sq(y \cdot Sq^{-1}w) + y], \\
&= \phi[Sq(\chi^*y \cdot Sq^{-1}\chi^*w) + \chi^*y], \\
&= \phi[Sq(\chi^*y \cdot Sq^{-1}(1/w)) + \chi^*y], \\
&= \phi[Sq\, x + x \cdot Sq^{-1}w], \\
&= \phi[Sq\, x + vx], \\
&= 0
\end{aligned}$$

(where $x = \chi^*y \cdot Sq^{-1}(1/w) = \chi^*y \cdot (1/Sq^{-1}w)$). **

This completes the analysis of the unoriented cobordism ring. Beginning the pattern which will be followed throughout, one wishes to know the relationship with other cobordism theories and the structure of the related bordism theory.

Relation to framed cobordism: The Hopf invariant

Recall that a framed manifold is a manifold together with an equivalence class of trivializations of the stable normal bundle. The cobordism corresponding is (B,f) cobordism with B_r a point and the cobordism groups Ω_n^{fr} are identified with $\lim_{r\to\infty} \pi_{n+r}(S^r,\infty)$. (Pontrjagin [101]).

The forgetful functor F which ignores framing defines a homomorphism $F_* : \Omega_*^{fr} \longrightarrow \mathfrak{N}_*$ and a relative cobordism semigroup $\Omega_n(F)$ (obtained by joining manifolds along common boundaries). As with any pair of (B,f) theories one then has an exact sequence

$$\begin{array}{ccc} \Omega_*^{fr} & \xrightarrow{F_*} & \mathfrak{N}_* \\ & \nwarrow \quad \swarrow & \\ & \Omega_*(F) & \end{array}$$

which is the homotopy exact sequence

$$\begin{array}{ccc} \lim_{r\to\infty} \pi_{*+r}(S^r,\infty) & \xrightarrow{F_*} & \lim_{r\to\infty} \pi_{*+r}(TBO_r,\infty) \\ & \nwarrow \qquad \swarrow & \\ & \lim_{r\to\infty} \pi_{*+r}(TBO_r,S^r,\infty) & \end{array}$$

where F_* is induced by the inclusion of a sphere which is the Thom space of the fiber over a base point in BO_r.

Making use of the calculation of $\mathfrak{N}_*$ one may analyze this relationship. One has:

<u>Proposition</u>: A framed manifold of positive dimension bounds in the unoriented sense; i.e. $F_* : \Omega_n^{fr} \longrightarrow \mathfrak{N}_n$ is the zero homomorphism if $n > 0$. Further $F_* : \Omega_0^{fr} = Z \longrightarrow \mathfrak{N}_0 = Z_2$ is epic.

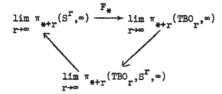

<u>Proof</u>: If M^n is framed, the stable normal bundle is trivial, so $w(\nu) = 1$. Thus for $n > 0$ all Stiefel-Whitney numbers of M are zero and M bounds. In dimension zero, M is a union of points (with signs = orientations) and $w_o(\nu)[M]$ is the cardinality of M modulo 2. **

The homotopy exact sequence then splits up into short exact sequences giving the diagrams

$$\begin{array}{ccccccccc}
 & & 0 \longrightarrow \mathcal{N}_n & \longrightarrow & \Omega_n(F) & \longrightarrow & \Omega^{fr}_{n-1} \longrightarrow 0 \\
 & & \downarrow q_n & & \downarrow r_n & & \downarrow s_n \\
\tilde{H}_n(\underset{\sim}{S};Z_2) & \longrightarrow & \tilde{H}_n(\underset{\sim}{TBO};Z_2) & \longrightarrow & \tilde{H}_n(\underset{\sim}{TBO},\underset{\sim}{S};Z_2) & \longrightarrow & \tilde{H}_{n-1}(\underset{\sim}{S};Z_2) \\
\| & & & & & & \| \\
0 & & & & & & 0
\end{array}$$

for $n - 1 > 0$, and

$$\begin{array}{ccccccccccc}
 & & 0 = \mathcal{N}_1 & \longrightarrow & \Omega_1(F) & \longrightarrow & \Omega^{fr}_o & \longrightarrow & \mathcal{N}_o \longrightarrow 0 \\
 & & \downarrow q_1 & & \downarrow r_1 & & \downarrow s_o & & \downarrow q_o \\
\tilde{H}_1(\underset{\sim}{S};Z_2) & \longrightarrow & \tilde{H}_1(\underset{\sim}{TBO};Z_2) & \longrightarrow & \tilde{H}_1(\underset{\sim}{TBO},\underset{\sim}{S};Z_2) & \longrightarrow & \tilde{H}_o(\underset{\sim}{S};Z_2) & \longrightarrow & \tilde{H}_o(\underset{\sim}{TBO};Z_2) & \longrightarrow & 0 \\
\| & & & & & & \| & & \| \\
0 & & & & & & Z_2 & & Z_2
\end{array}$$

in which the vertical maps are the Hurewicz homomorphisms (and notationally $\tilde{H}_n(\underset{\sim}{S};Z_2) = \lim_{r\to\infty} H_{n+r}(S^r,\infty;Z_2)$, etc.). Since $\mathcal{N}_n$ and $\tilde{H}_n(\underset{\sim}{TBO};Z_2)$ are Z_2 vector spaces with q_n monic, one has a splitting homomorphism $u_n : \tilde{H}_n(\underset{\sim}{TBO};Z_2) \longrightarrow \mathcal{N}_n$, and the homomorphism

$$\Omega_n(F) \xrightarrow{r_n} \tilde{H}_n(\underset{\sim}{TBO},\underset{\sim}{S};Z_2) \xrightarrow{\cong} \tilde{H}_n(\underset{\sim}{TBO};Z_2) \xrightarrow{u_n} \mathcal{N}_n$$

gives a splitting of the short exact sequences for $n > 1$. This defines in turn a homomorphism $v_n : \Omega^{fr}_{n-1} \longrightarrow \Omega_n(F)$ also splitting the sequence (for $n > 1$).

<u>Note</u>: One could choose a splitting by choosing for any framed manifold M^{n-1} a manifold V^n with $\partial V = M$ and $s_\omega(\nu)[V,\partial V] = 0$ if ω is non-dyadic. This corresponds to writing $\underset{\sim}{TBO}$ as a product $\Pi\ \underset{\sim}{K}(Z_2,n(\omega))$ for ω non-dyadic. Suppose the splitting has been chosen in this fashion.

Let M^{n-1} be a framed manifold with $M^{n-1} = \partial V^n$ and $s_\omega(\nu)[V,\partial V] = 0$ if ω is non-dyadic. The imbedding in some space H^{n+r} defines the map $\nu : (V,M) \longrightarrow (BO_r,*)$ (the framing being interpreted as a specific equivalence class of deformation of M to the base point). The Pontrjagin-Thom construction defines a map $(H^{n+r},R^{n+r-1}) \longrightarrow (TBO_r,T*)$ which will be considered as a map $f : (D^{n+r},S^{n+r-1}) \longrightarrow (TBO_r,S^r)$ representing the homotopy class corresponding to (V,M). Let X denote the two-cell complex formed by attaching D^{n+r} to S^r by the map $f : S^{n+r-1} \longrightarrow S^r$. One then has the diagram of cofibrations

$$\begin{array}{ccccc}
S^r & \xrightarrow{j} & X & \xrightarrow{\pi} & X/S^r = D^{n+r}/S^{n+r-1} \\
{\scriptstyle 1}\downarrow & & {\scriptstyle g}\downarrow & & {\scriptstyle \bar{f}}\downarrow \\
S^r & \xrightarrow{i} & TBO_r & \longrightarrow & TBO_r/S^r
\end{array}$$

in which $\bar{f}$ is just f on the quotient spaces and g is induced by f.

Recall that $H^*(X;Z_2)$ is a vector space over Z_2 with base $1 \in H^0(X;Z_2)$, $a \in H^r(X;Z_2)$ and $b \in H^{n+r}(X;Z_2)$ (the nonzero classes in these dimensions) with $j^*(a) = \iota \in \tilde{H}^r(S^r;Z_2)$ and $b = \pi^*(\iota')$ where $\iota' \in H^{n+r}(D^{n+r},S^{n+r-1};Z_2)$.

From the relationship between the Pontrjagin-Thom construction and characteristic numbers, one has $\bar{f}^*(w_\omega U) = w_\omega(\nu)[V,M] \cdot \iota'$ or $g^*(w_\omega U) = (w_\omega(\nu)[V,M]) \cdot b$. Writing TBO_r as $\Pi K(Z_2, r + n(\omega))$, ω non-dyadic corresponding to the $\mathcal{A}_2$ generators $s_\omega U$ one has $g^*(s_\omega U) = 0$ if $n(\omega) > 0$ (by dimension if $n(\omega) \neq n$, by choice for $n(\omega) = n$) so that the only possibly non-zero numbers are of the form $g^*(Sq^I U)$. If $I = (i_1, \ldots, i_r)$ with $r > 1$, $g^*(Sq^{i_r} U) = 0$. Thus the only nonzero characteristic number would be $g^*(Sq^n U) = g^*(w_n U) = w_n(\nu)[V,M] \cdot b$. Now $j^*g^*(U) = 1^*i^*(U) = \iota$ so $g^*(U) = a$. Thus $Sq^n a = w_n(\nu)[V,M] \cdot b$.

Following Steenrod [113], page 983, the element $H(f) \in Z_2$ for which $Sq^n a = H(f) \cdot b$ is called the <u>Hopf invariant of f</u>.

<u>Note</u>: For any framed M^{n-1} and any W with $\partial W = M$, one may form the characteristic number $w_n(\nu)[W,M]$. There is a closed manifold T with $s_\omega(\nu)[T] = s_\omega(\nu)[W,M]$ for all non-dyadic ω. Then $V = T \cup W$ satisfies the above conditions. Since $w_n(\nu)[T] = 0$ for any closed T, one has $w_n(\nu)[W,M] = w_n(\nu)[V,M] = H(f)$. One did not need the assumptions $s_\omega(\nu)[V,M] = 0$ to get that this is the Hopf invariant, but only to show that this is the only additional characteristic number arising.

Combining the above one has:

<u>Theorem</u>: For $n > 1$, $\Omega_n(F) \cong \mathcal{N}_n \oplus \Omega^{fr}_{n-1}$ and $\Omega_1(F) \cong Z = 2\Omega^{fr}_0$. If M^{n-1} is a closed framed manifold and $M = \partial V$, then the characteristic number $w_n(\nu)[V,M]$ coincides with the Hopf invariant of a map $f : S^{n+r-1} \to S^r$ representing the framed cobordism class of M. This is the only possibly nonzero homomorphism from $\Omega^{fr}_{n-1} \to Z_2$ obtainable from Stiefel-Whitney numbers.

From Adams' work [1] on the nonexistence of maps of Hopf invariant one, one knows that there is a map $f : S^{n+r-1} \to S^r$ with $H(f) \neq 0$ if and only if $n = 1,2,4$, or 8. In the above notation this may be phrased

Corollary: For $n \neq 1,2,4$, or 8 the image of $\mathcal{N}_n$ in $\tilde{H}_n(\underset{\sim}{TBO};Z_2)$ coincides with that of $\Omega_n(F)$. For $n = 1,2,4$, or 8 the image of $\mathcal{N}_n$ has codimension 1 in that of $\Omega_n(F)$. Equivalently, the homomorphism $w_n(\nu)[\quad] : \Omega_n(F) \longrightarrow Z_2$ is nontrivial if and only if $n = 1,2,4$, or 8.

Unoriented bordism : Steenrod representation

Let $\mathcal{T}$ denote the category of topological spaces and continuous maps and $F : \mathcal{B} \longmapsto \mathcal{T}$ the forgetful functor assigning to each differentiable manifold its underlying topological space. For any space X one may form the cobordism category $(\mathcal{B}/X,\tilde{\partial},\tilde{i})$, obtained by Construction I. This gives rise to a cobordism semigroup $\mathcal{N}_*(X)$ which was first defined by Atiyah [13] and which is called the bordism group of X.

Let (B,f) be the sequence of spaces and maps given by $B_r = X \times BO_r$ and $f_r : B_r \longrightarrow BO_r$ the projection on the second factor. A (B,f) structure on a manifold is then a $(BO,1)$ structure together with a homotopy class of maps into X. Since homotopic maps define the same class in $\mathcal{N}_*(X)$ one has induced a homomorphism $\Omega_*(B,f) \longrightarrow \mathcal{N}_*(X)$ which is clearly an isomorphism.

If $A \hookrightarrow X$ is a subspace, the inclusion map defines a functor $(\mathcal{B}/A,\tilde{\partial},\tilde{i}) \longmapsto (\mathcal{B}/X,\tilde{\partial},\tilde{i})$ and a map of sequences $(A \times BO,\pi_2) \longrightarrow (X \times BO,\pi_2)$ giving rise to relative bordism groups $\mathcal{N}_*(X,A)$. Note: If $f : (M,\partial M) \longrightarrow (X,A)$ is a map, the fact that $f|_{\partial M}$ factors through the inclusion of A into X may be thought of as an additional structure on the boundary. The standard joining together along common boundaries permits definition of the relative groups.

From the relative Pontrjagin-Thom construction one then has

<u>Theorem</u>: $\mathfrak{N}_n(X,A) = \lim_{r\to\infty} \pi_{n+r}(T(X \times BO_r), T(A \times BO_r), \infty)$,

$$= \lim_{r\to\infty} \pi_{n+r}((X/A)_\wedge TBO_r, \infty),$$

$$= H_n(X,A;\underset{\sim}{TBO}).$$

In particular, $\mathfrak{N}_*(X,A)$ is the homology theory defined by the ring spectrum $\underset{\sim}{TBO}$.

Since $\underset{\sim}{TBO}$ is a ring spectrum, $H_*(X,A;\underset{\sim}{TBO})$ is an $H_*(pt;\underset{\sim}{TBO}) = \mathfrak{N}_*$ module. If $f : (V,\partial V) \longrightarrow (X,A)$ represents a class in $\mathfrak{N}_*(X,A)$ and M is a closed manifold representing a class in $\mathfrak{N}_*$, then $f \circ \pi_1 : (V \times M, \partial V \times M = \partial(V \times M)) \longrightarrow (X,A)$ represents the product class.

The structure of $\mathfrak{N}_*(X,A)$ is given by:

<u>Theorem</u>: For every CW pair (X,A), $\mathfrak{N}_*(X,A)$ is a free graded $\mathfrak{N}_*$ module isomorphic to $H_*(X,A;Z_2) \otimes_{Z_2} \mathfrak{N}_*$.

<u>Proof</u>: Let $c_{n,i} \in H_n(X,A;Z_2)$ be an additive base with dual base $c^*_{n,i} \in H^n(X,A;Z_2)$. Applying the Künneth theorem $\tilde{H}^*((X/A)_\wedge\underset{\sim}{TBO};Z_2) = \lim_{r\to\infty} \tilde{H}^{*+r}((X/A)_\wedge TBO_r;Z_2)$ has a Z_2 base consisting of elements $c^*_{n,i} \otimes Sq^I(s_\omega U)$ for I admissible and ω nondyadic and in particular is a free $\mathcal{A}_2$ module on the classes $c^*_{n,i} \otimes s_\omega U$ since $Sq^I(c^*_{n,i} \otimes s_\omega U) = c^*_{n,i} \otimes Sq^I(s_\omega U)$ plus terms having second factor of lower degree.

One may then choose homotopy classes $a_{n,i} \in \pi_{n+r}((X/A)_\wedge TBO_r, \infty)$ (r large) for which $c^*_{n,j} \otimes U$ pulls back to a generator if $i = j$ and to zero otherwise. Applying the Pontrjagin-Thom procedure, the class $a_{n,i}$ is represented by a manifold $V^n_i \subset H^{n+r}$ with map $f : V^n_i \longrightarrow X$, $f(\partial V^n_i) \subset A$ and one has

$$S^{n+r} \longrightarrow T\nu_V/T\nu_{\partial V} \longrightarrow (V^n_i/\partial V^n_i)_\wedge T\nu_V \xrightarrow{f_\wedge T\nu} (X/A)_\wedge TBO_r \xrightarrow{c^*_{n,i}\wedge U} K(Z_2)_\wedge K(Z_2) \xrightarrow{m} K(Z_2)$$

so that $a^*_{n,i}(c^*_{n,j} \otimes U) = \delta_{ij}\,\iota_{n+r}$ is equivalent to $f^*(c^*_{n,j})[V^n_i,\partial V^n_i] = \delta_{ij}$ or $f_*[V^n_i,\partial V^n_i] = c_{n,i}$.

For any closed M, $(f \circ \pi_1)^*(c^*_{n,j} \otimes s_\omega)[V^n_i \times M,\partial V^n_i \times M] = \delta_{ij}s_\omega(\nu)[M]$, so that if $[M_\omega] \in \mathfrak{N}_*$ form a base, the classes $(V^n_i \times M_\omega,\partial V^n_i \times M_\omega;f \circ \pi_1)$ are a Z_2 base for the 2 primary part of the limit of the homotopy groups. Since $\mathfrak{N}_*(X,A)$ is 2 primary, these form a base of $\mathfrak{N}_*(X,A)$. This is precisely the assertion that the $(V^n_i, \partial V^n_i;f)$ form a base for $\mathfrak{N}_*(X,A)$ as $\mathfrak{N}_*$ module. **

This proof has several immediate consequences:

<u>Corollary</u>: The natural evaluation homomorphism $e : \mathfrak{N}_n(X,A) \longrightarrow H_n(X,A;Z_2)$ which sends the class represented by $f : (V,\partial V) \longrightarrow (X,A)$ into $f_*[V,\partial V]$ is an epimorphism.

This is often phrased: Every mod 2 homology class is Steenrod representable. (See problem 25 in Eilenberg [48]). This is, of course, very close to Poincaré's original concept of homology as given by submanifolds of a space.

<u>Corollary</u>: Unoriented bordism theory is determined by Z_2 cohomology characteristic numbers.

In particular, for each $x \in H^m(X,A;Z_2)$ and partition ω of $n - m$, one has a generalized "Stiefel-Whitney number" which is defined for a map $f : (V^n,\partial V^n) \longrightarrow (X,A)$ by $\{w_\omega(\tau) \cup f^*(x)\}[V,\partial V]$. Since the classes $x \otimes w_\omega$ form a base of $H^*((X/A) \times BO;Z_2)$ the associated characteristic numbers give a complete set of invariants.

It is clear from the free $\mathcal{A}_2$ module structure of $\tilde{H}^*((X/A)\wedge \underset{\sim}{TBO};Z_2)$ that all relations among these generalized Stiefel-Whitney numbers arise from the Wu relations.

References: In addition to Atiyah's paper [13], one may find a discussion of unoriented bordism in Conner and Floyd [36]. The Steenrod representability is due to Thom [127].

Chapter VII

Complex Cobordism

Historically the next cobordism problem to be completely solved was the cobordism of stably almost complex manifolds. This was defined and completely determined by Milnor [82] and by Novikov [93]. Specifically this is (B,f) cobordism in which $B_{2r} = B_{2r+1}$ is the classifying space BU_r for complex r-plane bundles. Since a complex vector bundle has a unique stable inverse, the objects are then manifolds with a chosen complex vector bundle structure on the normal or stable tangent bundle.

Since one has, essentially by definition, $\Omega_n^U \cong \lim_{r\to\infty} \pi_{n+2r}(TBU_r,\infty)$ the cobordism problem is ready for homotopy theoretic analysis. It is well known that the integral cohomology ring of $\mathbb{C}P(n)$ is the truncated polynomial ring on a 2 dimensional generator (make use of the multiplicative structure in the Serre spectral sequence of the fibration $S^1 \longrightarrow S^{2n+1} \longrightarrow \mathbb{C}P(n)$) and thus the integral cohomology ring $H^*(BU_r;Z)$ is the polynomial ring over Z on the universal Chern classes c_i (of dimension $2i$) with $1 \leq i \leq r$.

<u>Theorem</u>: The groups Ω_n^U are finitely generated and $\Omega_*^U \otimes Q$ is the rational polynomial ring on the cobordism classes of the complex projective spaces, the product corresponding to the Whitney sum of complex vector bundles.

<u>Proof</u>: By the Thom isomorphism theorem $\tilde{H}^n(\underset{\sim}{TBU};Z) = \lim_{r\to\infty} H^{n+2r}(TBU_r,\infty,Z)$ is torsion free with rank the number of partitions of m if $n = 2m$ and rank zero if n is odd. By the universal coefficient theorem, $\tilde{H}_n(\underset{\sim}{TBU};Z)$ is also torsion free of the same rank. Since TBU_r is $2r-1$ connected one has by Serre's theorem [107] that the Hurewicz homomorphism $\Omega_n^U \longrightarrow \tilde{H}_n(\underset{\sim}{TBU};Z)$ is an isomorphism modulo the class of finite groups. Thus Ω_n^U is finitely generated

and $\Omega_*^U \otimes Q$ has the same rank as a polynomial algebra on even dimensional generators. The Whitney sum of complex vector bundles gives a complex vector bundle structure to the normal bundle of the product of two stably almost complex manifolds, making Ω_*^U into a ring. Since $s_{(n)}(c(\nu))[\mathbb{C}P(n)] = -s_{(n)}(c(\tau))[\mathbb{C}P(n)] = -(n+1) \neq 0$ the monomials $\mathbb{C}P(n_1) \times \ldots \times \mathbb{C}P(n_r) = \mathbb{C}P(\omega)$ (for $\omega = (n_1, \ldots, n_r)$) are linearly independent in $\Omega_*^U \otimes Q$ (as in the unoriented case). Thus $\Omega_*^U \otimes Q$ is the polynomial ring on the cobordism classes of the complex projective spaces by dimension count. **

In order to study the torsion subgroup, one makes use of Z_p cohomology for all primes p. Since $H^*(BU_r;Z)$ is torsion free one has by the universal coefficient theorem that $H^*(BU_r;Z_p) \cong H^*(BU_r;Z) \otimes Z_p$ is the Z_p polynomial algebra on the mod p Chern classes c_i (reductions of the integral classes coincide with the direct definition using mod p cohomology).

In order to proceed one needs a knowledge of the operations in mod p cohomology. Briefly:

The mod p Steenrod algebra $\mathcal{A}_p$ for p an odd prime is the graded algebra defined by

$$(\mathcal{A}_p)^i = H^{n+i}(K(Z_p,n);Z_p) \qquad i < n.$$

Then:

a) $\mathcal{A}_p$ is the associative graded algebra over Z_p generated by symbols β of degree 1 and $\mathcal{P}^i$ of degree $2i(p-1)$ with all relations given by

$$\beta^2 = 0,$$

$$\mathcal{P}^a \mathcal{P}^b = \sum_{t=0}^{[a/p]} (-1)^{a+t} \binom{(p-1)(b-t)-1}{a-pt} \mathcal{P}^{a+b-t} \mathcal{P}^t$$

if $a < pb$, and

$$\mathcal{P}^a \beta \mathcal{P}^b = \sum_{t=0}^{[a/p]} (-1)^{a+t} \binom{(p-1)(b-t)}{a-pt} \beta \mathcal{P}^{a+b-t} \mathcal{P}^t$$
$$+ \sum_{t=0}^{[(a-1)/p]} (-1)^{a+t-1} \binom{(p-1)(b-t)-1}{a-pt-1} \mathcal{P}^{a+b-t} \beta \mathcal{P}^t$$

if $a \leq pb$. ($\mathcal{P}^0 = 1$).

b) For any pair (X,A) there is a natural pairing $\mathcal{A}_p \otimes H^*(X,A;Z_p) \longrightarrow H^*(X,A;Z_p)$ such that:

1') β is the Bockstein coboundary operator associated with the coefficient sequence $0 \longrightarrow Z_p \longrightarrow Z_{p^2} \longrightarrow Z_p \longrightarrow 0$,

2') $\beta(xy) = (\beta x)y + (-1)^{\dim x} x(\beta y)$,

and

1) $\mathcal{P}^i : H^n(X,A;Z_p) \longrightarrow H^{n+2i(p-1)}(X,A;Z_p)$ is additive;

2) $\mathcal{P}^0 u = u$ for all u,

$\mathcal{P}^i u = u^p$ if $\dim u = 2i$, and

$\mathcal{P}^i u = 0$ if $\dim u < 2i$; and

3) (Cartan formula) $\mathcal{P}^i(xy) = \sum_{j+k=i} \mathcal{P}^j x \cdot \mathcal{P}^k y$.

(See Steenrod and Epstein [114]).

One may define a diagonal map $\Delta : \mathcal{A}_p \longrightarrow \mathcal{A}_p \otimes \mathcal{A}_p$ by $\Delta(\beta) = \beta \otimes 1 + 1 \otimes \beta$, $\Delta(\mathcal{P}^i) = \sum_{j+k=i} \mathcal{P}^j \otimes \mathcal{P}^k$ which makes $\mathcal{A}_p$ into a connected Hopf algebra over Z_p (Milnor [79]).

One then has:

<u>Lemma</u>: Let p be any prime. The maps $TBU_r \wedge TBU_s \longrightarrow TBU_{r+s}$ obtained from the Whitney sum of vector bundles induce a diagonal map ψ on $\tilde{H}^*(\underset{\sim}{TBU};Z_p) = \lim_{r\to\infty} \tilde{H}^{*+2r}(TBU_r;Z_p)$ making it into a connected coalgebra over

Z_p with counit $U \in \tilde{H}^0(\underset{\sim}{TBU}; Z_p)$. Under the natural action of $\mathcal{A}_p$ on $\tilde{H}^*(\underset{\sim}{TBU}; Z_p)$ the Bockstein operator Q_0 ($Q_0 = \beta$ for p odd, $Q_0 = Sq^1$ for $p = 2$) is trivial and $\tilde{H}^*(\underset{\sim}{TBU}; Z_p)$ is a left module over the Hopf algebra $\mathcal{A}_p/(Q_0)$, the Steenrod algebra mod the two sided ideal generated by Q_0, with ψ a homomorphism of $\mathcal{A}_p/(Q_0)$ modules. Further, the homomorphism $\nu : \mathcal{A}_p/(Q_0) \longrightarrow \tilde{H}^*(\underset{\sim}{TBU}; Z_p) : a \longrightarrow a(U)$ is a monomorphism.

Proof: Since Q_0 has degree 1 while $\tilde{H}^*(\underset{\sim}{TBU}; Z_p)$ has only elements of even degree, the ideal (Q_0) acts trivially, making $\tilde{H}^*(\underset{\sim}{TBU}; Z_p)$ an $\mathcal{A}_p/(Q_0)$ module. To prove ν is monic one applies the splitting principle and calculation of the Thom class of a line bundle to write U formally as the product $V = x_1 x_2 \ldots$ of two dimensional classes x_i, $i \geq 1$. Letting $\mathcal{P}^i = Sq^{2i}$ if $p = 2$, the Adem formulae show that $\mathcal{A}_p/(Q_0)$ has a base consisting of operations $\mathcal{P}^I = \mathcal{P}^{i_1} \cdots \mathcal{P}^{i_r}$ with $i_\alpha \geq p i_{\alpha+1}$ (Note: For $p = 2$, $Sq^{2i+1} = Sq^1 Sq^{2i} \in (Q_0)$ so all odd terms vanish in the Adem formulae). Since $\mathcal{P}^i x^{p^s} = x^{p^s}, x^{p^{s+1}}$, or 0 as $i = 0$, p^s, or any other if $\dim x = 2$ (and if in addition $Sq^1 x = 0$ if $p = 2$; this holds for all x_i above since $x_i = c_1(\ell)$ for some line bundle ℓ), one may duplicate the proof of monicity of ν given in studying $\tilde{H}^*(\underset{\sim}{TBO}; Z_2)$ for this case. **

Corollary: $\tilde{H}^*(\underset{\sim}{TBU}; Z_p)$ is a free $\mathcal{A}_p/(Q_0)$ module.

One also has:

Lemma: Let X be a convergent spectrum such that $\tilde{H}^*(X; Z)$ has no p-primary torsion and such that $\tilde{H}^*(X; Z_p)$ is a free $\mathcal{A}_p/(Q_0)$ module. Then the homotopy of X has no p-primary torsion.

Remark: This was first proved by Milnor [82] using the Adams spectral sequence. Another proof was given by Brown and Peterson [30] by constructing a spectrum whose Z_p cohomology is free on one generator. The result given here is weaker than that of Brown and Peterson, but admits a reasonably elementary proof since one need not become involved with the k-invariants.

Proof: Let E_p be the 2 cell complex formed by attaching a 2 cell e_2 to a circle S^1 by a map of degree p, giving a cofibration

$$S^1 \longrightarrow E_p \longrightarrow S^2.$$

After smashing with X one has an exact sequence

$$\begin{array}{c} \pi_*(X) \xrightarrow{\ p\ } \pi_*(X) \xrightarrow{\ \rho\ } \pi_*(X \wedge E_p). \\ \uparrow\!\underline{\qquad\qquad\qquad\qquad\qquad\qquad}\!\downarrow \end{array}$$

Since X is a convergent spectrum the Hurewicz homomorphism $\pi_*(X) \longrightarrow \tilde{H}_*(X;Z)$ is an isomorphism modulo torsion and ρ maps $(\pi_*(X)/\text{Torsion}) \otimes Z_p \cong \tilde{H}_*(X;Z)\otimes Z_p$ monomorphically into a Z_p vector space $P_* \subset \pi_*(X \wedge E_p)$.

If the lemma is true then $P_* = \pi_*(X \wedge E_p)$, i.e. ρ is epic and $\pi_*(X \wedge E_p)$ is a Z_p vector space. If one can show directly that $P_* = \pi_*(X \wedge E_p)$ then ρ would be epic so multiplication by p would be monic on $\pi_*(X)$ - hence $\pi_*(X)$ could have no p-primary torsion. The remainder of the proof is devoted to showing $P_* = \pi_*(X \wedge E_p)$.

One needs a knowledge of the Steenrod algebra as described by Milnor [79].

Let $\mathcal{R}$ be the set of sequences of integers $(r_1, r_2, \ldots)$ such that $r_i \geq 0$ and $r_i = 0$ for all but a finite number of i. If $U, V \in \mathcal{R}$, $U - V \in \mathcal{R}$ is defined if $u_i \geq v_i$ for all i, and is equal to $(u_1 - v_1, u_2 - v_2, \ldots)$. $\Delta_j \in \mathcal{R}$ denotes the sequence with 1 in the j-th place and zeros elsewhere.

There exist elements Q_i and $\mathcal{P}^R$ in $\mathcal{A}_p$ for $i = 0,1,2,\ldots,$ and $R \in \mathcal{R}$ such that:

1) $\dim Q_i = 2p^i - 1$; $\dim \mathcal{P}^R = \dim R = \Sigma\, 2r_i(p^i - 1)$ if $R = (r_1, r_2, \ldots)$.

2) $\{Q_i\}$ generate a Grassmann subalgebra $\mathcal{A}_0$ of $\mathcal{A}_p$; i.e. $Q_iQ_i = 0$ and $Q_iQ_j + Q_jQ_i = 0$ for $i \neq j$.

3) $\mathcal{A}_p$ has a Z_p base given by $\{Q_o^{\varepsilon_0} \dots Q_i^{\varepsilon_i} \dots \mathcal{P}^R\}$ where $\varepsilon_i = 0$ or 1, and $\mathcal{P}^R Q_k - Q_k \mathcal{P}^R = \sum_{i=1}^{\infty} Q_{k+i}\mathcal{P}^{R-p^k\Delta_i}$.

Assertion: $\tilde{H}^*(X \wedge E_p;Z_p) \cong \tilde{H}^*(X;Z_p) \otimes Z_p\{u,Q_0u\}$, where $\{\quad\}$ denotes "vector space spanned by", dim $u = 1$, and is a free $\mathcal{A}_p/\mathcal{J}$ module where $\mathcal{J}$ denotes the two sided ideal generated by the elements Q_i, $i > 0$. If $\{x_\alpha\}$ is a base of $\tilde{H}^*(X;Z_p)$ as $\mathcal{A}_p/(Q_0)$ module, then $\{x_\alpha \otimes u\}$ is a base of $\tilde{H}^*(X \wedge E_p;Z_p)$ as $\mathcal{A}_p/\mathcal{J}$ module.

To see this, $H^*(E_p;Z_p) \cong Z_p\{1,u,Q_0u\}$ and so by the Künneth theorem $H^*(X \times E_p, X \times *;Z_p)$ is as asserted. Then $Q_0(x \otimes u) = x \otimes Q_ou$, $\mathcal{P}^R(x \otimes u) = \mathcal{P}^R x \otimes u$ gives the action of the Steenrod algebra. Since $Q_{k+1} = \mathcal{P}^{p^k} Q_k - Q_k \mathcal{P}^{p^k}$, while the operator Q_0 commutes with the $\mathcal{P}^R$ in $\tilde{H}^*(X \wedge E_p;Z_p)$, this is an $\mathcal{A}_p/\mathcal{J}$ module. A base of $\mathcal{A}_p/\mathcal{J}$ is given by the $Q_0^\varepsilon \mathcal{P}^R$, $\varepsilon = 0,1$, and the $\{\mathcal{P}^R x_\alpha \otimes Q_0^\varepsilon u\}$ are clearly a Z_p base for the cohomology.

For a sequence $R = (r_1,r_2,\dots)$ let $\ell(R) = \Sigma r_i$. Let V_s be the Z_p vector space generated by $R \in \mathcal{R}$ for which $\ell(R) = s$. Let $M_s = \mathcal{A}_p \otimes V_s$ and $d_s : M_s \longrightarrow M_{s-1}$ the $\mathcal{A}_p$ homomorphism of degree +1 given by

$$d_s(1 \otimes R) = \sum_{j=1}^{\infty} Q_j \otimes (R-\Delta_j).$$

Let $d_o : M_o \longrightarrow \mathcal{A}_p/\mathcal{J}$ by $d_0(1 \otimes (0,0,\dots)) = 1$. Just as in Brown and Peterson [30] one has:

Assertion: The sequence

$$\dots \longrightarrow M_s \xrightarrow{d_s} M_{s-1} \longrightarrow \dots \longrightarrow M_0 \xrightarrow{d_0} \mathcal{A}_p/\mathcal{J} \longrightarrow 0$$

is exact.

To see this, let B be the Grassmann algebra on the Q_i, $i > 0$. Then

$$\cdots \longrightarrow B \otimes V_s \xrightarrow{d_s} B \otimes V_{s-1} \longrightarrow \cdots \longrightarrow B \otimes V_0 \xrightarrow{d_0} Z_p \longrightarrow 0$$

with homomorphisms as above is exact, being the standard resolution of a field over a Grassmann algebra. Since $\mathcal{A}_p$ is a free B module and $\mathcal{A}_p \otimes_B Z_p \cong \mathcal{A}_p/\mathcal{J}$, tensoring with $\mathcal{A}_p$ gives the exact sequence.

Now let $\{x_\alpha\}$, $\alpha \in I$, be a base of $\tilde{H}^*(X;Z_p)$ as $\mathcal{A}_p/(Q_0)$ module and let T be the graded Z_p vector space on the $x_\alpha \otimes u$. Using the resolution constructed above for $\mathcal{A}_p/\mathcal{J}$ one can build a modified Postnikov tower for $X \wedge E_p$ using products of $K(Z_p,n)$'s. Specifically, one has a sequence of fibrations

$$\begin{array}{ccc} Y_{i+1} & & \\ \pi_i \downarrow & & \\ Y_i & \xrightarrow{f_i} & K(T \otimes V_i) \end{array}$$

induced from the path fibrations, with $Y_0 = X \wedge E_p$, $\tilde{H}^*(Y_i;Z_p) \cong T \otimes \text{image } d_i \cong T \otimes \ker d_{i-1}$, $\tilde{H}^*(K(T \otimes V_i);Z_p) = T \otimes M_i$ and f_i^* being given by $1 \otimes d_i$. (Note: For brevity of notation dimension shifts involved are ignored here). Thus the homotopy of $X \wedge E_p$ is built up by exact sequences from the Z_p vector spaces $T \otimes V_i$.

On the other hand, $P_* = \bigoplus_s (T \otimes V_s) \subset \pi_*(X \wedge E_p)$ so that by dimension count one must have $\pi_*(X \wedge E_p) = P_*$, proving the lemma. **

Remark: One may consider this as computing the Adams spectral sequence for $X \wedge E_p$. Since $\pi_*(X \wedge E_p)$ contains the vector space P_* isomorphic to the E_1 term, this spectral sequence must collapse and all extensions must be trivial.

Theorem: The group Ω_n^U is zero for n odd and for $n = 2m$ is free abelian of rank equal to the number of partitions of m. In addition, two stably almost complex manifolds are cobordant if and only if they have the same integral cohomology characteristic numbers.

Proof: Since $\tilde{H}^*(\underset{\sim}{TBU};Z)$ has no torsion and the mod p groups are free over $\mathcal{A}_p/(Q_0)$ for every prime p, the homotopy groups Ω_n^U are torsion free. Since the kernel of the Hurewicz homomorphism $\Omega_n^U \longrightarrow \tilde{H}_n(\underset{\sim}{TBU};Z)$ is a torsion group, while Ω_n^U is torsion free, complex cobordism is determined by integral cohomology characteristic numbers. **

In order to get the remaining structure use will be made of complex K theory. References for this are: Atiyah and Hirzebruch [18], Bott [24], Atiyah [15], and Husemoller [59]. Briefly, there is a multiplicative cohomology theory K^* indexed by the integers (positive and negative) for which $K^0(X)$ is the Grothendieck group of isomorphism classes of complex vector bundles over X. This cohomology theory is periodic of period 2, the periodicity isomorphism $p : K^i(X) \longrightarrow K^{i-2}(X)$ being given by multiplication by a generator $p(1) \in K^{-2}(pt) \cong \tilde{K}^0(S^2) \cong Z$.

For a complex vector bundle ξ over X one defines the Chern character of ξ, $ch(\xi) \in H^*(X;Q)$ by $ch(\xi) = \dim \xi + \sum_{i=1}^{\infty} (s_{(i)}(c(\xi)))/i!$. If $\xi = \ell_1 \oplus \ldots \oplus \ell_n$ is a sum of line bundles with $c_1(\ell_i) = x_i$, then $ch(\xi) = \sum_{i=1}^{n} e^{x_i}$. By primitivity of the s-classes, $ch(\xi \oplus \eta) = ch(\xi) + ch(\eta)$ so that ch extends to a homomorphism $K^0(X) \longrightarrow H^*(X;Q)$. Using the tensor product of bundles, which induces the product in $K^0(X)$ one has $ch(\xi \otimes \eta) = ch(\xi)\cdot ch(\eta)$ for if $\xi = \ell_1 \oplus \ldots \oplus \ell_n$, $\eta = m_1 \oplus \ldots \oplus m_p$ then $\xi \otimes \eta = \Sigma\, \ell_i \otimes m_j$, but $c_1(\ell_i \otimes m_j) = c_1(\ell_i) + c_1(m_j)$ so

$\mathrm{ch}(\xi \otimes \eta) = \Sigma\, e^{x_i+y_j} = (\Sigma\, e^{x_i})\cdot(\Sigma\, e^{y_j}) = \mathrm{ch}(\xi)\cdot\mathrm{ch}(\eta)$, and hence by the splitting principle in rational cohomology the formula holds for any pair of vector bundles. Thus $\mathrm{ch} : K^0(X) \longrightarrow H^*(X;Q)$ is a ring homomorphism.

For ξ a vector bundle over S^2, $\mathrm{ch}(\xi) = \dim \xi + c_1(\xi) \in H^*(S^2;Z)$ or $\mathrm{ch} : \tilde{K}^0(S^2) \longrightarrow H^2(S^2;Z)$. In particular, one may consider S^2 as $CP(1)$ and then $\mathrm{ch}(1-\lambda) = 1-e^{\alpha} = -\alpha$. Thus $\mathrm{ch} : \tilde{K}^0(S^2) \longrightarrow H^2(S^2;Z)$ is an isomorphism and one choose the generator $p(1) \in \tilde{K}^0(S^2)$ to be such that $\mathrm{ch}(p(1))[S^2] = 1$. (Simply take $p(1) = 1 - \lambda$. <u>Note</u>: This is the same as $\bar{\lambda} - 1$ on S^2, which is Bott's choice.)

One may then extend the Chern character to a ring homomorphism $\mathrm{ch} : K^*(X) \longrightarrow H^*(X;Q)$ so that $\mathrm{ch}(p(x)) = \mathrm{ch}\, x$. For $K^{2i}(X)$ one could use that as definition, leaving one to check the compatibility condition of commutativity in

$$\begin{array}{ccc}
K^0(X) & \xrightarrow{\quad\mathrm{ch}\quad} & H^*(X;Q) \\
p\downarrow & & \downarrow\Sigma^2 \\
K^{-2}(X) = \tilde{K}^0(S^2 \wedge X) & \xrightarrow{\mathrm{ch}} & \tilde{H}^*(S^2 \wedge X;Q) \cong H^*(X;Q)
\end{array}$$

which is immediate since $\mathrm{ch}(p(1)) = \iota \in H^2(S^2;Z)$, Σ^2 being suspension. It then suffices to define ch on $K^{-1}(X)$ and for this one takes

$$K^{-1}(X) \cong \tilde{K}^0(S^1 \wedge X) \xrightarrow{\mathrm{ch}} \tilde{H}^*(S^1 \wedge X;Q) \cong H^*(X;Q).$$

In order to apply K-theory to complex cobordism, let $\tilde{\alpha} = p^{-1}(\lambda-1)$ in $K^2(CP(n))$ where λ is the canonical line bundle. Then $K^*(CP(n))$ is the free $K^*(pt)$ module on $1, \tilde{\alpha},\ldots,\tilde{\alpha}^n$ and $\tilde{\alpha}^{n+1} = 0$, with $\tilde{\alpha}^n$ the image of $(-1)^n \iota$, $\iota \in \tilde{K}^{2n}(S^{2n})$ being the generator. Thus, by the general theory of characteristic classes one has defined K theoretic Chern classes, denoted $\gamma_i(\xi) \in K^{2i}(X)$, for ξ a complex vector bundle over X.

Note: Following Atiyah [12], if ξ is a vector bundle over X, let $\lambda_t(\xi) = \sum_{i=0}^{\infty} \lambda^i(\xi)\cdot t^i \in K^0(X)[[t]]$ where λ^i denotes the i-th exterior power. Then $\lambda_t(\xi \oplus \eta) = \lambda_t(\xi)\lambda_t(\eta)$ and one may extend λ_t to a homomorphism from $K^0(X)$ into $K^0(X)[[t]]$. Then $\lambda_{t/(1-t)}(\xi - \dim \xi) = \sum_{i=0}^{\infty} t^i \cdot p^i(\gamma_i(\xi))$. Thus the K-theory Chern classes are the Atiyah γ functions of stable vector bundles except for a periodicity factor. It should be noted that this does not coincide with the K-theory Chern classes of Bott [24] which use $1 - \bar{\lambda}$ rather than $\lambda - 1$, these being related by $p^{-1}(1 - \bar{\lambda}) = -c^*(\tilde{\alpha})$ where $c : \mathbb{C}P(V) \longrightarrow \mathbb{C}P(V)$ is given by complex conjugation.

The previous difficulty with c^* really exists in complex K-theory. Since $\lambda\bar{\lambda} = 1$, $\bar{\lambda} = 1/\lambda = 1/[1 + (\lambda - 1)] = 1 - (\lambda - 1) + (\lambda - 1)^2 + \ldots$ or $\bar{\lambda} - 1 = -(\lambda - 1) + (\lambda - 1)^2 + \ldots$. Thus $c^*(\tilde{\alpha}) = -\tilde{\alpha} + p(1)\cdot\tilde{\alpha}^2 + \ldots$. *If one could choose a generator* β *for which* $c^*(\beta) = -\beta$, *then* $\tilde{\alpha} = \beta + \sum_{i\geq 2} \phi_i \beta^i$ so $c^*(\tilde{\alpha}) = -\beta + \sum_{i\geq 2} \phi_i(-\beta)^i = -\tilde{\alpha} + 2 \sum_{i\equiv 0(2)} \phi_i\beta^i$ or $c^*(\tilde{\alpha}) \equiv -\tilde{\alpha} \bmod 2$. On the other hand $p(\bar{1}) \not\equiv 0\ (2)$ so $c^*(\tilde{\alpha}) \not\equiv -\tilde{\alpha} \bmod 2$. Thus one cannot choose generators well in complex K theory.

Combining the periodicity phenomenon with the Chern character permits the calculation of K-theoretic characteristic numbers. Specifically:

Lemma: Let M^n be a compact stably almost complex manifold and $x \in K^i(M,\partial M)$. Then

$$x[M,\partial M] = \begin{cases} 0 & \text{if } n-i \text{ is odd,} \\ \{ch(x)\cdot \mathscr{J}(M)\}[M,\partial M]\cdot p(1)^s \in K^{-2s}(pt) & \text{if } n-i = 2s, \end{cases}$$

where $\mathscr{J}(M) \in H^*(M;Q)$ is the class given by $\prod_{j=1}^{k} x_j/(e^{x_j} - 1)$ when $c_i(\tau(M))$ are expressed formally as the i-th elementary symmetric functions of two dimensional classes $x_1,\ldots,x_k$.

<u>Proof</u>: If λ is the canonical bundle over $\mathbb{C}P(n)$, then the Thom class $\tilde{U}_\lambda \in \tilde{K}^2(\mathbb{C}P(n+1))$ is $\tilde{\alpha}$ so $ch(\tilde{U}_\lambda) = ch\tilde{\alpha} = e^\alpha - 1$. Thus $ch(\tilde{U}_\lambda) = (\frac{e^\alpha - 1}{\alpha})U_\lambda$ where $U_\lambda = \alpha$ is the Z cohomology Thom class. If for any bundle ξ one defines $\mathscr{J}(\xi)$ to be $\prod_{j=1}^{k} x_j/(e^{x_j} - 1)$ when formally $c(\xi) = \prod_{j=1}^{k}(1+x_j)$, then, via the splitting principle and multiplicativity of Thom classes one has $ch(\tilde{U}_\xi) = \mathscr{J}(\xi)^{-1}\cdot U_\xi = \mathscr{J}(-\xi)\cdot U_\xi$ for all ξ.

If M is a stably almost complex manifold with $x \in K^i(M,\partial M)$ then one has

$$S^{n+2r} \xrightarrow{c} TN/TN' \xrightarrow{\emptyset} (M/\partial M) \wedge TN$$

for which

$$X[M,\partial M] = c^*\emptyset^*(x \otimes \tilde{U}) \in \tilde{K}^{i+2r}(S^{n+2r}) = K^{i-n}(pt).$$

This is zero if $i - n$ is odd, and if $n - i = 2s$, this is of the form $\theta\cdot p(1)^s = c^*\emptyset^*(x \otimes \tilde{U})$. Then $\theta = ch(c^*\emptyset^*(x \otimes \tilde{U}))[S^{n+2r}] = c^*\emptyset^*(chx\cdot ch\tilde{U})[S^{n+2r}] = c^*\emptyset^*(chx\cdot\mathscr{J}(M)\cdot U)[S^{n+2r}] = \{chx\cdot\mathscr{J}(M)\}[M,\partial M]$. **

In order to compute characteristic numbers for specific manifolds consider first $\mathbb{C}P(n)$. Since for $\mathbb{C}P(n)$ one has $(\tau \otimes 1) \cong (n+1)\xi$ where $\xi = \bar{\lambda}$, one has $\gamma_i(\tau) = \binom{n+1}{i}[p^{-1}(\xi-1)]^i$. Thus one needs the evaluation:

<u>Lemma</u>: $[p^{-1}(\xi-1)]^j[\mathbb{C}P(n)] = (-1)^{n-j}p(1)^{n-j}$.

<u>Proof</u>: By the previous lemma

$$[p^{-1}(\xi-1)]^j[\mathbb{C}P(n)] = [ch(\xi-1)]^j\mathscr{J}(\mathbb{C}P(n))[CP(n)]\cdot p(1)^{n-j}.$$

Then $c(\xi) = 1-\alpha$ and $(-\alpha)^n[\mathbb{C}P(n)] = 1$ so

$$[p^{-1}(\xi-1)]^j[\mathbb{C}P(n)] = (e^{-\alpha}-1)^j(\frac{-\alpha}{e^{-\alpha}-1})^{n+1}[CP(n)]\cdot p(1)^{n-j},$$

$$= \{\text{coefficient of } (-\alpha)^n \text{ in } (-\alpha)^{n+1}/(e^{-\alpha}-1)^{n+1-j}\} \cdot p(1)^{n-j},$$

$$= (1/2\pi i) \oint_{z=0} \frac{dz}{(e^z-1)^{n+1-j}} \cdot p(1)^{n-j},$$

$$= (1/2\pi i) \oint_{u=0} \frac{du}{u^{n+1-j}(1+u)} \cdot p(1)^{n-j}, \quad (u = e^z-1),$$

$$= (1/2\pi i) \oint_{u=0} \frac{\sum_{k=0}^{\infty} (-u)^k du}{u^{n+1-j}} \cdot p(1)^{n-j},$$

$$= (-1)^{n-j}p(1)^{n-j}. \quad **$$

In addition to the manifolds previously constructed it is convenient to consider the manifolds $H_{q^s,\ldots,q^s} \subset \mathbb{C}P(q^s) \times \ldots \times \mathbb{C}P(q^s)$ (with q factors) dual to the line bundle $\pi_1^*\xi \otimes \pi_2^*\xi \otimes \ldots \otimes \pi_q^*\xi$, where q is a prime. This is a hypersurface of degree $(1,\ldots,1)$ in the product of q copies of $\mathbb{C}P(q^s)$ and may be taken to be a projective algebraic variety. One has:

Lemma: The complex K theoretic characteristic numbers

$$s_\omega(\gamma(\tau))[H_{q^s,\ldots,q^s}]$$

are all congruent to zero modulo q if $n(\omega) \geq q^{s+1}-q$ unless $n(\omega) = q^{s+1}-q$ and ω refines $(q^s-1,\ldots,q^s-1)$ (q copies of q^s-1). In particular,

$$s_{(q^s-1,\ldots,q^s-1)}(\gamma(\tau))[H_{q^s,\ldots,q^s}] \equiv p(1)^{q-1}$$

modulo q.

Proof: Denote $H_{q^s,\ldots,q^s}$ simply by H, $\mathbb{C}P(q^s) \times \ldots \times \mathbb{C}P(q^s)$ simply by CP, and let $\xi_i = \pi_i^*(\xi)$ and $\beta_i = p^{-1}(\xi_i-1)$. Then (ignoring restriction homomorphisms in notation),

$$\gamma(H) = \prod_{j=1}^{q} (1 + \beta_j)^{q^s+1}/1 + p^{-1}(\xi_1 \otimes \ldots \otimes \xi_q - 1)$$

and H is dual to $p^{-1}(\xi_1 \otimes \ldots \otimes \xi_q - 1)$.

One has

$$\xi_1 \otimes \ldots \otimes \xi_q = [1+(\xi_1-1)] \otimes \ldots \otimes [1+(\xi_q-1)] = (1+p\beta_1) \otimes \ldots \otimes (1+p\beta_q)$$

or

$$p^{-1}(\xi_1 \otimes \ldots \otimes \xi_q-1) = \Sigma\beta_1 + p\Sigma\beta_1\beta_2 + \ldots + p^{k-1}\Sigma\beta_1 \ldots \beta_k + \ldots + p^{q-1}\beta_1 \ldots \beta_q$$

where $\Sigma\, \beta_1 \ldots \beta_k$ is the k-th elementary symmetric function in the β's. Thus letting v_i be the i-th elementary symmetric function of the β's,

$$\gamma(H) = (1 + v_1 + \ldots + v_q)^{q^s+1}/(1 + v_1 + pv_2 + \ldots + p^{q-1}v_q)$$

and H is dual to $v_1 + pv_2 + \ldots + p^{q-1}v_q$.

Noting that $p^{t-1}v_t = p(1)^{t-1}\cdot v_t$ any characteristic class $s_\omega(\gamma)(H)$ may be expressed as a polynomial with coefficients in $K^*(pt)$ in the variables $v_1,\ldots,v_q$ and the polynomial in question has degree in the v's or β's at least $n(\omega)$, the terms $p^{t-1}v_t$, $t > 1$, in the denominator giving rise to terms of degree greater than $n(\omega)$.

Thus $s_\omega(\gamma)(H)[H] = P_\omega(v_1,\ldots,v_q)\cdot(v_1 + pv_2 + \ldots + p^{q-1}v_q)[\mathbb{C}P]$ where P_ω is the polynomial just discussed. Now if σ is a permutation of $1,\ldots,q$, one has

$$p^{t-1}P_\omega\cdot\beta_{\sigma(1)} \cdots \beta_{\sigma(t)}[\mathbb{C}P] = p^{t-1}P_\omega\cdot\beta_1 \cdots \beta_t[\mathbb{C}P]$$

in $K^*(pt)$ by symmetry among the factors. Since $p^{t-1}v_t$ has precisely $\binom{q}{t}$ terms $p^{t-1}\beta_{\sigma(1)} \cdots \beta_{\sigma(t)}$, the numbers $P_\omega p^{t-1}v_t[\mathbb{C}P]$ are all divisible by q if $1 \leq t < q$. By exactly the same argument, the terms of P_ω involving $v_1,\ldots,v_{q-1}$ contribute zero in mod q characteristic numbers.

Thus the mod q characteristic numbers are the same as if $\gamma(H)$ was $(1 + v_1 + \dots + v_q)^{q^s+1}/(1 + p^{q-1}v_q)$ and H were dual to $p^{q-1}v_q$.

One then has $s_\omega(\gamma)[H] \equiv Q_\omega(v_1,\dots,v_q,p^{q-1}v_q)\cdot p^{q-1}v_q[\mathbb{C}P]$ where Q_ω is an integral polynomial of degree $n(\omega)$ in $v_1,\dots,v_q,p^{q-1}v_q$ (the latter having degree 1). Any monomial $a = v_1^{k_1}\dots v_q^{k_q}(p^{q-1}v_q)^{k_{q+1}}$ of Q_ω of degree $n(\omega)$ makes $a\cdot p^{q-1}v_q$ of degree $n(\omega) + q + k_{q+1}(q-1)$ in the β's, and since $\beta_i^{q^s+1} = 0$, $a\cdot p^{q-1}v_q = 0$ if $n(\omega) + q + k_{q+1}(q-1) > q^{s+1}$. If $n(\omega) > q^{s+1}-q$, this occurs so $s_\omega(\gamma)[H] \equiv 0 \bmod q$. For $n(\omega) = q^{s+1}-q$ this occurs if $k_{q+1} > 0$.

For characteristic numbers mod q of degree $q^{s+1}-q$, the numbers are the same as if $\gamma(H)$ were $\prod_{j=1}^{q} (1 + \beta_j)^{q^s+1}$ and H were dual to $p^{q-1}\beta_1\dots\beta_q$. Then $s_\omega(\gamma)(H)$ is the sum of monomials $\beta_1^{i_1}\dots\beta_q^{i_q}$ with $\omega = \omega_1 \cup \dots \cup \omega_q$, $\deg \omega_t = i_t$, $\Sigma\, i_t = q^{s+1}-q$ and to be nonzero one must have $i_1 = \dots = i_q = q^s-1$.

Thus for $s_\omega(\gamma)[H] \not\equiv 0(q)$ with $n(\omega) \geq q^{s+1}-q$ one must have $n(\omega) = q^{s+1}-q$ and ω refining $(q^s-1,\dots,q^s-1)$. In particular

$$s_{(q^s-1,\dots,q^s-1)}(\gamma)[H] = (q^s+1)^q\beta_1^{q^s}\dots\beta_q^{q^s}\cdot p(1)^{q-1}[\mathbb{C}P] = (q^s+1)^q\cdot p(1)^{q-1}$$

if $q^s-1 > 0$, i.e. $q^s > 1$ so $q^s \equiv 0\ (q)$ and this is $p(1)^{q-1}$. If $q^s-1 = 0$ so $s = 0$, then

$$s_{(q^s-1,\dots,q^s-1)}(\gamma)[H] = 1\cdot p^{q-1}\beta_1 \dots \beta_q[\mathbb{C}P] = p(1)^{q-1}. \quad **$$

Returning now to BU, recall from Chapter V that $H^*(BU;Z)$ is the formal power series ring over Z on the universal Chern classes c_i of dimension $2i$, and $H^*(BU;Q)$ is the rational power series algebra on these classes. One may consider $H_*(BU;Z) = \operatorname{Hom}(H^*(BU;Z);Z)$ as a subring of

$H_*(BU;Q) = \mathrm{Hom}(H^*(BU;Q);Q)$, being respectively the polynomial ring over Z or Q on classes α_i of dimension $2i$, where α_i is dual to $s_{(i)}(c)$ with respect to the base consisting of the $s_\omega(c)$. $(s_\omega(c)(\alpha_i) = \delta_{\omega,(i)})$.

Writing the Chern classes formally so that c_i is the i-th elementary symmetric function in variables x_j of dimension 2, let $s_\omega(e) \in H^*(BU;Q)$ be the s_ω symmetric function of the $e^{x_j}-1$ and let $\mathscr{S} \in H^*(BU;Q)$ be the product of the $x_j/(e^{x_j}-1)$. Under the diagonal homomorphism $\Delta : H^*(BU;Q) \longrightarrow H^*(BU;Q) \otimes H^*(BU;Q)$ one has $\Delta s_\omega(e) = \sum_{\omega'\omega''=\omega} s_{\omega'}(e) \otimes s_{\omega''}(e)$ and $\Delta\mathscr{S} = \mathscr{S} \otimes \mathscr{S}$.

Define a function $\rho : H_*(BU;Q) \longrightarrow Q[\beta_i]$ by $\rho(\alpha) = \sum_\omega s_\omega(e)\mathscr{S}[\alpha]\cdot\beta_\omega$ where for $\omega = (i_1,\ldots,i_r)$, $\beta_\omega = \beta_{i_1} \ldots \beta_{i_r}$. By the diagonal formulae this is a ring homomorphism. (<u>Note</u>: $e^x-1 = x +$ higher terms in x, so $s_\omega(e)\mathscr{S} = s_\omega(c) +$ higher terms, and the sums involved are finite.).

Let $B_n \subset H_n(BU;Q) = \mathrm{Hom}(H^n(BU;Q);Q)$ be the set of elements $\alpha \in H_n(BU;Q)$ with $\rho(\alpha) \in Z[\beta_i]$, and let $B_* = \oplus_n B_n \subset H_*(BU;Q)$. B_* is a subring of $H_*(BU;Q)$. Since $H_{2k}(BU;Z) = \{\alpha \in H_{2k}(BU;Q) | s_\omega(c)[\alpha] \in Z \text{ if } n(\omega) = k\}$ and for $u \in B_{2k}$, $n(\omega) = k$ gives $s_\omega(c)[u] = s_\omega(e)\mathscr{S}[u] \in Z$, one has $B_{2k} \subset H_{2k}(BU;Z)$. Trivially $B_{2k+1} \subset H_{2k+1}(BU;Z)$ since both groups are zero. Let $\rho_q : B_* \longrightarrow Z_p[\beta_i]$ be the composition of $\rho : B_* \longrightarrow Z[\beta_i]$ and reduction mod q, q a prime.

If M^n is a closed stably almost complex manifold, let $\tau(M) : H^*(BU;Q) \longrightarrow Q$ be the homomorphism which sends x to the value of the tangential characteristic class $x(\tau)$ on the fundamental homology class of M. This defines a ring homomorphism $\tau : \Omega_*^U \longrightarrow H_*(BU;Q)$ which was previously shown to be monic. Since for any bundle η one has $s_\omega(e)(\eta) = ch(s_\omega(\gamma(\eta)))$ one has for all ω

$$\{s_\omega(e)\mathscr{S}\}[\tau M] = ch(s_\omega(\gamma(\tau)))\cdot\mathscr{S}(M)[M],$$

$$= \begin{cases} 0 & \text{if } n-n(\omega) \text{ is odd,} \\ s_\omega(\gamma(\tau))[M]\cdot p(1)^{-t} & \text{if } n-n(\omega) = 2t, \end{cases}$$

which is integral and thus $\tau M \in B_n$. Thus one has inclusions $\tau\Omega_n^U \subset B_n \subset H_n(BU;Z)$.

Definition: If $P \in Z_q[\beta_1,\ldots]$ is any polynomial in variables β_i, P is said to have largest monomial $\beta_{i_1} \ldots \beta_{i_r}$ if

(1) the coefficient of $\beta_{i_1} \ldots \beta_{i_r}$ in P is nonzero, and

(2) if the coefficient of $\beta_{j_1} \ldots \beta_{j_s}$ is nonzero, with $\beta_{j_1} \ldots \beta_{j_s} \neq \beta_{i_1} \ldots \beta_{i_r}$ then either

(a) $j_1 + \ldots + j_s < i_1 + \ldots + i_r$, or

(b) $j_1 + \ldots + j_s = i_1 + \ldots + i_r$ and $s > r$.

(Note: A polynomial need not have a largest monomial).

If $P,Q \in Z_q[\beta_1,\ldots]$ have largest monomials β_ω and $\beta_{\omega'}$, then $P\cdot Q$ has largest monomial $\beta_\omega\cdot\beta_{\omega'} = \beta_{\omega\cup\omega'}$. If $P_i \in Z_q[\beta_1,\ldots]$, $i = 1,\ldots,n$, are polynomials having distinct largest monomials, then the polynomials P_i are linearly independent over Z_q.

Proposition: There exist almost complex manifolds M_i^p of dimension $2i$ for each prime p and each integer i such that $\rho_p(\tau M_i^p)$ has largest monomial β_i if $i + 1 \neq p^s$ for any s or $[\beta_{p^{s-1}-1}]^p$ if $i + 1 = p^s$ for some s.

Proof: If $i + 1 \not\equiv 0\ (p)$ let $M_i^p = \mathbb{C}P(i)$. Then $\{s_{(i)}(e)\mathscr{S}\}[\tau M_i^p] = s_{(i)}(c)[\mathbb{C}P(i)] = i + 1 \not\equiv 0\ (p)$.

If $i + 1 \equiv 0\ (p)$ but $i + 1 \neq p^s$ for any s, then one may write $i + 1 = p^r(pu+v)$ with $r > 0$ and $0 < v < p$. If $u = 0$, $v > 1$ and let $M_i^p = H_{p^r, p^r(v-1)}$, for which $\{s_{(i)}(e)\mathcal{S}\}[\tau M_i^p] = s_{(i)}(c)[\tau M_i^p] = -\binom{p^r v}{p^r} \not\equiv 0\ (p)$. If $u > 0$, let $M_i^p = H_{p^r v, p^{r+1} u}$, and then $\{s_{(i)}(e)\mathcal{S}\}[\tau M_i^p] = s_{(i)}(c)[\tau M_i^p] = -\binom{p^r(pu+v)}{p^r v} \not\equiv 0\ (p)$.

If $i + 1 = p^s$ for some s, let $M_i^p = H_{p^{s-1}, \ldots, p^{s-1}}$ (p subscripts). Then $\{s_\omega(e)\mathcal{S}\}[\tau M_i^p] = s_\omega(\gamma(\tau))[M_i^p]$ which is zero mod p if $n(\omega) \geq p^s - p$ unless $n(\omega) = p^s - p$ and ω refines $(p^{s-1}-1, \ldots, p^{s-1}-1)$. **

<u>Corollary</u>: If $\omega = (i_1, \ldots, i_r)$ let $M_\omega^p = M_{i_1}^p \times \ldots \times M_{i_r}^p$. Then for each prime p and each integer n, the polynomials $\rho_p(\tau M_\omega^p) = \rho_p(\tau M_{i_1}^p) \ldots \rho_p(\tau M_{i_r}^p)$ in $Z_p[\beta_1, \ldots]$ with $\omega \in \pi(n)$ are linearly independent.

<u>Proof</u>: The polynomials $\rho_p(\tau M^p)$ with $\omega \in \pi(n)$ have distinct largest monomials. **

<u>Lemma</u>: Let $\mathcal{R}_*$ be a graded subring of the graded polynomial ring $Z[\alpha_1, \alpha_2, \ldots]$, degree $\alpha_i = i$, and suppose that for each prime p there are elements $c_i^p \in \mathcal{R}_i$, $i \geq 1$, with $(\mathcal{R}_*/p\mathcal{R}_*) \cong Z_p[c_i^p]$. Then $\mathcal{R}_*$ is the integral polynomial ring on classes $b_i \in \mathcal{R}_i$, $i \geq 1$. If $\mathcal{S}_* \subset \mathcal{R}_*$ is a subring containing all of the c_i^p then $\mathcal{S}_* = \mathcal{R}_*$.

<u>Note</u>: All rings here are assumed to have the unit 1. To say that $\mathcal{R}_*$ is a graded subring means that the homogeneous components of elements of $\mathcal{R}_*$ themselves belong to $\mathcal{R}_*$.

Proof: Let $\mathcal{L}_* = Z[\alpha_1,\ldots]$. Since $\mathcal{A}_n \subset \mathcal{L}_n$, $\mathcal{A}_n$ is a free abelian group of rank at most $|\pi(n)|$ (the number of partitions of n). Since $\mathcal{A}_n \otimes Z_p = (\mathcal{A}_*/p\,\mathcal{A}_*)_n$ has dimension $|\pi(n)|$ over Z_p, $\mathcal{A}_n$ has rank exactly $|\pi(n)|$.

Let $a'_\omega = \Sigma\, \lambda'^{\omega}_{\omega'}\alpha_{\omega'}$, $\omega,\omega' \in \pi(n)$, $\lambda'^{\omega}_{\omega'} \in Z$ be any base of $\mathcal{A}_n$. Applying the usual triangularization process (as for integral matrices) one may form from $\{a'_\omega\}$ a new base $a_\omega = \Sigma\, \lambda^{\omega}_{\omega'}\alpha_{\omega'}$, ω, $\omega' \in \pi(n)$, $\lambda^{\omega}_{\omega'} \in Z$ in which $\lambda^{\omega}_{(n)} = 0$ if $\omega \neq (n)$. ($\lambda^{(n)}_{(n)}$ is the greatest common divisor of the $\lambda'^{\omega}_{(n)}$). For each n, let $b_n \in \mathcal{A}_n$ be any one of the $a_{(n)}$ obtained in this way. Since $\lambda^{(n)}_{(n)} \neq 0$ by rank, one may solve inductively to write α_i as a rational polynomial in the b_j $(j \leq i)$, and hence a base of $\mathcal{A}_n$ is given by b_n and elements $a_\omega = \Sigma\mu^{\omega}_{\omega'}b_{\omega'}$, ω, $\omega' \in \pi(n)$; $\omega,\omega' \neq (n)$, $\mu^{\omega}_{\omega'} \in Q$.

Suppose inductively that in dimensions less than n, $\mathcal{A}_*$ is the integral polynomial ring on the classes b_i, $i < n$. Let L be the free group generated by the a_ω and M the free group generated by the b_ω, $\omega \in \pi(n)$, $\omega \neq (n)$. Since M consists entirely of decomposable elements of $\mathcal{L}_n$, $M \subset L$, and since they have the same rank, the index of M in L is finite.

Let p be any prime. Since $\mathcal{A}_*$ is the integral polynomial ring on the b_i in dimensions less than n, c_i^p is an integral polynomial in the b's of degree i (if $i < n$), so $c_\omega^p \in M$ for all $\omega \neq (n)$, $\omega \in \pi(n)$. Thus the image of M in $\mathcal{A}_n \otimes Z_p$ has the same rank as the image of L (equal to $|\pi(n)|-1$) and the index of M in L cannot be divisible by p.

Since this is true for all primes $M = L$. Hence $\mathcal{A}_n$ has a base consisting of the b_ω, $\omega \in \pi(n)$. Hence $\mathcal{A}_*$ is the integral polynomial ring on the classes b_i, $i < n + 1$ in dimensions less than $n + 1$, and by induction $\mathcal{A}_* = Z[b_1,b_2,\ldots]$.

To show that $\mathcal{S}_* = \mathcal{Q}_*$, $\mathcal{S}_n \subset \mathcal{Q}_n$ is a free abelian group and for any prime p, $\mathcal{S}_n$ maps onto $\mathcal{Q}_n \otimes Z_p$, so the rank of $\mathcal{S}_n$ is $|\pi(n)|$ and the index of $\mathcal{S}_n$ in $\mathcal{Q}_n$ is not divisible by p. Thus $\mathcal{S}_n = \mathcal{Q}_n$ for each n. **

Theorem: Ω^U_* is the integral polynomial ring on classes x_i of dimension 2i. A stably almost complex manifold M^{2i} may be taken to be the 2i-dimensional generator if and only if

$$s_{(i)}(c(\tau))[M^{2i}] = \begin{cases} \pm 1 & \text{if } i+1 \neq p^s \text{ for any prime } p \\ \pm p & \text{if } i+1 = p^s \text{ for some prime } p \text{ and integer } s. \end{cases}$$

Proof: Let $Z[\alpha_1, \ldots] = H_*(BU;Z)$, $\mathcal{Q}_* = B_*$, and $\mathcal{S}_* = \tau\Omega^U_*$ in the lemma, and let $c^p_i = \tau M^p_i$ as defined above. Then $\mathcal{S}_* = \mathcal{Q}_*$ is the integral polynomial ring on classes b_i of dimension 2i. Further, a generator is characterized by its s-class. Under reduction mod p one has $\Omega^U_* \otimes Z_p = Z_p[b_i] = Z_p[c^p_i]$, so that $b_i = x\cdot c^p_i + u + pv$ with $x \in Z$, $x \not\equiv 0(p)$, u decomposable, $u,v \in \Omega^U_{2i}$. Thus $s_{(i)}(c)[b_i] \equiv x\cdot s_{(i)}(c)[c^p_i]$ mod p, giving $s_{(i)}(c)[b_i] \equiv 0$ mod p if $i+1 = p^s$ and $s_{(i)}(c)[b_i] \not\equiv 0\ (p)$ if $i+1 \neq p^s$. Thus if $i+1 \neq p^s$ for any p, $s_{(i)}(c)[b_i]$ is not divisible by any prime so must be ±1. If $i+1 = p^s$, then p is unique so $s_{(i)}(c)[b_i]$ is divisible by only the prime p. Since

$$s_{p^s-1}(c)[H_{p^{s-1},\ldots,p^{s-1}}] =$$

$$\{\sum_{i=1}^{p} (p^{s-1}+1)(\pi_i^*\bar{\alpha})^{p^s-1} - (\sum_{i=1}^{p} \pi_i^*\bar{\alpha})^{p^s-1}\}\cdot(\sum_{i=1}^{p} \pi_i^*\bar{\alpha})[\mathbb{C}P] = -(\sum_{i=1}^{p} \pi_i^*\bar{\alpha})^{p^s}[\mathbb{C}P] =$$

$$-\binom{p^s}{p^{s-1}}\binom{p^{s-1}(p-1)}{p^{s-1}} \cdots \binom{p^{s-1}}{p^{s-1}}$$

is not divisible by p^2, one must have $s_{(i)}(c)[b_i] = \pm p$. (Note: This is the same type of computation as in Chapter V. Here $s \geq 1$ so $p^s-1 > p^{s-1}$ and $(\pi_i^*\bar{\alpha})^{p^s-1} = 0$ provided $p \neq 2$ or $s > 1$. For $p = 2$, $s = 1$, this becomes $\{\sum_1^2 2\pi_i^*\bar{\alpha} - \sum_1^2 \pi_i^*\bar{\alpha}\}(\sum_1^2 \pi_i^*\bar{\alpha})[\mathbb{C}P]$ which changes the sign but not the divisibility by p^2.). **

Theorem: All relations among the integral cohomology Chern numbers of closed stably almost complex manifolds come from complex K theory. Specifically, if $\phi : H^n(BU;Q) \longrightarrow Q$ is a homomorphism, there is a closed stably almost complex manifold M^n with $\phi(x) = x(\tau)[M^n]$ for all $x \in H^n(BU;Q)$ if and only if ϕ sends the n-dimensional component of each $s_\omega(e)\mathcal{J}$ into an integer.

Proof: This is the fact that $B_* = \tau\Omega_*^U$ proved above. **

Remarks: 1) This may also be phrased: The image of the Hurewicz homomorphism $\pi_{2k+2N}(TBU_N) \longrightarrow \tilde{K}_{2k+2N}(TBU_N)$ is a direct summand (N large with respect to k) (Hattori [52]).

2) The completeness of these relations was conjectured by Atiyah and Hirzebruch [19].

Now suppose $i + 1 = p^s$ for some prime p (unique) and write $b_i = xc_i^p + u + pv$, $x \in Z$, $x \not\equiv 0$ (p), u decomposable, $u,v \in \Omega_{2i}^U$. One may then replace b_i by $b_i' = b_i - u = xc_i^p + pv$ giving another acceptable generator. For $\omega \in \pi(i)$, this gives

$$s_\omega(c)[b_i'] = xs_\omega(\gamma(\tau))[M_i^p] + ps_\omega(c)[v]$$

which is divisible by p. Thus one has:

Theorem: One may choose generators $x_i \in \Omega_{2i}^U$ such that if $i + 1 = p^s$, all integral cohomology Chern numbers of x_i will be divisible by p.

Remark: That this is possible was first noted by Conner and Floyd [36], section 41, who called such generating manifolds "Milnor manifolds".

This gives the following relationship with integral homology pointed out by Joel M. Cohen [32].

Corollary: There exist polynomial generators x_i, $i \geq 1$, of Ω_*^U and z_i, $i \geq 1$, of $H_*(BU;Z)$ such that $\tau x_i = m_i \cdot z_i$ where m_i is p if $i + 1 = p^s$ for some prime p and is 1 otherwise. If $\omega = (i_1, \ldots, i_r)$, let $m_\omega = m_{i_1} \ldots m_{i_r}$. Then $H_{2k}(BU;Z)/\tau\Omega_{2k}^U$ is the direct sum of the cyclic groups $Z/m_\omega Z$ for $\omega \in \pi(k)$.

Proof: Choose x_i to be the classes of Milnor manifolds with $s_{(i)}(c)[x_i] = m_i$. If $i + 1 = p^s$, $\tau x_i \in H_{2i}(BU;Z)$ maps to zero in $H_{2i}(BU;Z_p)$ so is divisible by p and uniquely so since $H_{2i}(BU;Z)$ is torsion free. Let $z_i = (1/m_i)\cdot\tau x_i$ in $H_{2i}(BU;Z)$. Since $s_{(i)}(c)[z_i] = 1$, z_i is an acceptable generator for $H_*(BU;Z)$. **

One also has the result of Milnor (see Hirzebruch [54] or Thom [129]):

Theorem: Every class $x \in \Omega_n^U$ contains a non-singular algebraic variety (not necessarily connected) if $n > 0$.

Proof: Let $\mathcal{U}_* \subset \Omega_*^U$ be the set of cobordism classes represented by non-singular algebraic varieties. $\mathcal{U}_*$ is closed under sums (disjoint unions) and products, but not necessarily under additive inverses. (Note: If one could sensibly interpret $-1 \in \Omega_0^U$ as a variety inverses would exist trivially).

Now $\mathcal{U}_*$ contains the classes of the $\mathbb{C}P(n)$ and H_{n_1,n_2} which generate Ω_*^U (Note: $s_{(p^s-1)}(H_{p^{s-1},p^s-p^{s-1}}) = -\binom{p^s}{p^{s-1}}$ if $p^{s-1} > 1$ while

$s_{(p-1)}(CP(p-1)) = p$ so these are not divisible by p^2; hence $\{\mathbb{C}P(n), H_{n_1,n_2}\}$ generate $\Omega_*^U \otimes Z_p$ for all primes p, and hence the subring they generate coincides with Ω_*^U).

If one can show that there exist classes $x_i, x_i' \in \mathcal{U}_{2i}$ with $s_{(i)}(c)[x_i] = m_i$ and $s_{(i)}(c)[x_i'] = -m_i$ then one is done, for then suppose inductively that $\mathcal{U}_{2j} = \Omega_{2j}^U$ if $j < k$. If $x \in \Omega_{2k}^U$ then $s_{(i)}(c)[x] = tm_i$, $t \in Z$, and if $t > 0$, $x = tx_i + v$, if $t < 0$, $x = |t|x_i' + v$, where v is decomposable and hence $v \in \mathcal{U}_*$ (inductively). Thus also $x \in \mathcal{U}_*$.

For $\mathcal{U}_{2i}$, $i \geq 1$, one has $s_{i+1}[\mathbb{C}P(i)] = i + 1 > 0$. Let $M_i \subset \mathbb{C}P(i+1)$ be the hypersurface defined by $\sum_{j=0}^{i+1} z_j^{t+1} = 0$, $t \geq 1$, where $(z_0,\ldots,z_{i+1})$ give local coordinates. The derivative of $u = \sum z_j^{t+1}$ with respect to z_k is $(t+1)z_k^t$, not all partials can vanish simultaneously, and M_i is a non-singular hypersurface. Letting

$$f : \mathbb{C}P(i+1) \xrightarrow{\Delta} \prod_{j=1}^{t+1} \mathbb{C}P(i+1) \xrightarrow{g} \mathbb{C}P((t+1)(i+2)-1)$$

be the composition of the diagonal $\Delta(z) = (z,\ldots,z)$ and the map given in local coordinates by $u_{j_1\ldots j_{t+1}} = z_{j_1}^{(1)} \ldots z_{j_{t+1}}^{(t+1)}$, f is transverse regular on the hyperplane section $\sum u_{j\ldots j} = 0$ with preimage M_i. Since $g^*(\xi) = \xi \otimes\ldots\otimes \xi$, $f^*(\xi) = \xi^{t+1}$ and M_i is dual to ξ^{t+1}. Thus $c(M_i) = (1+\bar{\alpha})^{i+1}/(1+(t+1)\bar{\alpha})$ and $s[M_i] = (t+1)[i+1-(t+1)^i]$ and this is negative if $1 \leq i < t$.

Consider $A_{2k} = \{x \in Z | x = s_{(k)}(c)[u]$ for some $u \in \mathcal{U}_{2k}\}$. If $x,y \in A_{2k}$, then $x+y \in A_{2k}$. The above constructions show that A_{2k} contains both positive and negative elements. Let p be the least positive element of A_{2k} and n the largest negative element of A_{2k}. Then $p+n = 0$ (if $p+n > 0$, $p > p+n > 0$ contradicts the choice of p; if $p+n < 0$,

$n < p+n < 0$ contradicts the choice of n). If $q \in A_{2k}$, $q = tp + s$ with $t, s \in Z$, $0 \leq s < p$, but then $s = q + (-t)p \in A_{2k}$ if $t < 0$ and s $s = q + tn \in A_{2k}$ if $t > 0$ and since $s < p$, $s = 0$. Thus A_{2k} is the set of multiples of p. Since the greatest common divisor of the elements of A_{2k} is m_k, one has $p = m_k$, $n = -m_k$. **

Open question: (Hirzebruch [54]) Which classes of Ω_*^U contain connected non-singular algebraic varieties?

Remarks: 1) One may obtain the polynomial structure and the completeness of K-theoretic relations in other ways. The proof given here is based on Stong [117] and is simplified based upon Conner and Floyd [41], Chapter III. One may prove the polynomial structure using the Adams spectral sequence as was done by Milnor and Novikov (an exposition appears in Conner and Floyd [36], section 41 - for a similar situation) and then use the proof for completeness of relations given by Hattori [52].

2) If one uses the Bott choice for Chern classes with $p^{-1}(1-\bar{\lambda})$ as generator for $K^*(\mathbb{C}P(n))$ one obtains a different orientation class $U'_\xi \in \tilde{K}^*(T\xi)$ for which $\operatorname{ch} U'_\xi = T(\xi)^{-1}U_\xi$ where $T(\eta)$ is the universal Todd class given by $\Pi(x_i/1-e^{-x_i})$ if $c(\eta) = \Pi(1+x_i)$. The classes U'_ξ and $\tilde{U}_\xi$ are related by $U'_\xi = \det\bar{\xi}\cdot\tilde{U}_\xi$ (det being the determinant bundle), and $\det\bar{\xi}$ is an invertible element in K^0(Base space) which restricts to 1 at each point. The literature is very confused in that the choices of Chern classes and orientation are frequently made with opposite conventions. The choice made here was intended to keep the Atiyah Chern classes and keep a consistent universal orientation, avoiding complex conjugation whenever possible.

Relation to framed cobordism : The Adams invariant $e_{\mathbb{C}}$.

Just as with unoriented cobordism one has a forgetful functor F from the category of framed manifolds to that of stably almost complex manifolds giving a homomorphism of cobordism groups and a relative group, denoted $\Omega_*^{U,fr}$. As with any pair of (B,f) theories, the sequence

$$\begin{array}{ccc} \Omega_*^{fr} & \longrightarrow & \Omega_*^{U} \\ & \nwarrow \quad \swarrow & \\ & \Omega_*^{U,fr} & \end{array}$$

is exact. One then has:

Proposition: A framed manifold of positive dimension bounds a stably almost complex manifold; i.e. $F_* : \Omega_n^{fr} \longrightarrow \Omega_n^{U}$ is the zero homomorphism if $n > 0$. Further $F_* : \Omega_0^{fr} \longrightarrow \Omega_0^{U}$ is an isomorphism.

Proof: For $n > 0$, Ω_n^U is torsion free while Ω_n^{fr} is a finite group so $F_* = 0$. For $n = 0$ both groups are isomorphic to Z, given by oriented points. **

The homotopy exact sequences on the Thom spaces then split into short exact sequences giving the diagrams:

$$\begin{array}{ccccccccc} 0 & \longrightarrow & \Omega_n^U & \longrightarrow & \Omega_n^{U,fr} & \longrightarrow & \Omega_{n-1}^{fr} & \longrightarrow & 0 \\ & & \downarrow{\scriptstyle k_n} & & \downarrow{\scriptstyle h_n} & & \downarrow & & \\ 0 = \tilde{H}_n(\underset{\sim}{S};Z) & \longrightarrow & \tilde{H}_n(\underset{\sim}{TBU};Z) & \xrightarrow{\cong} & H_n(\underset{\sim}{TBU},\underset{\sim}{S};Z) & \longrightarrow & \tilde{H}_{n-1}(\underset{\sim}{S};Z) = 0 & & \end{array}$$

for $n-1 > 0$, and

$$\begin{array}{ccccccc}
0 = \Omega_1^U & \longrightarrow & \Omega_1^{U,fr} & \longrightarrow & \Omega_0^{fr} & \xrightarrow{\cong} & \Omega_0^U \\
\downarrow & & \downarrow & & \downarrow \cong & & \downarrow \cong \\
0 = \tilde{H}_1(TBU;Z) & \longrightarrow & H_1(TBU,S;Z) & \longrightarrow & \tilde{H}_0(S;Z) & \xrightarrow{\cong} & \tilde{H}_0(TBU;Z) \\
 & & \| & & \| & & \| \\
 & & 0 & & Z & & Z
\end{array}$$

where the vertical arrows are the Hurewicz homomorphisms.

For n odd, $\Omega_n^U = 0$ so if $n > 1$, $\Omega_n^{U,fr} \cong \Omega_{n-1}^{fr}$ which is a finite group. Also $H_n(TBU,S;Z) = 0$ and no information is available using Chern numbers.

For n even, $n > 0$, $h_n(\Omega_n^{U,fr})$ is a free abelian group, containing the subgroup $k_n(\Omega_n^U) \cong \Omega_n^U$. Thus $h_n(\Omega_n^{U,fr})$ has rank equal to the number of partitions of $n/2$ and contains $k_n(\Omega_n^U)$ as a subgroup of finite index.

Let $\alpha \in \Omega_{2k}^{U,fr}$ be represented by a stably almost complex manifold V^{2k} with a compatible framing of $\partial V = M^{2k-1}$. Let $\tau : (V,M) \longrightarrow (BU,*)$ be a map classifying the stable tangent bundle of V, the framing of M being interpreted as a specific equivalence class of deformation of M to the base point. One then has defined Chern numbers $\tau^*(c_\omega)[V,M]$ which completely determine $h_{2k}(\alpha)$.

In order that $h_{2k}(\alpha) \in k_{2k}(\Omega_{2k}^U)$ it is necessary and sufficient that $\tau^*(s_\omega(e)\mathscr{S})[V,M] \in Z$ for all ω. Since $s_\omega(e) = ch(s_\omega(\gamma))$ given by the K theoretic Chern classes $\gamma_i \in K^{2i}(BU,*)$ for $i > 0$, $\tau^*(s_\omega(e)\mathscr{S})[V,M] = \tau^* s_\omega(\gamma)[V,M]\cdot p(1)^{n(\omega)-k} \in Z$ where $\tau^* s_\omega(\gamma) \in K^{2n(\omega)}(V,M)$ for $n(\omega) > 0$. Thus one has:

<u>Theorem</u> (Conner and Floyd [41]): A necessary and sufficient condition that a stably almost complex manifold with framed boundary have the same Chern numbers as a closed stably almost complex manifold is that the $\mathscr{S}$ class be integral.

Since the homomorphism $\mathscr{T} : \Omega_{2k}^{U,fr} \longrightarrow Q : \alpha \longrightarrow \tau^{*}\mathscr{T}[V,M]$ sends Ω_{2k}^{U} into Z one has defined a quotient homomorphism $E : \Omega_{2k-1}^{fr} \longrightarrow Q/Z$. One has the result of Conner and Floyd [41]:

Theorem: The homomorphism $E : \Omega_{2k-1}^{fr} \longrightarrow Q/Z$ coincides with the Adams invariant $e_C : \lim\limits_{s\to\infty} \pi_{2k-1+s}(S^s) \longrightarrow Q/Z$.

Remarks: The Adams invariant is defined in Adams [4]. The proof given here is due to P. S. Landweber.

Proof: Let (V,M) be imbedded in $(H^{2k+2r}, R^{2k-1+2r})$ with complex normal bundle trivialized over M, defining the normal map $\nu : (V,M) \longrightarrow (BU_r,*)$ (trivialization determining the deformation of M to a point). Applying the Pontrjagin-Thom construction defines a map $f : (D^{2k+2r}, S^{2k+2r-1}) \longrightarrow (TBU_r, T*)$ and as with the unoriented case one has a diagram of cofibrations

$$\begin{array}{ccccc} S^{2r} & \xhookrightarrow{j} & X & \longrightarrow & X/S^{2r} = D^{2k+2r}/S^{2k+2r-1} \\ {\scriptstyle 1}\downarrow & & {\scriptstyle g}\downarrow & & {\scriptstyle \bar{f}}\downarrow \\ S^{2r} & \xhookrightarrow{i} & TBU_r & \longrightarrow & TBU_r/S^{2r} \end{array}$$

where X is the two cell complex formed by attaching D^{2k+2r} to S^{2r} by the map $f : S^{2k+2r-1} \longrightarrow S^{2r} = T*$, g and $\bar{f}$ being induced by f. In particular the class of f in the stable homotopy of spheres is the element corresponding to $[M] \in \Omega_{2k-1}^{fr}$.

The cohomology groups of X are free abelian with base $1 \in H^0(X;Z)$, $a \in H^{2r}(X;Z)$, and $b \in H^{2k+2r}(X;Z)$ characterized by $j^{*}(a) = \iota \in \tilde{H}^{2r}(S^{2r};Z)$ and $b = \pi^{*}(\iota')$ with $\iota' \in H^{2k+2r}(D^{2k+2r}, S^{2k+2r-1};Z)$.

To define the Adams invariant of the class of f one chooses any element $u \in \tilde{K}(X)$ with $ch(u) = a + \phi\cdot b$, $\phi \in Q$ (possible from the Atiyah-Hirzebruch [18] spectral sequence for K theory) and lets $e_C([f]) = \phi$ in Q/Z.

From the relation between the Thom homomorphism and characteristic numbers one has $\tau^*\mathscr{S}[V,M]\cdot\iota' = \bar{f}^*(\mathscr{S}^{-1}\cdot U)^{2k+2r}$ so $\tau^*\mathscr{S}[V,M]\cdot b = g^*(\mathscr{S}^{-1}U)^{2k+2r}$, $U \in H^{2r}(TBU_r;Z)$ being the Thom class. On the other hand $\mathscr{S}^{-1}U = ch(\tilde{U})$ where $\tilde{U} \in \tilde{K}^{2r}(TBU_r)$ is the K-theoretic Thom class. Since $j^*g^*U = i^*U = \iota$, $g^*U = g^*(\mathscr{S}^{-1}U)^{2r} = a$ and one has $ch(g^*p(1)^r\tilde{U}) = g^*(ch\tilde{U}) = a + \tau^*\mathscr{S}[V,M]\cdot b$ with $g^*p(1)^r\tilde{U} \in \tilde{K}(X) = \tilde{K}^0(X)$. Thus $e_C([f]) = \tau^*\mathscr{S}[V,M] = E(\alpha)$. **

From Adams' computations with e_C one has:

Corollary: The homomorphism $\mathscr{S} : \Omega_{2k}^{U,fr} \longrightarrow Q$ maps precisely onto the integral multiples of the numbers:

a) $1/d_{2t}$ for $\Omega_{8t}^{U,fr}$,

b) 1 for $\Omega_{8t-2}^{U,fr}$,

c) $1/d_{2t-1}$ for $\Omega_{8t-4}^{U,fr}$,

d) $1/2$ for $\Omega_{8t-6}^{U,fr}$,

where $d_{2t} = a_{2t}$, $2d_{2t+1} = a_{2t+1}$ and a_n is the denominator of $B_n/4n$, B_n being the n-th Bernoulli number.

For facts concerning Bernoulli numbers, see Milnor [86] or Adams [3]. That the Bernoulli numbers enter into the result is not surprising since

$$x/(e^x-1) = \sum_{t=0}^{\infty} \beta_t \frac{x^t}{t!}$$

where

$$\beta_{2s} = (-1)^{s-1}B_s,\ \beta_1 = -\frac{1}{2},\ \text{and}\ \beta_{2s+1} = 0\ \text{if}\ s > 0.$$

Relation to unoriented cobordism

The relation of complex cobordism to unoriented cobordism was completely explored by Milnor [87]. As with any pair of (B,f) theories one has the homomorphism $F_* : \Omega_*^U \longrightarrow \mathfrak{N}_*$ which is obtained by ignoring the complex structure.

Proposition: Let M^n be a stably almost complex manifold. Then the Stiefel-Whitney classes w_{2i+1} of M are zero and the classes w_{2i} are the mod 2 reductions of the Chern classes c_i. In particular, all Stiefel-Whitney numbers of M having an odd Stiefel-Whitney class as a factor must be zero.

Proof: This follows at once from the change of fields theorem of Chapter V. **

Proposition: A closed manifold M^n has all Stiefel-Whitney numbers with an odd class as a factor zero if and only if there is a manifold M' with M cobordant to $M' \times M'$.

Proof: If $M \sim M' \times M'$, M has the same Stiefel-Whitney numbers as $M' \times M'$. Under the comultiplication $\Delta(w_i) = \sum_{j+k=i} w_j \otimes w_k$ so

$$w_{i_1} \cdots w_{i_r}[M' \times M'] = \sum_{j_\alpha + k_\alpha = i_\alpha} w_{j_1} \cdots w_{j_r}[M'] \cdot w_{k_1} \cdots w_{k_r}[M'].$$

If $J = (j_1, \ldots, j_r)$, $K = (k_1, \ldots, k_r)$ and $J \neq K$, then the terms $w_J[M']w_K[M']$ and $w_K[M']w_J[M']$ are paired, having the same value and so add to zero in Z_2. In particular, if any i_α is odd, this pairs all terms and the number $w_{i_1} \cdots w_{i_r}[M' \times M']$ is zero. If every i_α is even, then $w_I[M' \times M'] = (w_{I/2}[M'])^2 = w_{I/2}[M']$.

Suppose all numbers of M^n divisible by an odd class are zero. If n is odd M^n must bound; this being interpreted as being a product vacuously. If $n = 2k$, consider the homomorphism $\phi : H^k(BO;Z_2) \longrightarrow Z_2$ with $\phi(w_{i_1} \dots w_{i_r}) = w_{2i_1} \dots w_{2i_r}[M]$ obtained by composing the doubling homomorphism $\psi : H^*(BO;Z_2) \longrightarrow H^*(BO;Z_2)$ which sends w_i into w_{2i} with the evaluation on M. Let $\lambda : H^*(BO;Z_2) \longrightarrow H^*(BO;Z_2)$ by sending w_i into $\sum_{j+k=i} w_j w_k$. λ is induced by maps of spaces $BO_r \longrightarrow BO_{2r}$ classifying $\gamma_r \oplus \gamma_r$, so λ commutes with all Sq^i. Further $\lambda(w_{2i+1}) = 0$, $\lambda(w_{2i}) = w_i^2$ so the kernel of λ is precisely the ideal generated by elements of odd degree. The composition $\lambda \cdot \psi : H^*(BO;Z_2) \longrightarrow H^*(BO;Z_2)$ is the homomorphism given by $x \longrightarrow x^2$.

Then $\lambda(Sq^{2i}\psi x) = Sq^{2i}(\lambda\psi x) = Sq^{2i}(x^2) = (Sq^i x)^2 = \lambda(\psi Sq^i x)$ and $\lambda(v\psi x) = v \cdot v \cdot \lambda\psi(x) = v^2 . x^2 = (vx)^2 = \lambda\psi(vx)$, where $v = 1 + v_1 + \dots$ is the Wu class. Comparing terms of equal degree $\lambda(v_{2i}\psi x) = \lambda\psi(v_i x)$. Thus for all $x \in H^{k-i}(BO;Z_2)$, $\psi(Sq^i x + v_i x) + Sq^{2i}\psi x + v_{2i}\psi x$ is in the kernel of λ, hence vanishes on the fundamental class of M. Then $\phi(Sq^i x + v_i x) = \{Sq^{2i}\psi(x) + v_{2k}\psi(x)\}[M] = 0$. Thus there is a manifold M' with $\phi(x) = x[M']$ for all $x \in H^k(BO;Z_2)$, and $w_I[M] = w_I[M' \times M'] = w_{I/2}[M']$ for all I, so $M \sim M' \times M'$. **

One then has:

<u>Theorem</u>: The homomorphism $F_* : \Omega_*^U \longrightarrow \mathfrak{N}_*$ has image $\mathfrak{N}_*^2 = \{x^2 | x \in \mathfrak{N}_*\}$, i.e. precisely those classes for which the Stiefel-Whitney numbers having an odd degree factor are zero. Further, one may find generators b_i of Ω_*^U and $x_i (i \neq 2^s-1)$ of $\mathfrak{N}_*$ for which $F_*(b_i) = x_i^2$ if $i \neq 2^s-1$ and $F_*(b_{2^s-1}) = 0$.

<u>Proof</u>: Since $x \longrightarrow x^2$ is a homomorphism in an algebra over Z_2 one need only map onto generators. Let $x_{2i} = [RP(2i)]$, $b'_{2i} = [CP(2i)]$, and for $i = 2^p(2q+1)-1$, let $x_i = [H_{2^{p+1}q,2^p}(R)]$, $b'_i = [H_{2^{p+1}q,2^p}(\mathbb{C})]$ while for $i = 2^s-1$, let $b'_i = [H_{2^{s-1},2^{s-1}}(\mathbb{C})]$.

It has already been shown that $\mathfrak{N}_* = Z_2[x_i]$, $\Omega^U_* \otimes Z_2 = Z_2[b'_i]$. The characteristic number computations give $c_\omega[b'_i] = w_{2\omega}[b'_i]$ (mod 2) by exactly the same formulae as the computations of $w_\omega[x_i]$. Hence $F_*(b'_i) = x_i^2$. Further all Chern numbers (= K-theoretic numbers) of b'_{2^s-1} are even, so $F_*(b'_{2^s-1}) = 0$. Since one may choose $b_i \in \Omega^U_*$ which generate and which reduce to b'_i mod 2, the result is clear. **

If one then considers the relative group $\Omega^{0,U}_* = \pi_*(\underset{\sim}{TBO},\underset{\sim}{TBU})$ one has an exact triangle

$$\begin{array}{ccc} \Omega^U_* & \xrightarrow{F_*} & \mathfrak{N}_* \\ & \partial \nwarrow \quad \swarrow d & \\ & \Omega^{0,U}_* & \end{array} .$$

Since $\Omega^U_{2k+1} = 0$, this gives rise to an exact sequence

$$\begin{array}{ccccccccc} 0 = \Omega^U_{2n+1} & \xrightarrow{F_{2n+1}} & \mathfrak{N}_{2n+1} & \xrightarrow{d_{2n+1}} & \Omega^{0,U}_{2n+1} & \xrightarrow{\partial_{2n+1}} & \Omega^U_{2n} & \xrightarrow{F_{2n}} & \mathfrak{N}_{2n} \\ \downarrow & & \downarrow & & \downarrow & & \downarrow & & \downarrow \\ 0 = \tilde{H}_{2n+1}(\underset{\sim}{TBU}) & \longrightarrow & \tilde{H}_{2n+1}(\underset{\sim}{TBO}) & \longrightarrow & H_{2n+1}(\underset{\sim}{TBO},\underset{\sim}{TBU}) & \longrightarrow & \tilde{H}_{2n}(\underset{\sim}{TBU}) & \longrightarrow & \tilde{H}_{2n}(\underset{\sim}{TBO}) \end{array}$$

$$\begin{array}{ccccc} \mathfrak{N}_{2n} & \xrightarrow{d_{2n}} & \Omega^{0,U}_{2n} & \xrightarrow{\partial_{2n}} & \Omega^U_{2n-1} = 0 \\ \downarrow & & \downarrow & & \downarrow \\ \tilde{H}_{2n}(\underset{\sim}{TBO}) & \longrightarrow & H_{2n}(\underset{\sim}{TBO},\underset{\sim}{TBU}) & \longrightarrow & \tilde{H}_{2n-1}(\underset{\sim}{TBU}) = 0 \end{array}$$

in which the vertical arrows are the Hurewicz homomorphisms. From the knowledge of F_{2n} one may decompose this still further.

First, image $F_{2n} = \mathcal{N}_n^2$, so $\Omega_{2n}^{O,U} = \mathcal{N}_{2n}/\mathcal{N}_n^2$ is a Z_2 vector space (of known dimension). Further, this group is clearly detected by the Stiefel-Whitney numbers with an odd factor.

Next, kernel F_{2n} is the ideal in Ω_*^U generated by the elements b_{2^s-1} (letting $b_0 = 2$) and since this is a free abelian group, $\Omega_{2n+1}^{O,U} \cong \mathcal{N}_{2n+1} \oplus \ker F_{2n}$, $\mathcal{N}_{2n+1}$ being precisely the torsion subgroup. The torsion free part of this group, $\Omega_{2n+1}^{O,U}/\text{Torsion}$, may be characterized by mapping into Ω_{2n}^U, so that the class is determined by the Chern numbers of the boundary. The torsion subgroup is $d_{2n+1}\mathcal{N}_{2n+1}$ and is detected by Stiefel-Whitney numbers. In mod 2 homology $\tilde{H}_{2n}(\underset{\sim}{TBU}) \longrightarrow \tilde{H}_{2n}(\underset{\sim}{TBO})$ is monic (the cohomology map being epic), so that in fact the Hurewicz map defines a splitting

$$\Omega_{2n+1}^{O,U} \longrightarrow H_{2n+1}(\underset{\sim}{TBO},\underset{\sim}{TBU};Z_2) \cong \tilde{H}_{2n+1}(\underset{\sim}{TBO};Z_2) \longrightarrow \mathcal{N}_{2n+1},$$

the latter map being a projection.

The interesting question is how much of $\Omega_*^{O,U}$ may be detected by Z_2 cohomology characteristic numbers. For this one has:

<u>Proposition</u>: Under the product of manifolds $\Omega_*^{O,U}$ is an Ω_*^U module. Writing $\Omega_*^U = Z[b_i]$ and letting $b_0 = 2$, $\Omega_*^{O,U}$ is generated over Ω_*^U by the elements

$$\alpha_{2^{s+1}-1} \in \Omega_{2^{s+1}-1}^{O,U} \quad \text{with} \quad \partial_{2^{s+1}-1}\alpha_{2^{s+1}-1} = b_{2^s-1} \quad (s \geq 0)$$

and

$$d_*(x_{i_1} \cdots x_{i_r}) \in \Omega_{i_1+\dots+i_r}^{O,U} \qquad 0 < i_1 < \dots < i_r$$

where $\mathcal{M}_* = Z_2[x_i]$, the complete set of relations being given by

$$b_{2^s-1} \cdot d_*(x_{i_1} \ldots x_{i_r}) = 0$$

and

$$b_{2^t-1}\alpha_{2^{s+1}-1} = b_{2^s-1}\alpha_{2^{t+1}-1}.$$

In addition, the kernel of the Hurewicz homomorphism $\Omega_*^{O,U} \longrightarrow H_*(\underset{\sim}{TBO},\underset{\sim}{TBU};Z_2)$ is precisely the submodule consisting of multiples of the b_{2^s-1} (i.e. the image is the free $\mathcal{M}_*^2$ module on the classes $\alpha_{2^{s+1}-1}$ and $d_*(x_{i_1} \ldots x_{i_r})$).

Proof: The first part is obvious. Since $\partial(b_{2^t-1}\alpha_{2^{s+1}-1}) = \partial(b_{2^s-1}\alpha_{2^{t+1}-1}) = b_{2^t-1}b_{2^s-1}$, one must have $b_{2^t-1}\alpha_{2^{s+1}-1} - b_{2^s-1}\alpha_{2^{t+1}-1} = u$ in the torsion subgroup, but all Stiefel-Whitney numbers of u are zero since those of the b's are, and hence $u = 0$. Further the b_{2^s-1} having zero mod 2 numbers implies that the submodule $\Sigma b_{2^s-1}\Omega_*^{O,U}$ is annihilated by the mod 2 Hurewicz homomorphism. To see that this is the entire kernel one considers $\tilde{H}^*(\underset{\sim}{TBO}) \longrightarrow \tilde{H}^*(\underset{\sim}{TBU})$ (using Z_2 coefficients unless otherwise noted), which maps a free $\mathcal{A}_2$ module onto a free $\mathcal{A}_2/(Q_0)$ module. By a good choice of generators this may be written $(T \oplus S) \otimes \mathcal{A}_2 \longrightarrow T \otimes \mathcal{A}_2/(Q_0)$ where T,S are Z_2 vector spaces (by proper choice of characteristic numbers S to detect cokernel F_* and T to detect image F_*). Writing $TBO = K(T) \times K(S)$ one may project onto $K(T)$, splitting $K(S)$ out of the problem.

Letting $X = TBU$, one has $f : X \longrightarrow K(T)$ with cohomology map $T \otimes \mathcal{A}_2 \longrightarrow T \otimes \mathcal{A}_2/(Q_0)$ and one wishes to know how much of the homotopy of $K(T)/X$ is detected by mod 2 cohomology. Since $\tilde{H}^*(X;Z)$ is torsion free the classes of T are the reduction of integral classes, and letting $\bar{T}$ be a free abelian group in $\pi_*(X)$ (direct summand) with $\bar{T} \otimes Z_2 \cong T$, π^+ a complementary summand, one has the diagram

$$
\begin{array}{ccccc}
\Sigma X & \longrightarrow & \Sigma K(\bar{T}) & \longrightarrow & G \\
\downarrow & & \downarrow & & \downarrow \\
\Sigma X & \longrightarrow & \Sigma K(\bar{T}) & \longrightarrow & G \\
\downarrow & \searrow^{\Sigma f} & \downarrow & & \downarrow \\
X \wedge E_2 & \longrightarrow & \Sigma K(T) & \longrightarrow & F
\end{array}
$$

with homotopy and cohomology diagrams

$$
\begin{array}{ccccccc}
\pi^+ & \longrightarrow & \pi^+ \oplus \bar{T} & \longrightarrow & \bar{T} & \xrightarrow{0} & \pi^+ \\
\downarrow & & \downarrow 2 & & \downarrow 2 & & \downarrow 2 \\
\pi^+ & \longrightarrow & \pi^+ \oplus \bar{T} & \longrightarrow & \bar{T} & \xrightarrow{0} & \pi^+ \\
\downarrow & & \downarrow & \searrow & \downarrow & & \downarrow \\
\pi^+ \otimes Z_2 & \longrightarrow & \pi^+ \otimes Z_2 \oplus T & \longrightarrow & T & \xrightarrow{0} & \pi^+ \otimes Z_2
\end{array}
$$

$$
\begin{array}{ccc}
T \otimes \mathcal{A}_2/(Q_0) & \longleftarrow & T \otimes \mathcal{A}_2/\mathcal{A}_2 Sq^1 \\
\uparrow 0 & & 0 \uparrow \\
T \otimes \mathcal{A}_2/(Q_0) & \longleftarrow & T \otimes \mathcal{A}_2/\mathcal{A}_2 Sq^1 \\
\uparrow & \nwarrow & \uparrow \\
T \otimes \mathcal{A}_2/\mathcal{J} & \longleftarrow & T \otimes \mathcal{A}_2 \\
\uparrow Q_0 & & Sq^1 \uparrow \\
T \otimes \mathcal{A}_2/(Q_0) & \longleftarrow & T \otimes \mathcal{A}_2/\mathcal{A}_2 Sq^1 .
\end{array}
$$

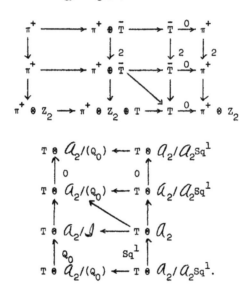

From the analysis of spectra of the type of X, one knows that the summand $T \otimes V_1$ of $\pi^+ \otimes Z_2 = \pi_*(F)$ is detected by mod 2 cohomology ($T \otimes$ image d_1), while the quotient ΣT of $\Sigma\bar{T} = \pi_*\Sigma^2 K(\bar{T})$ is detected by $T \otimes 1 \subset T \otimes \mathcal{A}_2/\mathcal{A}_2 Sq^1$; i.e. in $H^*(K(T)/X)$ there is a summand mapping to $T \otimes Sq^1 \oplus T \otimes d_1(V_1) \subset T \otimes \mathcal{A}_2 = H^*(K(T))$, and detecting homotopy classes of the form $\Sigma T \oplus T \otimes V_1 \subset \pi_*(K(T)/X)$.

Thus under the Hurewicz homomorphism $\Omega_*^{O,U} \longrightarrow H_*(\underset{\sim}{TBO},\underset{\sim}{TBU};Z_2)$ the image is at least as large as $S \oplus \Sigma T \oplus T \otimes V_1$. Since this has exactly the right dimension, the result is proven. **

Note: By the analysis above, one may if desired detect $\alpha_{2^{s+1}-1}$ mod 2 by the Stiefel-Whitney class $w_{2^{s+1}-1}(\nu)$, corresponding to $Sq^{2^{s+1}-1}U$ in $H^*(\underset{\sim}{TBO},\underset{\sim}{TBU};Z_2)$, these numbers annihilating the image of $\mathcal{N}_*$. In particular, α_1 comes from $\Omega_1^{O,fr}$ and is related to the Hopf invariant, corresponding to $\Omega_0^{fr} \cong \Omega_0^U$. The other classes α_i do not come via framed cobordism.

Complex Bordism

Corresponding to the forgetful functor from stably almost complex manifolds to topological spaces one has defined relative bordism groups $\Omega_*^U(X,A) \cong \lim_{r\to\infty} \pi_{*+2r}((X/A) \wedge TBU_r) = H_*(X,A;\underset{\sim}{TBU})$. The product of manifolds makes $\Omega_*^U(X,A)$ a module over Ω_*^U. These modules have been studied by Conner and Floyd [35], [37], or [39] by analysis of the spectral sequence from $H_*(X,A;\Omega_*^U)$ to $\Omega_*^U(X,A)$. One can also obtain these results as was done for Ω_*^U.

Theorem: For every CW pair (X,A), $\Omega_*^U(X,A) \otimes Q$ is a free $\Omega_*^U \otimes Q$ module isomorphic to $H_*(X,A;Q) \otimes_Q (\Omega_*^U \otimes Q)$.

Proof: $\pi_*((X/A) \wedge TBU) \longrightarrow H_*((X/A) \wedge TBU;Z)$ is an isomorphism modulo torsion. **

Theorem: If (X,A) has no torsion in its integral homology then $\Omega_*^U(X,A)$ is a free Ω_*^U module isomorphic to $H_*(X,A;Z) \otimes \Omega_*^U$. In particular, the evaluation homomorphism $e : \Omega_*^U(X,A) \longrightarrow H_*(X,A;Z)$ is epic. If $\{x_i\}$ is a homogeneous base of $H_*(X,A;Z)$ and $f_i : (M^i,\partial M^i) \longrightarrow (X,A)$ is a map of a stably almost complex manifold into (X,A) with $f_{i*}([M,\partial M]) = x_i$, then $\Omega_*^U(X,A)$ is the free Ω_*^U module with base the classes of the (M^i,f_i).

Proof: Since (X,A) has no p-torsion, Q_0 acts trivially in $H^*(X,A;Z_p)$, making $\tilde{H}^*((X/A) \wedge \underset{\sim}{TBU};Z_p)$ a free $\mathcal{A}_p/(Q_0)$ module on the classes $\{x_i^* \otimes u_\alpha\}$, where x_i^* are a base dual to the x_i mod p and $\{u_\alpha\}$ is a base of $\tilde{H}^*(\underset{\sim}{TBU};Z_p)$ as $\mathcal{A}_p/(Q_0)$ module. Thus $\Omega_*^U(X,A)$ is torsion free and maps monomorphically into $\tilde{H}_*((X/A) \wedge \underset{\sim}{TBU};Z)$. Further, the map $\Omega_*^U(X,A) \xrightarrow{e} H_*(X,A;Z) \to H_*(X,A;Z)\otimes Z_p$ is epic for each prime p, so the index of image e in $H_*(X,A;Z)$ is not divisible by p. Thus e is epic. Choose classes (M^i,f_i) mapping to x_i. By the Atiyah-Hirzebruch [18] spectral sequence for K-theory, there exist elements $z_i \in K^*(X,A)$ with $ch(z_i) = x_i^* +$ higher terms, $\{x_i^*\}$ being dual to the x_j. Let $B_* \subset H_*((X/A) \times BU;Z)$ be the ring of homogeneous elements x for which $ch(z_i)s_\omega(e)\mathcal{J}[x] \in Z$ for all i and ω. If $f : (M,\partial M) \longrightarrow (X,A)$ then $ch(z_i)\cdot s_\omega(e)\mathcal{J}((f \times \tau)_*[M,\partial M]) = f^*z_i \cdot s_\omega(\gamma(\tau))[M,\partial M]$ which is integral, so $\Omega_*^U(X,A) \longrightarrow H_*((X/A) \times BU;Z) : (M,F) \longrightarrow (f \times \tau)_*[M,\partial M]$ maps into B_*. The elements $b_\omega \cdot (M^i,f_i)$ have linearly independent mod p characteristic numbers $ch(z_i)s_\omega(e)\mathcal{J}$ and hence B_* is the image of the free Ω_*^U module on the $\{(M^i,f_i)\}$. In particular the $\{(M^i,f_i)\}$ generate $\Omega_*^U(X,A)$ freely as Ω_*^U module. **

Corollary: If (X,A) has no torsion in its integral homology, then integral cohomology characteristic numbers determine cobordism class in $\Omega_*^U(X,A)$. Further, all relations among these numbers come from K-theory.

Briefly, for $f : (M,\partial M) \longrightarrow (X,A)$ one has generalized Chern numbers $f^*(x)\cdot c_\omega(\tau)[M,\partial M]$ for $x \in H^*(X,A;Z)$. All relations among these are given by $ch(f^*z)\cdot s_\omega(e)\mathscr{J}[M,\partial M] \in Z$ for $z \in K^*(X,A)$.

This result may be modified slightly to give results in the presence of limited torsion.

Theorem: If (X,A) has no torsion which is p-primary for $p \in P'$ (P' a set of primes) and $Q' \subset Q$ is the set of rational numbers which have denominators relatively prime to the elements of P' (when expressed in lowest terms) then $\Omega_*^U(X,A) \otimes Q'$ is a free $\Omega_*^U \otimes Q'$ module isomorphic to $H_*(X,A;Z) \otimes \Omega_*^U \otimes Q'$. In particular, the cokernel of the evaluation is finite, of order not divisible by p for any $p \in P'$.

Proof: One proceeds exactly as above using only the primes p belonging to P', showing that $\Omega_*^U(X,A)$ has no p torsion and that coker(e) is finite of order prime to p. One then chooses classes $x_i \in H_*(X,A)$ which freely generate $H_*(X,A) \otimes Q'$ and can select maps (M^i,f_i) which realize $n_i x_i$, $1/n_i \in Q'$. To prove that these generate $\Omega_*^U(X,A) \otimes Q'$ one may assume that (X,A) is a finite complex (to prove freeness up through dimension n one may restrict to the $n + 1$ skeleton of X and A. This introduces no new torsion and induces isomorphisms of $\Omega_*^U(X,A)$ in dimensions less than or equal to n). Noting that all differentials in the Atiyah-Hirzebruch [18] spectral sequence have finite order, which cannot be p primary for any $p \in P'$, one may find elements $z_i \in K^*(X,A)$ for which $ch(z_i)$ has least component of degree equal to $\dim x_i$ and with $ch(z_i)[x_i] = m_i \in Z$, $1/m_i \in Q'$ (and annihilating all other x_j of the same or lower dimension). One lets $B_* \subset H_*((X/A) \times BU;Z)$ be defined by $ch(z_i)s_\omega(e)\mathscr{J}[x] \in Z$ and proceeds as above to prove $\Omega_*^U \otimes Z\{(M^i,f_i)\} \subset \Omega_*^U(X,A) \subset B_*$ having index of order prime to p for each $p \in P'$. **

Corollary: If (X,A) has no p primary torsion in its homology then $\Omega_*^U(X,A) \otimes Z_p$ is a free $\Omega_*^U \otimes Z_p$ module isomorphic to $H_*(X,A) \otimes \Omega_*^U \otimes Z_p$.

Corollary: If (X,A) has no p primary torsion in its homology for all $p \in P'$, then generalized Chern numbers determine bordism class up to torsion of order prime to all such p. Further, all p primary relations among these numbers follow from the K-theory of the finite skeleta of (X,A).

Note: It has been assumed throughout (implicity) that (X,A) has finite type. Notice that the K-theory of (X,A) may be zero, while that of its finite skeleta is not. In the above one takes a large skeleton to determine the relations (which are independent of the skeleton chosen) with respect to the prime p and its powers. See Hodgkin [56] in which it is shown that inverse limit K-theory vanishes for many spaces (homotopy class of maps theory factors through inverse limit theory for characteristic numbers). The important point is that the K-theory is not closely related to skeletal decomposition, while homology and bordism are.

It is to be noted that the results concerning the p-primary situation are valid for spaces with p torsion provided one remains below the dimension in which that torsion occurs, since one may restrict to a skeleton.

Landweber [66] has examined the homomorphism $\Omega_*^U(X) \longrightarrow H_*(X;Z)$ for $X = K(Z_p,n)$, $K(Z,n)$, or $BU(2q,\ldots,\infty)$ (the connective cover of BU) in the stable range, the interest being entirely in torsion elements. He completely determines the image, but this does not determine the bordism since there are nontrivial extensions involved.

Chapter VIII

σ_1 - Restricted Cobordism

Let K be one of the fields $\mathbb{R}$ or $\mathbb{C}$. If μ is an n-dimensional K vector bundle, the determinant bundle of μ, $\det\mu$, is the K line bundle $\Lambda_K^n(\mu)$ given by the n-fold exterior power over K of the bundle μ. If μ' has dimension n', then $\Lambda_K^{n+n'}(\mu \oplus \mu') \cong \Lambda_K^n(\mu) \otimes_K \Lambda_K^{n'}(\mu')$ so the determinant takes Whitney sums to products. Combining this with the fact that $\det\rho = \rho$ if ρ is a line bundle, one has $\det(\mu \oplus 1) \cong \det\mu \otimes 1 = \det\mu$, extending the determinant to stable K vector bundles.

For any integer $r \geq 1$ one may form a cobordism category of manifolds with "$P(K^r)$ structure" as follows:

1) An object consists of:

a) A compact manifold M with a chosen K vector bundle structure on its stable tangent bundle (equivalently normal bundle; i.e. a (BG,g) manifold, where $G = O$ or U);

b) A map $f : M \longrightarrow P(K^r)$; and

c) An equivalence of $f^*(\xi)$ with the determinant bundle of the K-tangent bundle τ of M (i.e. a bundle isomorphism of K line bundles). Note: For $r = 1$, ξ is the trivial line bundle and the equivalence is a trivialization.

2) A map $\phi : (M',f') \longrightarrow (M,f)$ is an imbedding ϕ with trivialized normal bundle for which the K tangent bundles are compatible (a (BG,g) map) such that $f' = f\cdot\phi$ with the equivalence given by restriction.

3) The boundary functor assigns to M its boundary with inner normal trivialization to define the induced structure, and the 'inclusion' natural transformation is the inclusion map with inner normal trivialization.

The cobordism semigroup corresponding will be denoted $\mathcal{W}_*(K,r)$.

Letting M be a (BG,g) manifold with $\nu : M \longrightarrow BG$ defining the normal structure and letting $\psi : BG \longrightarrow P(K^\infty)$ be a map with $\psi^*(\lambda) = \det \mathcal{U}$ where λ is the canonical bundle over $P(K^\infty)$ and $\mathcal{U}$ is the universal stable bundle over BG, one has $(\psi\cdot\nu)^*(\xi) \cong \det\tau$. Any two maps obtained in this way are homotopic and one has a canonical choice of homotopy defined by the isomorphism of (BG,g) structures for different imbeddings and choice of homotopy for the maps ψ. If $f : M \longrightarrow P(K^r)$ is any map, an equivalence of $f^*(\xi)$ with $\det\tau$ may be interpreted as a homotopy of the maps f and $\psi\cdot\nu$. Thus a "$P(K^r)$ structure" on M may be interpreted as a deformation of the canonical map $\psi\cdot\nu$ into $P(K^r)$.

Interpreting a homotopy as a cobordism, it is clear that within cobordism only the homotopy class of the equivalence matters and also that homotopic maps f give isomorphic families of structures (the isomorphism depending on the choice of homotopy). Further, a structure on M defines by projection a structure on $M \times I$ with the "opposite end" defining an inverse to M.

In order to make this more precise and to determine this cobordism category as a (B,f) theory, one constructs a classifying space as follows. Let $\rho : BG \times P(K^r) \longrightarrow P(K^\infty)$ be a map for which $\rho^*(\lambda) = (\det \mathcal{U}) \otimes \xi$. Let $BK^{(r)}$ be the total space of the induced fibration of the sphere $S(K^\infty)$ over the projective space $P(K^\infty)$, giving

$$\begin{array}{ccc} BK^{(r)} & \xrightarrow{\bar{\rho}} & S(K^\infty) \\ {\scriptstyle\pi}\downarrow & & \downarrow \\ BG \times P(K^r) & \xrightarrow{\rho} & P(K^\infty) \end{array}$$

and let $\theta : BK^{(r)} \longrightarrow BG$ be the composition of π and the projection on BG.

Being given a manifold (M,f) with $P(K^r)$ structure with $\nu : M \longrightarrow BG$ the normal map, $\nu \times f : M \longrightarrow BG \times P(K^r)$ pulls $(\det \mathcal{U}) \otimes \xi$ back to $(\det \nu) \otimes f^*(\xi) \cong (\det \nu) \otimes (\det \tau)$ which is trivial. Thus $\nu \times f$ lifts to $BK^{(r)}$, the choice of lifting being equivalent to the choice of homotopy of $\rho \cdot (\nu \times f)$ to a point map or to the choice of equivalence of $f^*(\xi)$ and $\det \tau$. (<u>Note</u>: π is a principal G_1 bundle.)

Conversely if $\bar{\nu} : M \longrightarrow BK^{(r)}$ is a lifting of the normal map $\nu : M \longrightarrow BG$, the composition $f = \pi_2 \cdot \pi \cdot \bar{\nu}$ where π_2 projects on $P(K^r)$ maps M into $P(K^r)$ with $\det \nu \otimes f^*(\xi)$ trivialized ($S(K^\infty)$ is the sphere bundle of λ and the pullback of λ to it is naturally trivialized) and the trivialization may be interpreted as an equivalence of $\det \tau$ and $f^*(\xi)$.

Letting $BK_n^{(r)}$ denote the pullback of $BK^{(r)}$ over BG_n, one has

<u>Theorem</u>: $\mathcal{W}_n(K,r) \cong \lim_{s \to \infty} \pi_{n+ks}(TBK_s^{(r)}, \infty)$, where $k = \dim_R K$.

The interest in these cobordism theories is primarily that they provide intermediate levels between the "unoriented" theories $(r = \infty)$ and the "oriented" theories $(r = 1)$. Briefly, one has:

1) For $r = \infty$, the space $BK^{(r)}$ may be identified with BG by means of $BG \xrightarrow{1 \times \psi} BG \times P(K^\infty)$ for then $\rho \cdot (1 \times \psi)^*(\lambda) = (\det \mathcal{U}) \otimes (\overline{\det \mathcal{U}})$ which is trivial. In fact, if $\dim M = n$ then the classifying map $\psi \cdot \nu$ for $\det \tau$ sending M into $P(K^\infty)$ may be deformed into the n-skeleton and the homotopy given by two different deformations may be pushed into the $(n+1)$-skeleton giving a unique $P(K^r)$ structure provided $(n+1) \leq k(r-1)$. Thus for $r \geq (n+1)/k + 1$, $\mathcal{W}_n(K,r) = \mathcal{W}_n(K,\infty)$ is the "unoriented" cobordism group $\mathfrak{N}_n$ or Ω_n^U. This will be denoted Ω_*^G in this chapter.

2) For $r = 1$, a "$P(K^r)$ structure" on M is a trivialization of $\det \tau$. If τ is represented as an n-plane bundle with an inner product, each fiber

V becomes an n-dimensional inner product space. This extends to an inner product on the graded algebra $\Lambda(V) = \Sigma_0^n \Lambda^n(V)$ by letting $\Lambda^j V$ be orthogonal to $\Lambda^k V$ if $j \neq k$ and setting

$$\langle X,Y \rangle = \det |\langle x_i, y_j \rangle|$$

if $X = x_1 \wedge \dots \wedge x_s$, $Y = y_1 \wedge \dots \wedge y_s$. Giving the exterior power bundles these inner products, a trivialization of $\det\tau$ may be thought of as choosing (continuously) a unit vector in the n-th exterior power of each fiber. Thus the structure group of τ is reduced to those linear transformations of V fixing a unit vector in $\Lambda^n(V)$. If $T : V \longrightarrow V$, the induced transformation on $\Lambda^n(V)$ is multiplication by the determinant of T, and the transformations of determinant one are the special orthogonal or special unitary groups.

<u>Note</u>: The groups $\mathcal{W}_*(K,1)$ will be denoted Ω_*^{SG}, and the space $BK^{(1)}$ is denoted BSG. Although the main reason for interest in the $\mathcal{W}_*(K,r)$ is the calculation of the $r = 1$ case, there will be few results in this chapter directly concerned with that calculation.

3) The first analysis of the groups $\mathcal{W}_*(K,r)$ for $r \neq 1,\infty$ was by Wall [130] for the case $K = \mathbb{R}$. Making use of the case $r = 2$, which may be thought of as "σ_1 - spherical" cobordism, he exploited the various inter-relationships to determine the 2 primary structure of Ω_*^{SO}. Additional material may be found in Atiyah [13] and Wall [133]. The complex case was studied by Conner and Floyd [39] patterned closely on the work of Wall (but using the methods of Atiyah). It has been noted by Novikov in conjunction with his study of the Adams spectral sequence with "unoriented" cobordism coefficients (Novikov [96]) that the $\mathcal{W}_*(K,2)$ arise naturally in the calculation of the "oriented" theories.

The relationship of "$P(K^2)$" theory and "σ_1 - spherical" theory may be seen as follows.

If Z_K denotes Z_2 for $K = R$ or Z for $K = \mathbb{C}$ then the first characteristic class $\sigma_1(M) = \sigma_1(\tau) \varepsilon H^k(M;Z_K)$ coincides with the characteristic class $\sigma_1(\det\tau)$. (To see this one has $\sigma_1(\rho) = \sigma_1(\det\rho)$ if ρ is a line bundle and as mentioned in discussing $H_{m,n}$ in Chapter V, $\sigma_1(a \otimes b) = \sigma_1(a) + \sigma_1(b)$ if a,b are line bundles, so that applying the splitting principle and induction establishes $\sigma_1(\tau) = \sigma_1(\det\tau)$ for general bundles).

Since $P(K^2) = S^k$ with $\sigma_1(\xi) = \iota \varepsilon H^k(S^k;Z_K)$ one has for an object (M,f) with "$P(K^2)$" structure a map $f : M \longrightarrow S^k$ with $f^*(\iota) = \sigma_1(M)$. Thus $\sigma_1(M)$ is <u>spherical</u>. Conversely, being given a manifold M for which $\sigma_1(M)$ is spherical there is a map $f : M \longrightarrow P(K^2)$ with $f^*(\iota) = \sigma_1(M)$. Since $P(K^\infty) = K(Z_K,k) = BG_1$, equivalence classes of K line bundles are determined by the first Z_K-characteristic class. Hence there is an equivalence of $f^*(\xi)$ and $\det\tau$.

4) The general case $r > 2$ is not of special interest. Results in the real case have been obtained by J. B. Minkus and by C. T. C. Wall, but nothing appears in the literature. The use of the Atiyah bordism approach makes the computation an exercise.

Semi - geometric methods : $\mathcal{W}_*(K,2)$

Let (X,A) be any pair, and define a homomorphism $\Phi : \Omega_*^G(X,A) \longrightarrow \mathcal{W}_*(K,2)(X,A)$ as follows. If $\alpha \varepsilon \Omega_*^G(X,A)$ choose a map $g : (M,\partial M) \longrightarrow (X,A)$ representing α and let $\psi\cdot\nu : M \longrightarrow P(K^\infty)$ be the map inducing $\det\tau$. By compactness of M, there is an integer Q such that $\psi\cdot\nu(M) \subset P(K^Q)$. Let $\beta : P(K^Q) \times P(K^2) \longrightarrow P(K^N)$ be the usual imbedding

($N = 2Q$, given in local coordinates by $u_{ij} = x_i y_j$) so that $\theta = \beta\cdot(\psi\cdot\nu \times 1) : M \times P(K^2) \longrightarrow P(K^N)$ classifies $\det\tau \otimes \xi$; i.e. $\theta^*(\xi) = \det\tau \otimes \xi$. By means of a homotopy θ may be deformed so that $\theta|\partial M \times P(K^2)$ is transverse regular on $P(K^{N-1})$, and then to make θ transverse regular on $P(K^{N-1})$ keeping $\partial M \times P(K^2)$ fixed. Then $\theta^{-1}(P(K^{N-1})) = L \subset M \times P(K^2)$ with $\partial L = L \cap (\partial M \times P(K^2))$. The tangent bundle of L is isomorphic to the pullback of $\tau_M \oplus \tau_{P(K^2)} - \det\tau_M \otimes \xi$, giving an isomorphism $\det(\tau_L) \cong \det\tau_M \otimes \xi^2 \otimes (\det\tau_M)^{-1} \otimes \xi^{-1} = \xi$. Thus the composition $f : L \hookrightarrow M \times P(K^2) \longrightarrow P(K^2)$ defines a "$P(K^2)$" structure on L. The composition $\phi : (L,\partial L) \hookrightarrow (M \times P(K^2), \partial M \times P(K^2)) \xrightarrow{\pi} (M,\partial M) \xrightarrow{g} (X,A)$ then gives $((L,f),\phi)$ which is an $\mathcal{W}_*(K,2)$ bordism element of (X,A). That this defines a homomorphism $\Phi : \Omega_*^G(X,A) \longrightarrow \mathcal{W}_*(K,2)(X,A)$ is an easy consequence of transverse regularity (think of different homotopies as cobordisms and for a cobordism $G : (V,U) \longrightarrow (X,A)$ with $\partial V = M \cup (-M') \cup U$, first apply transverse regularity on $U \times P(K^2)$ keeping the boundary fixed, and then keeping $\partial V \times P(K^2)$ fixed make the map transverse regular on all of $V \times P(K^2)$).

Lemma: Let $f : M \longrightarrow P(K^2)$ be a differentiable map. Then $f \times 1 : M \times P(K^2) \longrightarrow P(K^2) \times P(K^2)$ is transverse regular on $H_{1,1} = \{(x,y) \in P(K^2) \times P(K^2) | x_0y_0 + x_1y_1 = 0\}$.

Proof: Let $\mu : P(K^2) \longrightarrow P(K^2) : (y_0,y_1) \longrightarrow (-y_1,y_0)$. Then μ is a differentiable involution ($\mu^2 = 1$). Then $f \times 1$ is transverse regular on $H_{1,1}$ if and only if $(1 \times \mu)\cdot(f \times 1)$ is transverse regular on $(1 \times \mu)(H_{1,1}) = \{(x,y)|x_0y_1 = x_1y_0\}$ which is the diagonal Δ in $P(K^2) \times P(K^2)$. This is the case if and only if $(1 \times \mu)\cdot(f \times 1)\cdot(id_M \times \mu)$ is transverse regular on Δ, and so it suffices to prove $f \times 1$ is transverse regular on Δ. Writing $(\tau_{M\times P(K^2)})_{(m,x)} = (\tau_M)_m \oplus (\tau_{P(K^2)})_x$ for $(f \times 1)(m,x) \in \Delta$, $(f \times 1)_*$ maps

$(0 \times (\tau_{P(K^2)})_x)$ onto $0 \times (\tau_{P(K^2)})_x \subset (\tau_{P(K^2)})_x \oplus (\tau_{P(K^2)})_x$ which is transverse to $(\tau_\Delta)_{(x,x)}$. Thus $f \times 1$ is transverse to Δ, completing the proof. **

(I am indebted to W. Browder for the above proof, which considerably simplifies my own proof.)

One then has:

Proposition: The composition

$$\mathcal{W}_*(K,2)(X,A) \xrightarrow{F_*} \Omega_*^G(X,A) \xrightarrow{\Phi} \mathcal{W}_*(K,2)(X,A)$$

is the identity, where F_* is the homomorphism induced by the forgetful functor which ignores "$P(K^2)$" structure.

Proof: Let $((M,f),\phi)$ represent a class in $\mathcal{W}_*(K,2)(X,A)$ with $f : M \longrightarrow P(K^2)$ a differentiable map having $f^*(\xi) \cong \det\tau$, and $\phi : (M,\partial M) \longrightarrow (X,A)$. Let $\theta : M \times P(K^2) \longrightarrow P(K^4)$ be the composite

$$M \times P(K^2) \xrightarrow{f\times 1} P(K^2) \times P(K^2) \xrightarrow{\beta} P(K^4)$$

where $\beta((x_0,x_1),(y_0,y_1)) = (x_0y_0,x_0y_1,x_1y_0,x_1y_1)$. Then $\theta^*(\xi) \cong \det\tau \otimes \xi$. Further, β is transverse regular on the subspace $P(K^3)$ given by $u_0 + u_3 = 0$, with preimage $H_{1,1}$ and $f \times 1$ is transverse regular on $H_{1,1}$ by the lemma, so that θ (and its restriction to $\partial M \times P(K^2)$) is transverse regular on $P(K^3)$.

Then $L = \theta^{-1}(P(K^3)) = \{(m,\mu f(m)) | m \in M\}$ with the composite $v : L \longrightarrow M \times P(K^2) \longrightarrow M$ a diffeomorphism. The map $f' : L \longrightarrow M \times P(K^2) \longrightarrow P(K^2)$ may be considered as $\mu f : M \longrightarrow P(K^2)$ and since μ is induced by $K^2 \longrightarrow K^2 : (a,b) \longrightarrow (-b,a)$ which is a rotation through $90°$, μ is easily homotoped to the identity, with a chosen isomorphism of ξ and $\mu^*\xi$. The tangent bundle of L is the pullback of $\tau_M \oplus \tau_{P(K^2)} - \det\tau_M \otimes \xi$ and $\tau_{P(K^2)} \oplus 1 \cong \xi \oplus \xi$ or equivalently $\tau_{P(K^2)} \cong \xi \otimes \xi$, while $\det\tau_M \cong \xi$, so that τ_L is isomorphic to τ_M.

Thus with only "universal identifications" $((L,f'),\phi\cdot v)$ coincides with $((M,f),\phi)$, and $\Phi F_* = 1$. **

Corollary: $\mathcal{W}_*(K,2)(X,A)$ is a direct summand of $\Omega_*^G(X,A)$ for every pair (X,A).

Remark: From this one has: $\mathcal{W}_*(K,2)(X,A)$ is isomorphic (via F_*) to the subset of $\Omega_*^G(X,A)$ consisting of those classes which are represented by a manifold-map (M,g) for which $\sigma_1(M)$ is spherical. This is Wall's original definition [130], having eliminated dependence on the choice of map to $P(K^2)$ and bundle equivalence.

Proposition: For any pair (X,A) the diagram

$$\begin{array}{ccc} \Omega_*^G(X,A) & \xrightarrow{\ \Phi\ } & \mathcal{W}_*(K,2)(X,A) \\ & {\scriptstyle e}\searrow \quad \swarrow{\scriptstyle e} & \\ & H_*(X,A;Z_K) & \end{array}$$

commutes. In particular, a Z_K homology class is representable by a manifold with "$P(K^2)$" structure if and only if it is representable by an "unoriented" manifold.

Proof: It suffices to show that for every M, the map $v : L \longrightarrow M \times P(K^2) \xrightarrow{\pi} M$ has $v_*[L,\partial L] = [M,\partial M]$. This is basically a consequence of Poincaré-Lefschetz duality. Letting $n = \dim M = \dim L$, $H^n(M,\partial M;Z_K)$ is a free Z_K module (by duality) and hence is isomorphic to

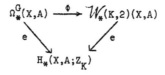

$\mathrm{Hom}(H_n(M,\partial M;Z_K),Z_K)$ (universal coefficient theorem) and thus it suffices to show $v^*(x)[L,\partial L] = x[M,\partial M]$ for all $x \in H^n(M,\partial M;Z_K)$. Then

$$v^*(x)[L,\partial L] = \pi^*(x)\cdot(\pi^*\sigma_1(M) + \pi_2^*\iota)[M \times P(K^2),\partial M \times P(K^2)],$$

$$= \pi^*(x)\cdot\pi_2^*(\iota)[(M,\partial M) \times P(K^2)]$$

since $x\cdot\sigma_1(M)$ has dimension greater than that of M, but this is then equal to

$$x[M,\partial M]\cdot\iota[P(K^2)] = x[M,\partial M]. \quad **$$

In order to determine $\mathcal{W}_*(K,2)(X,A)$, it is standard to use an exact sequence of Atiyah [13] (The proof here being due to Wall [133]). For this one needs to generalize the notion of submanifold dual to a line bundle.

Let M^n be a compact (BG,ρ) manifold and σ a K-line bundle over M. Let $h : M \longrightarrow P(K^\infty)$ with $h^*(\xi) \cong \sigma$, and by compactness, $h : M \longrightarrow P(K^S)$ for some large S. One may then deform $h|\partial M$ to be transverse regular to $P(K^{S-s})$ and keeping the map fixed on ∂M continue this deformation to make the map h transverse regular on $P(K^{S-s})$. Then $h^{-1}(P(K^{S-s})) = N$ is a submanifold of M of codimension $s\cdot k$ with normal bundle in M isomorphic to $s\cdot\sigma$ (as real vector bundles). For any $0 \leq t \leq s$ one may give this normal bundle the K vector bundle structure given by $t\cdot\bar{\sigma} + (s-t)\cdot\sigma$, and this gives N a (BG,ρ) structure.

N is known as "the" submanifold dual to $t\bar{\sigma} + (s-t)\sigma$. The manifold N is of course not unique, but is well defined up to choice of various homotopies used in making h transverse regular. Two such transverse regular maps being homotopic, one may make the homotopy $H : M \times I \longrightarrow P(K^S)$ transverse regular, keeping ends fixed to define a (BG,ρ) submanifold V of $M \times I$. The map

$V \hookrightarrow M \times I \xrightarrow{\pi} M$ gives a (BG,ρ) bordism of the two representatives. Thus the class of N in $\Omega_*^G(M,\partial M)$ is well-defined.

Proposition: There is a homomorphism

$$d : \Omega_*^G(X,A) \longrightarrow \Omega_*^G(X,A)$$

of degree $-2k$ which sends the class of $f : (M,\partial M) \longrightarrow (X,A)$ into the class of the composite

$$\tilde{f} : (N,\partial N) \xrightarrow{i} (M,\partial M) \xrightarrow{f} (X,A)$$

where N is the submanifold dual to $\det\tau_M \oplus (\overline{\det\tau_M})$. Further, the sequence

$$0 \longrightarrow \mathcal{W}_*(K,2)(X,A) \xrightarrow{F_*} \Omega_*^G(X,A) \xrightarrow{d} \Omega_*^G(X,A) \longrightarrow 0$$

is exact.

Proof: One first needs to see that d is well defined, but if $H : W \to X$, $\partial W = M \cup T \cup (-M')$ and $H|_M = f$, $H|_{M'} = f'$, $H(T) \subset A$, $\partial T = \partial M \cup (-\partial M')$, then $\det\tau_W$ restricts to $\det\tau_M$ on M and $\det\tau_{M'}$ on M'. Thus a submanifold of W dual to $\det\tau_W \oplus (\overline{\det\tau_W})$ gives a cobordism of the representatives defined by M and M'.

Since the construction may be performed separately in each summand of a disjoint union, d is clearly a homomorphism.

To prove exactness of the sequence one has:

1) F_* is monic by the previous results.

2) $d \cdot F_* = 0$, for if $f : (M,\partial M) \longrightarrow (X,A)$ and $\det\tau_M$ is induced by a map into $P(K^2)$, then $(N,\partial N)$ is the preimage of $P(K^0)$ which is empty.

3) If $d(\alpha) = 0$, α being represented by $f : (M,\partial M) \longrightarrow (X,A)$, let $h : M \longrightarrow P(K^s)$, s large, with $h^*(\xi) \cong \det\tau_M = \mu$ be made transverse regular

on $P(K^{s-2})$ with inverse image N dual to $\mu \oplus \bar{\mu}$, and also on $P(K^{s-1})$ with inverse image L dual to μ. [First make it transverse regular on $P(K^{s-2})$ giving N as inverse image. A neighborhood T of N is then mapped by a bundle map into a neighborhood S of $P(K^{s-2})$. By a deformation of h on $M-T$ one may push $M-T$ out of interior(S). Since $h|_T$ is transverse regular on $P(K^{s-1})$ one may make a small homotopy of h fixed on T to get h transverse regular on $P(K^{s-1})$ and by a proper choice of "small" assure that $h(M-T)$ does not intersect $P(K^{s-2})$].

Since $d(\alpha) = 0$, $\tilde{f} : (N,\partial N) \longrightarrow (X,A)$ bounds and there is a map $\tilde{F} : \tilde{N} \longrightarrow X$, $\tilde{N}$ a (BG,ρ) manifold, $\partial\tilde{N} = N \cup P/(\partial N \cong \partial P)$, $\tilde{F}|_N = \tilde{f}$, $\tilde{F}(P) \subset A$.

The normal bundle of N in L is $(\overline{\det\tau_M})|_N$, but $\det\tau_N \otimes \det\nu = (\det\tau_M)|_N$, where $\nu = \mu \oplus \bar{\mu}$ is the normal bundle of N in M. $\det\nu = \det\mu \otimes \det\bar{\mu} = \mu \otimes \bar{\mu}$ is trivial so the normal bundle of N in L is $\overline{\det\tau_N}$.

Let U be the manifold formed from $L \times I$ and D, the disc bundle of $(\overline{\det\tau_{\tilde{N}}})$, by identifying the part of D over N with a tubular neighborhood of $N \times 1$ in $L \times 1$. Let $U' \subset U$ be the subset $L \times I \cup \tilde{N}$, where $\tilde{N}$ is the zero section of D.

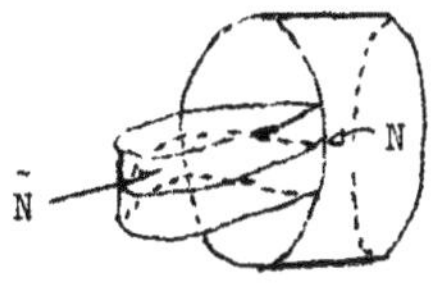

Since N has a neighborhood of the form $N \times [1,2)$ in $\tilde{N}$, there is a strong deformation retraction of U onto U' projecting D onto $\tilde{N}$ over $\tilde{N} - N \times [1,2)$ and collapsing $D|_{N\times[1,2)}$ onto $D|_{N\times 1} \cup N \times [1,2)$ by pushing out radially from the sphere bundle of $2D|_{N\times 2}$.

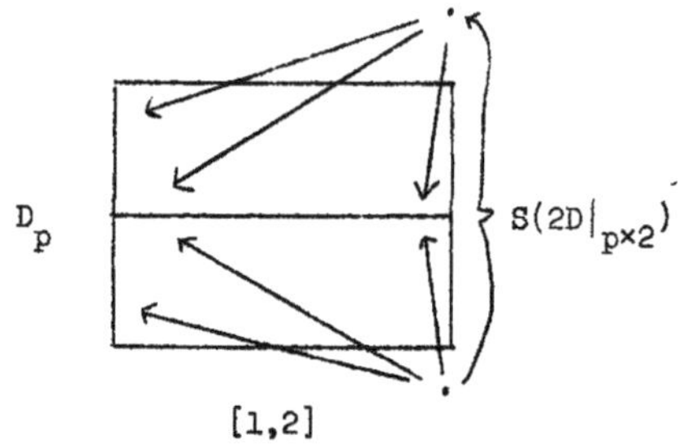

The tangent bundle of D is the pullback from $\tilde{N}$ of $\tau_{\tilde{N}} \oplus \overline{\det\tau_{\tilde{N}}}$ (If $p : X \longrightarrow Y$ is a differentiable vector bundle, $\tau_X \cong p^*X \oplus p^*\tau_Y$, p^*X being the bundle along the fibers = kernel of the differential, and its orthogonal complement in some metric being identifiable under the differential with $p^*\tau_Y$). Thus U admits a (BG,ρ) structure coinciding with that of $L \times I$ and D (agreeing on $D|_{N\times 1}$). Further $f\cdot\pi_1 : L \times I \longrightarrow X$ and $\tilde{F} : \tilde{N} \longrightarrow X$ agree on $N \times 1$ and so define a map $U' \longrightarrow X$. Composition with the retraction gives a map $\tilde{F} : U \longrightarrow X$ extending the map on U'.

The boundary of U has three pieces: $L \times 0$, $\partial L \times I \cup D|_p$, and $L \times 1 - (\text{nbhd } N) \times 1 \cup (\text{sphere bundle of } D) = L'$

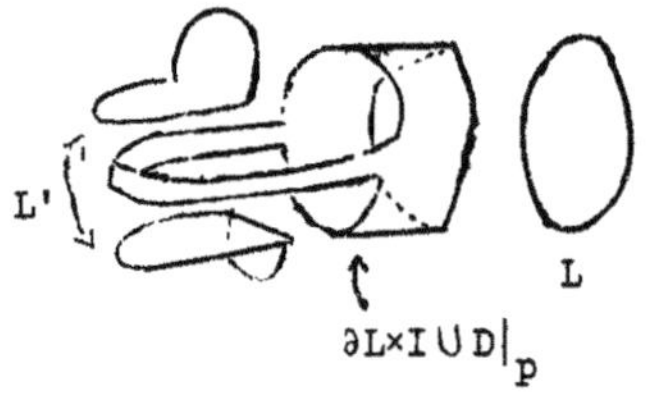

and $\partial L \times I \cup D|_p$ maps into A, giving a cobordism of $f|_L$ with $\tilde{F}|_{L'}$.

Since s is large, one may assume that $\det\tau_{\tilde{N}}$ is induced by a map of $\tilde{N}$ into $P(K^{s-2})$ agreeing on N with $h|_N$, and hence this extends to a bundle map $D \longrightarrow P(K^{s-1})$ sending D to a tubular neighborhood of $P(K^{s-2})$ and agreeing with $h \cdot \pi_1 : L \times I \longrightarrow P(K^{s-1})$ where both are defined. This gives a map $\tilde{H} : U \longrightarrow P(K^{s-1})$ which is transverse regular on $P(K^{s-2})$ (with inverse image $\tilde{N} \cup N \times I$). Let $\tilde{\mu}$ denote the bundle over U induced from ξ by $\tilde{H}$.

Thus $f \times h : (L, \partial L) \longrightarrow (X \times P(K^{s-1}),\ A \times P(K^{s-1}))$ is (BG, ρ) cobordant to the map $\tilde{F} \times \tilde{H}|_{L'} : (L', \partial L') \longrightarrow (X \times P(K^{s-1}),\ A \times P(K^{s-1}))$ and the line bundle $\tilde{\mu}$ over L' induced by that over $P(K^{s-1})$ is trivial $(\xi|_{P(K^{s-1})-P(K^{s-2})}$ is trivial).

Now the normal bundle of L in M is $\det\tau_M|_L \cong \tilde{\mu}_L$ and one may attach the disc bundle E of $\tilde{\mu}$ to $M \times [-1,0]$ by identifying $E|_{L\times 0}$ to a tubular neighborhood of $L \times 0$ in $M \times 0$ to form a manifold W.

As before, W retracts to $M \times [-1,0] \cup U$, where U is the zero section of E; W admits a (BG,ρ) structure given by that of $M \times [1,0]$ and E (agreeing on the intersection) and $\tilde{H} : U \longrightarrow P(K^{s-1})$ extends to a bundle map of E to a tubular neighborhood of $P(K^{s-1})$ in $P(K^s)$, hence giving a map $H : W \longrightarrow P(K^s)$.

The map H is transverse regular on $P(K^{s-1})$ with inverse image $U \cup L \times [-1,0]$, and on $P(K^{s-2})$ with inverse image $\tilde{N} \cup N \times [0,1] \cup N \times [-1,0]$ and the map H classifies $\det\tau_W$. (This is clear on $M \times [-1,0]$, while on E, $\det\tau_E \cong \pi^*\tilde{\mu} \otimes \pi^*\det\tau_U$ but on the part of U over $\tilde{N}$ one has $\det\tau_U = p^*\det\tau_{\tilde{N}} \otimes \overline{p^*\det\tau_{\tilde{N}}}$ - which is trivial, while on $L \times I$ one has $\det\tau_U = \det\tau_{L\times I}$ which is trivial since the normal bundle of L in M is $\det\tau_M$ and hence $\det\tau_L \otimes \det\tau_M = \det\tau_M$ on L.)

Thus one has a cobordism of $f : (M, \partial M) \longrightarrow (X,A)$ to $f' : (M', \partial M') \to (X,A)$ with N' empty and $\det\tau_{M'}|_{L'}$ trivial.

Now let $h' : M' \longrightarrow P(K^s)$ be transverse regular on $P(K^{s-1})$ with preimage L' and $h'^*(\xi)|_{L'}$ trivial. Since $P(K^s) = T(\xi)$ and $T(\xi) - P(K^{s-1})$ is contractible one may homotope h' to coincide with the map $M' \xrightarrow{c} T\nu' \xrightarrow{Th''} T\xi$, where ν' is the normal bundle of L' in M' and $h'' : L' \longrightarrow P(K^{s-1})$ classifies $h'^*(\xi)|_{L'}$. Since this bundle is trivial, h'' may be deformed to a point map, so h' is homotopic to a map into the Thom space of a point, i.e. a k sphere.

Thus the class of $f : (M,\partial M) \longrightarrow (X,A)$ is in the image of F_*.

4) Finally, d is epic.

To begin, one considers a differentiable K vector bundle ξ over a manifold M. $p : P(\xi) \longrightarrow M$ is then a differentiable bundle and $\tau_{P(\xi)} \cong p^*\tau_M \oplus \theta$ where $\theta = \text{kernel}(p^*)$ is the bundle tangent to the fibers.

The bundle θ is the quotient of the bundle tangent to the fibers of $S(\xi)$ by the action of S^{k-1} or

$$\{(x,y) \in S(\xi) \times E(\xi) | p(x) = p(y);\ x \perp y\}/(x,y) \sim (tx,ty).$$

If λ is the canonical line bundle over $P(\xi)$, $E(\lambda)$ may be identified with pairs $(x,s) \in S(\xi) \times K$ representing sx in the line $[x]$ where $(x,s) \sim (tx,s\bar{t})$, and $E(\bar{\lambda})$ is pairs $(x,s) \in S(\xi) \times K$ with $(x,s) \sim (tx,ts)[(x,s) \rightarrow (x,\bar{s})$ is a conjugate linear isomorphism].

Then $E(\bar{\lambda} \otimes p^*\xi)$ is pairs $([(x,s)],y) \in E(\bar{\lambda}) \times E(\xi)$ with $p(x) = p(y)$ and $([(x,ts)],y) \sim ([(x,s)],ty)$ or equivalently pairs $(x,y) \in S(\xi) \times E(\xi)$ with $(tx,ty) \sim (x,y)$ via $([(x,s)],y) \longrightarrow (x,sy)$. Thus $\theta \oplus 1 \cong \bar{\lambda} \otimes p^*\xi$ for θ is the orthogonal complement of the section $x \longrightarrow (x,x)$.

Now suppose $\xi = \xi' \oplus 1$, and then $P(\xi') \subset P(\xi)$ with normal bundle given by $\bar{\lambda}$, since $\theta \oplus 1 = \bar{\lambda} \otimes p^*(\xi) = \bar{\lambda} \otimes p^*(\xi') \oplus \bar{\lambda} \otimes 1$. $P(\xi')$ is the quotient of $S(\xi') \subset S(\xi) \subset E(\xi') \times K$, and the complement of a tubular neighborhood of $P(\xi')$ is the quotient of $(E(\xi') \times S^{k-1}) \cap S(\xi)$, which is the image of $M \times 1$. Over this subspace λ has a section, given by $m \longrightarrow [(0_m,1)]$. Thus λ is induced from the Thom space of the normal bundle of

$P(\xi')$. Finally then $P(\xi')$ is the submanifold of $P(\xi)$ dual to λ (or $\bar{\lambda}$, depending on choice of structure).

Now let $f : (M,\partial M) \longrightarrow (X,A)$ be any map and let $\mu = \det\tau_M$ (with τ_M the stable K tangent bundle) and $U = P(\mu \oplus 2)$, $V = P(\mu \oplus 1)$, $W = P(\mu)$ with projections $\pi : U \longrightarrow M$, $\pi' : V \longrightarrow M$, $\pi'' : W \longrightarrow M$. Let τ_U be given the stable K vector bundle structure of $\pi^*\tau_M \oplus \lambda \otimes \pi^*(\bar{\mu}) \oplus \bar{\lambda} \oplus \lambda$ (isomorphic as real bundles with $\pi^*\tau_M \oplus \bar{\lambda} \otimes_K \pi^*(\mu \oplus 2)$), and τ_V as $\pi'^*\tau_M \oplus \lambda \otimes \pi'^*(\bar{\mu}) \oplus \bar{\lambda}$, τ_W as $\pi''^*\tau_M \oplus \lambda \otimes \pi''^*(\bar{\mu})$.

Then $f\cdot\pi : (U,\partial U) \longrightarrow (X,A)$ with $d([U,f\cdot\pi])$ represented by $f\cdot\pi'' : (W,\partial W) \longrightarrow (X,A)$, since $\det\tau_U$ is $\pi^*\mu \otimes \lambda \otimes \pi^*\bar{\mu} \otimes \bar{\lambda} \otimes \lambda \cong \lambda$. Now $\pi'' : W \longrightarrow M$ is a diffeomorphism and $\pi''^*\mu = \lambda$, so $\lambda \otimes \pi''^*(\bar{\mu}) = 1$ and π'' is an isomorphism of (BG,ρ) manifolds. **

This determines $\mathcal{W}_*(K,2)(X,A)$ as the kernel of the homomorphism d. In particular, if $f : (M,\partial M) \longrightarrow (X,A)$ represents a class in $\Omega_*^G(X,A)$ with $(N,\partial N)$ dual to $\det\tau_M \oplus (\overline{\det\tau_M})$, then in Z_K cohomology, the characteristic class of N is

$$\sigma(N) = \sigma(M)/(1 - \sigma_1{}^2(M))$$

and $(N,\partial N)$ is dual to $\sigma_1(M)^2$. For $x \in H^*(X,A;Z_K)$,

$$\{x\cdot\sigma_\omega(N)\}[N,\partial N] = \{P_\omega(\sigma_i(M))\cdot\sigma_1(M)^2\cdot x\}[M,\partial M]$$

where P_ω is some integral polynomial and $P_\omega(\sigma_i(M)) = \sigma_\omega(M)$ plus terms with $\sigma_1(M)$ as a factor. Thus all characteristic numbers of $((N,\partial N),f)$ are zero if and only if all characteristic numbers of $((M,\partial M),f)$ with a factor $\sigma_1(M)^2$ are zero. [Use induction on the number of σ_1 factors in σ_ω]. Hence if Z_K-characteristic numbers determine bordism class in (X,A) one can characterize $\mathcal{W}_*(K,2)(X,A)$ in terms of numbers.

More generally, let $\mathcal{W}^!_*(K,r)(X,A) \subset \Omega^G_*(X,A)$ denote the set of cobordism classes $[M,g]$ for which all generalized Z_K characteristic numbers

$$\sigma_\omega(\tau_M) \cup g^*(x)[M,\partial M],$$

$x \in H^*(X,A;Z_K)$, which have a factor σ_1^r are zero. Since $\sigma_1(\xi)^r = 0$ in $P(K^r)$ one has $F_*\{\mathcal{W}_*(K,r)(X,A)\} \subset \mathcal{W}^!_*(K,r)(X,A)$. One then has:

Proposition: If $\alpha \in \mathcal{W}^!_*(K,2)(X,A)$ then $\Phi(\alpha)$ and α have the same generalized Z_K characteristic numbers. In particular, if $H_*(X,A;Z_K)$ is a free Z_K module, then $\mathcal{W}^!_*(K,2)(X,A)$ is equal to $F_*\mathcal{W}_*(K,2)(X,A)$.

Remarks: 1) The difficulty lies entirely in the case $K = \mathbb{C}$, where Z characteristic numbers do not determine bordism class for spaces with torsion.

2) That the following proof was valid in the complex case was pointed out to me by Wall. This proof was used in Stong [118].

Proof: Let $f : (M,\partial M) \longrightarrow (X,A)$ with $\{\sigma_1^2 \cdot \sigma_\omega \cdot f^*(x)\}[M,\partial M] = 0$ for all $x \in H^*(X,A;Z_K)$ and $M' \subset M \times P(K^2)$ dual to $\sigma_1(M) + \bar{\alpha}$, with $\bar{\alpha} \in H^k(P(K^2);Z_K)$ the usual generator, $\bar{\alpha}[P(K^2)] = 1$, and let

$$g : M' \hookrightarrow M \times P(K^2) \xrightarrow{\pi_1} M \xrightarrow{f} X.$$

Then $\sigma(M') = \sigma(M)\cdot(1 + \bar{\alpha})^2/\{1 + \bar{\alpha} + \sigma_1(M)\}$ while $g^*(x) = f^*(x) \otimes 1 = f^*(x)$ dropping useless $\otimes$'s and restrictions.

Now $\bar{\alpha}^2 = 0$ and mod σ_1^2 one has

$$\begin{aligned}(1 + \bar{\alpha})^2/(1 + \bar{\alpha} + \sigma_1) &= (1 + 2\bar{\alpha})(1 - \bar{\alpha} - \sigma_1 + 2\bar{\alpha}\sigma_1),\\ &= 1 + (\bar{\alpha} - \sigma_1).\end{aligned}$$

Thus $\sigma_i(M') = \sigma_i(M) + (\bar\alpha - \sigma_1)\sigma_{i-1}(M) \bmod \sigma_1^2$ or $\sigma_\omega(M') = \sigma_\omega(M) + (\bar\alpha - \sigma_1)u_\omega + \sigma_1^2 v_\omega$ where u_ω, v_ω are polynomials in $\bar\alpha$ and the $\sigma_i(M)$. Then

$$\{\sigma_\omega \cdot g^*(x)\}[M',\partial M'] = \{\sigma_\omega(\bar\alpha+\sigma_1)f^*(x) + (\bar\alpha^2-\sigma_1^2)u_\omega f^*(x) + \sigma_1^2(\bar\alpha+\sigma_1)v_\omega f^*(x)\}\cdot$$
$$\cdot[M,\partial M] \times [P(K^2)]$$

and deleting numbers with σ_1^2, which are zero, this is

$$\{\sigma_\omega \cdot f^*(x)\}(\bar\alpha+\sigma_1)[M,\partial M] \times [P(K^2)].$$

Since $\{\sigma_\omega \cdot f^*(x)\cdot\sigma_1\} \otimes 1$ evaluates to zero, this is

$$\{\sigma_\omega \cdot f^*(x)\}[M,\partial M]\cdot\bar\alpha[P(K^2)] = \{\sigma_\omega \cdot f^*(x)\}[M,\partial M]$$

so that $((M,\partial M),f)$ and $((M',\partial M'),g)$ have the same characteristic numbers. **

In order to compute $\mathcal{W}_*(K,2)$, it is convenient to give this group an algebraic structure. For $K = \mathbb{R}$, this is very easy since $P(\mathbb{R}^2) = S^1$ is an abelian group. Letting $m : S^1 \times S^1 \longrightarrow S^1 : (z,w) \longrightarrow z\cdot w$ be multiplication of complex numbers of norm 1, $m^*\xi$ is a line bundle restricting to ξ on $S^1 \times 1$ and $1 \times S^1$ so $m^*\xi = \pi_1^*\xi \otimes \pi_2^*\xi$. Thus if $f : M \longrightarrow S^1$, $g : M' \longrightarrow S^1$ pull ξ back to $\det\tau_M$ and $\det\tau_{M'}$, then $\tau_{M\times M'} \cong \tau_M \oplus \tau_{M'}$ so

$$M \times M' \xrightarrow{f\times g} S^1 \times S^1 \xrightarrow{m} S^1$$

realizes $\det(\tau_{M\times M'})$. This gives immediately:

<u>Proposition</u>: $\mathcal{W}_*(\mathbb{R},2)(X,A)$ is a free $\mathcal{W}_*(\mathbb{R},2)$ module isomorphic to $H_*(X,A;Z_2) \otimes \mathcal{W}_*(\mathbb{R},2)$ and $\mathcal{W}_*(\mathbb{R},2)$ is a Z_2 subalgebra of $\mathcal{N}_*$.

Proof: From the multiplication on S^1, $\mathcal{W}_*(R,2)(X,A)$ is a $\mathcal{W}_*(R,2)$ module and since generators of $H_*(X,A;Z_2)$ may be taken to be images of $\mathcal{W}_*(R,2)(X,A)$ classes, a dimension count using the Atiyah sequence of (X,A) and a point suffices to give the free module structure. **

For $K = \mathbb{C}$, $\mathcal{W}_*(K,2)$ is not a subring of Ω_*^G, for $P(\mathbb{C}^2)$ has c_1 spherical (by a map of degree 2 of S^2 into itself) while $c_1^2[P(\mathbb{C}^2) \times P(\mathbb{C}^2)] = (2\alpha_1 + 2\alpha_2)^2[P(\mathbb{C}^2) \times P(\mathbb{C}^2)] = 8 \neq 0$. Noting that 2 is the only prime dividing 8, it is not surprising that a best possible result is:

Proposition: $\mathcal{W}_*(K,2) \otimes Z_2 \subset \Omega_*^G \otimes Z_2$ is a Z_2 subalgebra. In fact, if $a,b \in \mathcal{W}_*(K,2)$ then

$$\Phi(a\cdot b) = a\cdot b + 2[V^{2k}]\cdot\partial a\cdot\partial b$$

where $[V^{2k}] = [P(K^2) \times P(K^2)] - [P(K^3)]$, and if $M \in a$, then ∂a is represented by the submanifold of M dual to $\det\tau_M$.

Proof: Since $\mathcal{W}_*(K,2)$ is a direct summand of Ω_*^G, $\mathcal{W}_*(K,2) \otimes Z_2 \subset \Omega_*^G \otimes Z_2$. If $p,q \in \mathcal{W}_*(K,2) \otimes Z_2$, represented by $x,y \in \Omega_*^G$, then $x = a + 2u$, $y = b + 2v$, with $a,b \in \mathcal{W}_*(K,2)$, $u,v \in \Omega_*^G$. Then $p\cdot q \in \Omega_*^G \otimes Z_2$ is represented by $a\cdot b$ or $\Phi(a\cdot b) \in \mathcal{W}_*(K,2)$ from the formula. Thus it suffices to prove the formula for $\Phi(a\cdot b)$.

To prove the formula, note that $P(K^\infty)$ has Z_K homology a free Z_K module and hence $\Omega_*^G(P(K^\infty))$ is a free Ω_*^G module with base given by the inclusion maps $i : P(K^r) \hookrightarrow P(K^\infty)$, $r \geq 1$. Let $x_j = (P(K^{j+1}),i)$.

Let $\Delta : \Omega_*^G(P(K^\infty)) \longrightarrow \Omega_*^G(P(K^\infty))$ by sending (M,f) to $(N,f\cdot j)$ where $j : N \hookrightarrow M$, N being dual to $f^*(\xi)$. Δ is clearly an Ω_*^G module homomorphism and $\Delta x_j = x_{j-1}$ since the normal bundle of $P(K^j)$ in $P(K^{j+1})$ is ξ.

Let $\varepsilon : \Omega_*^G(P(K^\infty)) \longrightarrow \Omega_*^G : (M,f) \longrightarrow [M]$ be the augmentation. Then ε is an Ω_*^G module homomorphism. Let $\mu : \Omega_*^G \longrightarrow \Omega_*^G(P(K^\infty))$ by sending $[M]$ into (M,f) where $f^*(\xi) = \det\tau_M$. If $P(K^\infty) \times P(K^\infty) \longrightarrow P(K^\infty)$ classifies the tensor product $\xi \otimes \xi$, one has induced a multiplication in $\Omega_*^G(P(K^\infty))$. Since $\det\tau_{M\times M'} = \det\tau_M \otimes \det\tau_{M'}$, μ is a ring homomorphism with this product.

If $x \in \mathcal{W}_*(K,2)$, then μx comes from $\Omega_*^G(P(K^2))$ so $\mu x = \alpha x_0 + \beta x_1$, $\alpha,\beta \in \Omega_*^G$. Then $x = \varepsilon\mu\, x = \alpha + \beta(\varepsilon x_1)$, and $\partial x = \varepsilon\Delta\mu x = \varepsilon(\beta x_0) = \beta\varepsilon x_0 = \beta$.

For any $c \in \Omega_*^G$, $\Phi(c) = \varepsilon\Delta(\mu(c)\cdot x_1)$. Thus if $a,b \in \mathcal{W}_*(K,2)$, $\mu a = \alpha x_0 + \alpha' x_1$, $\mu b = \beta x_0 + \beta' x_1$ and

$$\Phi(ab) = \varepsilon\Delta(\alpha\beta x_1 + (\alpha\beta' + \beta\alpha')x_1^2 + \alpha'\beta' x_1^3).$$

Now $\varepsilon\Delta x_1^2 = \varepsilon x_1$ for the submanifold $H_{1,1}$ of $P(K^2) \times P(K^2)$ dual to $\xi \otimes \xi$ is $P(K^2)$. $[\sigma(H) = \{(1+x)^2(1+y)^2\}/(1+x+y)$ and H is dual to $x + y$, so $\sigma_1[H] = (x+y)^2[P(K^2) \times P(K^2)] = 2 = \sigma_1[P(K^2)]]$. Also $\varepsilon\Delta x_1^3 = 3\varepsilon x_1^2 - 2\varepsilon x_2$, for if $H \subset P(K^2) \times P(K^2) \times P(K^2)$ is dual to $\xi \otimes \xi \otimes \xi$, then

$$\sigma(H) = \frac{(1+x)^2(1+y)^2(1+z)^2}{1 + x + y + z} = 1 + (x+y+z) + 2(xy+xz+yz)$$

and H is dual to $x+y+z$, so

$$\sigma_1^2[H] = \sigma_2[H] = 2(xy+xz+yz)(x+y+z)[P(K^2) \times P(K^2) \times P(K^2)],$$
$$= 6$$

with

$$\sigma_1^2[P(K^3)] = 9, \quad \sigma_2[P(K^3)] = 3; \quad \sigma_1^2[P(K^2)^2] = 8, \quad \sigma_2[P(K^2)^2] = 4.$$

Thus

$$\begin{aligned}\Phi(ab) &= \alpha\beta\varepsilon x_0 + (\alpha\beta' + \beta\alpha')\varepsilon x_1 + \alpha'\beta'(3\varepsilon x_1^2 - 2\varepsilon x_2), \\ &= (\alpha + \alpha'\varepsilon x_1)(\beta + \beta'\varepsilon x_1) + \alpha'\beta'(2\varepsilon x_1^2 - 2\varepsilon x_2), \\ &= ab + 2[V^{2k}]\, \partial a\cdot\partial b. \quad **\end{aligned}$$

For later use, one also has:

Lemma: If $a,b \in \mathcal{W}_*(K,2)$, then

$$\partial(a\cdot b) = a\cdot\partial b + b\cdot\partial a - [P(K^2)]\cdot\partial a\cdot\partial b.$$

Proof: Let $\mu a = \alpha x_0 + \alpha' x_1$, $\mu b = \beta x_0 + \beta' x_1$. Then

$$\begin{aligned}
\partial(ab) &= \varepsilon\Delta\mu(ab) = \varepsilon\Delta(\mu a\cdot\mu b),\\
&= \varepsilon\Delta(\alpha\beta x_0 + (\alpha\beta' + \beta\alpha')x_1 + \alpha'\beta' x_1^2),\\
&= (\alpha\beta' + \beta\alpha')\varepsilon x_0 + \alpha'\beta'\ \varepsilon\ x_1,\\
&= (\alpha + \alpha'\ \varepsilon x_1)\beta' + (\beta + \beta'\ \varepsilon\ x_1)\alpha' - \alpha'\beta'\ \varepsilon\ x_1,\\
&= a\cdot\partial b + b\cdot\partial a - (\varepsilon x_1)\partial a\cdot\partial b. \quad **
\end{aligned}$$

Theorem: One may choose generators x_i, $i \neq 2^s-1$, of $\mathcal{N}_*$ and b_i of Ω_*^U so that

$$\mathcal{W}_*(\mathbb{R},2) = Z_2[x_j, (x_{2^s})^2 \mid j \neq 2^t, 2^t-1],$$

and

$$\mathcal{W}_*(C,2) \otimes Z_2 = Z_2[b_j, (b_{2^{s+1}})^2 + c_{2^{s+2}} \mid s \geq 0,\ j \neq 2^{t+1},\ t \geq 0]$$

where $c_{2^{s+2}}$ belongs to the ideal in Ω_*^U generated by the b_{2^t-1}, having the further property that b_j maps to $x_j^{\ 2}$ in $\mathcal{N}_*$ if $j \neq 2^t-1$, and b_{2^s-1} maps to zero.

Proof: First, any elements b_j and $(b_{2^{s+1}})^2 + c_{2^{s+2}}$ of the given form generate a polynomial subalgebra of $\Omega_*^U \otimes Z_2$ as is easily verified by filtering by powers of the ideal generated by the b_{2^t-1}. Letting Q be one of the given polynomial algebras above, $Q = Z_2[y_j \mid j \neq 2, 2^s-1]$ or $Q = Z_2[y_j \mid j \neq 2]$

respectively. Thus $\Omega_*^G \otimes Z_2$ has the same rank over Z_2 as $Q[z]$ where $\dim z = 2k$. From the exact sequence

$$0 \longrightarrow \mathcal{W}_*(K,2) \otimes Z_2 \longrightarrow \Omega_*^G \otimes Z_2 \xrightarrow{d} \Omega_*^G \otimes Z_2 \longrightarrow 0$$

it follows that Q has the same rank as $\mathcal{W}_*(K,2) \otimes Z_2$ (d has degree $-2k$). Since $\mathcal{W}_*(K,2) \otimes Z_2$ is closed under multiplication, it suffices to construct the generators.

Define manifolds M_i as follows

1) If $i = 2^t$, let $M_i = KP(2^t)$.

2) If i is odd, not of the form 2^t-1, let $i = 2^p(2q+1)-1$, $p,q \geq 1$ and let $M_i \subset KP(2^p) \times KP(2^{p+1}q)$ be dual to $\sigma_1 = (2^p+1)\bar{\alpha}_1 + (2^{p+1}q+1)\bar{\alpha}_2$.

3) If i is even, not a power of 2, let $i = 2^p(2q+1)$, $p,q \geq 1$ and let $M_i \subset KP(1) \times KP(2^p) \times KP(2^{p+1}q)$ be dual to $\bar{\alpha}_1 + (2^p+1)\bar{\alpha}_2 + (2^{p+1}q+1)\bar{\alpha}_3$.

4) If $i = 2^{t+1}-1$, $t \geq 1$, let $M_i \subset KP(2^t) \times KP(2^t)$ be dual to $\sigma_1 = (2^t+1)(\bar{\alpha}_1 + \bar{\alpha}_2)$.

The manifolds M_i provide acceptable generators for the cobordism ring mod 2 since one has:

1) $\sigma(M_i) = (1 + \bar{\alpha})^{2^t+1}$ so $s[M_i] = 2^t+1$ which is odd for $t > 0$ and nonzero mod 4 if $t = 0$. Further $M_1^{\mathbb{C}}$ has $c_1 = 2\bar{\alpha}$ spherical by a map of degree 2, and $M_1^{\mathbb{C}}$ bounds in $\mathcal{U}_*$.

2) $\sigma(M_i) = (1 + \bar{\alpha}_1)^{2^p+1}(1 + \bar{\alpha}_2)^{2^{p+1}q+1}/[1 + (2^p+1)\bar{\alpha}_1 + (2^{p+1}q+1)\bar{\alpha}_2]$ has $\sigma_1 = 0$, hence spherical and $s[M_i] = -[(2^p+1)\bar{\alpha}_1 + (2^{p+1}q+1)\bar{\alpha}_2]^{i+1}[KP] =$
$-(2^p+1)^{2^p}(2^{p+1}q+1)^{2^{p+1}q}\binom{2^p(2q+1)}{2^p}$ which is odd.

3) $\sigma(M_i) = (1+\bar{\alpha}_1)^2(1+\bar{\alpha}_2)^{2^p+1}(1+\bar{\alpha}_3)^{2^{p+1}q+1}/[1+\bar{\alpha}_1+(2^p+1)\bar{\alpha}_2+(2^{p+1}q+1)\bar{\alpha}_3]$

so $\sigma_1 = \bar{\alpha}_1$ is spherical and $s[M_i] = -[\bar{\alpha}_1+(2^p+1)\bar{\alpha}_2+(2^{p+1}q+1)\bar{\alpha}_3]^{i+1}[KP]$

$= -(i+1)\bar{\alpha}_1\cdot\{(2^p+1)\bar{\alpha}_2+(2^{p+1}q+1)\bar{\alpha}_3\}^i[KP] = -(2^p(2q+1)+1)(2^p+1)^{2^p}(2^{p+1}q+1)^{2^{p+1}q}$.

$\binom{2^p(2q+1)}{2^p}$ which is odd.

4) $\sigma(M_i) = (1+\bar{\alpha}_1)^{2^t+1}(1+\bar{\alpha}_2)^{2^t+1}/1+(2^t+1)(\bar{\alpha}_1+\bar{\alpha}_2)$ has $\sigma_1 = 0$, hence spherical and $s[M_i] = -(2^t+1)^{2t+1}\binom{2^{t+1}}{2^t}$ which is nonzero mod 4. Further, by symmetry in $\bar{\alpha}_1,\bar{\alpha}_2$, all numbers of M_i are even (the dual class $\bar{\alpha}_1 + \bar{\alpha}_2$ gives equal terms with each summand $\bar{\alpha}_i$) so $M_i^{\mathbb{C}}$ and $M_i^{\mathbb{R}}$ are zero in $\mathcal{N}_*$.

One may let $x_i = [M_i^{\mathbb{R}}]$ for $i \neq 2^s-1$ and let b_i be a generator reducing to $[M_i^{\mathbb{C}}]$ mod 2. Then it suffices to show that $(x_{2^s})^2$ and $(b_{2^{s+1}})^2 + c_{2^{s+2}}$ belong to $\mathcal{W}_*(K,2) \otimes Z_2$.

Let $N \subset KP(1) \times KP(2^{t+1}) \times KP(2^{t+1})$, $t \geq 0$, be dual to $\bar{\alpha}_1 + (2^{t+1}+1)(\bar{\alpha}_2+\bar{\alpha}_3)$. Then $\sigma_1(N) = \bar{\alpha}_1$ is spherical so $[N] \in \mathcal{W}_*(K,2)$. Since $RP(2^{t+1})^2 \sim \mathbb{C}P(2^{t+1})$ which has $w_1 = 0$, $N^R = \Phi(RP(2^{t+1})^2) \sim RP(2^{t+1})^2$. Since $[N^{\mathbb{C}}] = [N^{\mathbb{R}}]^2$ in $\mathcal{N}_*$, $N^{\mathbb{C}} - \mathbb{C}P(2^{t+1})^2$ maps to zero in $\mathcal{N}_*$ and hence belongs to the ideal generated by 2 and the b_{2^s-1}. Thus $[N^{\mathbb{C}}] = [\mathbb{C}P(2^{t+1})]^2 + [N^{\mathbb{C}} - \mathbb{C}P(2^{t+1})^2]$ has the desired form. **

<u>Corollary</u>: Under the natural homomorphism $F_* : \Omega_*^U \longrightarrow \mathcal{N}_*$, the direct summand $\mathcal{W}_*(\mathbb{C},2)$ maps precisely onto the squares of elements in $\mathcal{W}_*(\mathbb{R},2)$.

<u>Relation between $\mathcal{W}_*(K,2)$ and Ω_*^{SG} : Semi-geometric methods.</u>

The importance of the groups $\mathcal{W}_*(K,2)$ is their relationship with the groups Ω_*^{SG}. This is expressed in the fashion:

<u>Theorem</u>: For every pair (X,A) there is an exact sequence

$$\begin{array}{ccc} \Omega_*^{SG}(X,A) & \xrightarrow{\ t\ } & \Omega_*^{SG}(X,A) \\ & {}_{\partial}\nwarrow \quad \swarrow_{\rho} & \\ & \mathcal{W}_*(K,2)(X,A) & \end{array}$$

in which ρ considers an oriented G manifold as a $\mathcal{W}_*(K,2)$ manifold (degree 0), ∂ sends (M,f) into $(N,f\cdot j)$ where $j : N \hookrightarrow M$ is the inclusion of the submanifold dual to $\det\tau_M$, and $\det\tau_N$ is trivialized via the identification $\det\tau_N \otimes \det\tau_M \cong \det(\tau_N \oplus \nu) \cong \det\tau_M$ where ν is the normal bundle of N in M (degree $-k$), and t is the homomorphism of degree $k-1$ obtained by multiplication by a fixed class $[S^{k-1}, \mathcal{O}] \in \Omega_*^{SG}$, $\mathcal{O}$ being an SG structure.

<u>Remarks</u>: The first proof of this type theorem was due to Rohlin [104] who showed $\Omega_*^{SO} \xrightarrow{2} \Omega_*^{SO} \xrightarrow{\rho} \mathcal{T}_*$ was exact. This proof was improved by Dold [45]. Wall [130] proved exactness of $\Omega_*^{SO} \longrightarrow \Omega_*^{SO} \longrightarrow \mathcal{W}_*(\mathbb{R},2) \longrightarrow \Omega_*^{SO}$, with an improved proof in Wall [133]. The bordism analog of Rohlin's result was proved in Conner and Floyd [36] giving exactness of $\Omega_*^{SO}(X,A) \longrightarrow \Omega_*^{SO}(X,A) \longrightarrow \mathcal{T}_*(X,A)$. The exactness of the complex sequence was proved by Conner and Floyd [39] using a modification of a proof due to Atiyah [13] for the real case.

<u>Proof</u>: 1) $\partial\rho = 0$. If $f : (M,\partial M) \longrightarrow (X,A)$ and $\det\tau_M$ is trivialized, then N is empty, so represents zero.

2) If $\partial x = 0$, then let $f : (M,\partial M) \longrightarrow (X,A)$ represent x. Let $j : N \hookrightarrow M$ be dual to $\det\tau_M$ by making the map $h : M \longrightarrow P(K^2)$ transverse on a point $P(K^1)$. Thus $\det\tau_M|_N$ is trivialized. Letting $L : U \longrightarrow X$ be a

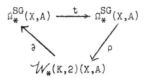

map of an oriented G manifold into X, $\partial U = N \cup (-P)/(\partial N \cong \partial P)$, $L|_N = f \cdot j$, $L(P) \subset A$, let V be formed from $M \times I$ and D, where D is the disc bundle of the trivial K line bundle over U, by identifying $D|_N$ with a tubular neighborhood of $N \times 1$ in $M \times 1$. Exactly as before, V has a BG-manifold structure, V maps into X extending $f \cdot \pi_1$ and L by means of a retraction and $\det\tau_V$ is induced by the map into $P(K^2)$ sending $M \times I$ by $h \cdot \pi_1$ and $D = U \times D^k$ by the map into a disc neighborhood of $h(N)$. This gives a cobordism of $f : (M, \partial M) \longrightarrow (X, A)$ to a map for which N is empty. Since $P(K^2) - pt$ is contractible, this gives a trivialization of the determinant bundle, and so x is represented by an SG manifold.

3) If $\rho y = 0$ with y represented by $f : (M, \partial M) \longrightarrow (X, A)$ with M oriented, then there is a BG manifold U such that $\partial U = M \cup (-P)/(\partial M \cong \partial P)$, a map $F : U \longrightarrow X$ extending f and sending P into A, and a map $h : U \longrightarrow P(K^2)$ with $h^*(\xi) = \det\tau_U$ sending M into a point $q \in P(K^2)$ defining the trivialization of $\det\tau_M$. Let $u \in P(K^2)$ be some other point and deform h to be transverse regular to u, using a deformation which keeps M fixed. Let $L = h^{-1}(u)$. Then $L \subset U$ is a submanifold with trivialized normal bundle, ∂L being contained in P. Let $L \times D^k$ be a tubular neighborhood of L mapped into the disc D^k with center u (by projection) under h, where $D^k \subset P(K^2)$ does not contain q, and let $W = U - \text{interior}(L \times D^k)$. Since ξ is trivial over $P(K^2) - u$, W has an SG structure given by this trivialization. By a homotopy of F one may assume $F|_{L \times D^k}$ coincides with the composition of projection on L and $F|_L$, since the neighborhood $L \times D^k$ may be deformed into L. Thus $F|_W : W \longrightarrow X$ gives an SG cobordism from $f : (M, \partial M) \longrightarrow (X, A)$ to a map $g : (L \times S^{k-1}, \partial L \times S^{k-1}) \longrightarrow (X, A)$ which factors through the projection onto L. The trivialization of the normal bundle of L in U and of $\det\tau_U|_L$ gives L an SG structure, and each fiber

S^{k-1} is given the SG structure obtained as follows: Let $\phi : D^k \longrightarrow P(K^2)$ be an imbedding and give S^{k-1} the SG structure obtained by trivializing the stable normal bundle $\phi^*\xi$ by deforming S^{k-1} to a point in $P(K^2) - \phi(0)$. Thus $y = tz$, where z is represented by $F|_L : (L,\partial L) \longrightarrow (X,A)$.

Note: In Wall's proof it is not known that $\det\tau_U$ is induced by a map into $P(R^2)$, hence he gets $M \sim V$ where V double covers $L = h^{-1}(u)$, $h : U \longrightarrow P(K^\infty)$. It is then necessary to show $V \sim 2L$.

4) $\rho t = 0$ for $\rho'([S^{k-1},\sigma]\cdot(M,f)) = \rho'([S^{k-1},\sigma])\cdot\rho'((M,f))$ where $\rho' : \Omega_*^{SG}(X,A) \longrightarrow \Omega_*^G(X,A)$ is reduction, while $[S^{k-1},\sigma]$ bounds in Ω_*^G. Thus $\rho' tx = 0$, but the map $\mathcal{W}_*(K,2)(X,A) \longrightarrow \Omega_*^G(X,A)$ is monic and hence $\rho tx = 0$.

5) If $f : (M,\partial M) \longrightarrow (X,A)$ is a $\mathcal{W}_*(K,2)$ map and $h : M \longrightarrow P(K^2)$ is transverse regular on $P(K^1)$ with $j : N \hookrightarrow M$ the submanifold $h^{-1}(P(K^1))$, let D^k be a neighborhood of $P(K^1)$ and $N \times D^k$ a tubular neighborhood of N mapped by projection onto D^k under h. Deform f so that $f|_{N\times D^k}$ coincides with the composition of projection on N and $f|_N$. Let $W = M - \text{interior}(N\times D^k)$ and then trivialize $\det\tau_W$ by deforming $h|_W$ to a point in $P(K^2) - P(K^1)$. Then $(W,f|_W)$ gives a cobordism in $\Omega_*^{SG}(X,A)$ of $[S^{k-1},\sigma]\cdot(N,f|_N)$ and the empty map.

6) If $f : (M,\partial M) \longrightarrow (X,A)$ is an SG bordism element and $f\cdot\pi : [S^{k-1},\sigma]\cdot(M,\partial M) \longrightarrow (X,A)$ represents zero, let $F : U \longrightarrow X$, $\partial U = M \times S^{k-1} \cup (-P)$, $F|_{M\times S^{k-1}} = f\cdot\pi$, $F(P) \subset A$ be a cobordism to zero. Beginning with a point map $H : U \longrightarrow P(K^2)$ (into $P(K^2) - D^k$, D^k a neighborhood of $P(K^1)$) trivializing $\det\tau_U$, one may homotope H in $P(K^2) - D^k$ to coincide with the standard map $M \times S^{k-1} \longrightarrow S^{k-1} \longrightarrow \partial D^k$ (using a tubular neighborhood). Let W be formed from U and $M \times D^k$ by joining the copies of $M \times S^{k-1}$ and extend $F : U \longrightarrow X$ by $f\cdot\pi_M$ on $M \times D^k$ to define

$F' : W \longrightarrow X$. The usual map $M \times D^k \longrightarrow D^k$ and H fit together to define a map $h : W \longrightarrow P(K^2)$ inducing $\det \tau_W$. Since $\partial W = P \cup \partial M \times D^k$ is mapped into A, $F' : (W, \partial W) \longrightarrow (X,A)$ is a $\mathcal{W}_*(K,2)$ bordism element in (X,A). Since $h : W \longrightarrow P(K^2)$ is transverse on $P(K^1)$, $\partial(W,F') = (M,f)$, and thus (M,f) is in the image of ∂. **

Relation to bordism groups

Rather than continue to attack the structure of the oriented cobordism groups, which will be relegated to later chapters, it seems better to study the Atiyah [13] approach to the above sequence.

Proposition: For $r > 1$, $\mathcal{W}_n(K,r)(X,A) \cong \tilde{\Omega}^{SG}_{n+k}(P(K^{r+1}) \wedge (X/A))$.

Proof: Let $f : BSG \times P(K^r) \longrightarrow BK^{(r)}$ classify $\mathcal{U} \oplus \xi$, $p : BK^{(r)} \to P(K^r)$ classify $\det \nu$ and $q : BK^{(r)} \longrightarrow BSG$ classify $\nu - \det \nu$. If $g = (q \times p)\cdot\Delta : BK^{(r)} \longrightarrow BSG \times P(K^r)$, then fg and gf are both homotopic to 1 since they classify the universal bundles. Thus $BK^{(r)}$ is identifiable to $BSG \times P(K^r)$ and the universal bundle to $\mathcal{U} \oplus \xi$. The Thom space of ξ is $P(K^{r+1})$ so $TBK^{(r)}_s$ is equivalent to $TBSG_{s-1} \wedge T\xi = TBSG_{s-1} \wedge P(K^{r+1})$ in the limit. Thus

$$\begin{aligned}\mathcal{W}_n(K,r)(X,A) &\cong \lim_{s\to\infty} \pi_{n+ks}(TBK^{(r)}_s \wedge (X/A)),\\ &\cong \lim_{s\to\infty} \pi_{n+ks}(TBSG_{s-1} \wedge P(K^{r+1}) \wedge (X/A)),\\ &\cong \tilde{\Omega}^{SG}_{n+k}(P(K^{r+1}) \wedge (X/A)). \quad **\end{aligned}$$

One has a cofibration sequence

$$S^k = P(K^2) \longrightarrow P(K^3) \longrightarrow S^{2k} = P(K^3)/P(K^2) \xrightarrow{a} \Sigma S^k = S^{k+1}$$

and smashing this with (X/A) and applying $\tilde{\Omega}_m^{SG}(\ \ )$ gives an exact sequence

$$\begin{array}{ccccccc}
\tilde{\Omega}_m^{SG}(S^k \wedge (X/A)) & \longrightarrow & \tilde{\Omega}_m^{SG}(P(K^3) \wedge (X/A)) & \longrightarrow & \tilde{\Omega}_m^{SG}(S^{2k} \wedge (X/A)) & \longrightarrow & \cdots \\
\wr\| & & \wr\| & & \wr\| & & \\
\Omega_{m-k}^{SG}(X,A) & \longrightarrow & \mathcal{W}_{m-k}(K,2)(X,A) & \longrightarrow & \Omega_{m-2k}^{SG}(X,A) & \longrightarrow & \cdots
\end{array}$$

where the outer identifications are suspension isomorphisms. This gives the Rohlin-Wall exact sequence. One may then recognize the homomorphism t as multiplication by the framed cobordism class represented by $a : S^{2k} \longrightarrow S^{k+1}$ which suspends the attaching map used to form $P(K^3)$ by attaching a $2k$ cell to S^k.

In order to obtain the Atiyah sequence, one first needs two results:

<u>Lemma 1</u>: $P(K^{m+n})/P(K^m)$ is the Thom complex of the bundle $m\xi$ over $P(K^n)$.

<u>Proof</u>: Let $f : S^{kn-1} \times D^{km} \longrightarrow S^{k(m+n)-1} : (x,y) \longrightarrow (\sqrt{1-|y|^2}\cdot x,y)$ where $S^{kn-1} = \{x \in K^n \,\big|\, |x| = 1\}$, $D^{km} = \{y \in K^m \,\big|\, |y| \leq 1\}$, and $S^{k(m+n)-1} = \{(x,y) \in K^n \times K^m \,\big|\, |x|^2 + |y|^2 = 1\}$. If $t \in K$, $|t| = 1$, then $f(tx,ty) = t\cdot f(x,y)$ so f is compatible with the usual action of S^{k-1}. If $(u,z) \in S^{k(m+n)-1}$ and $u \neq 0$, then $f^{-1}(u,z) = \{(\sqrt{1-|z|^2}\,(u/|u|),z)\}$. This gives a homeomorphism $\overline{f} : T(m\xi) \longrightarrow P(K^{m+n})/P(K^m)$ where $P(K^m)$ is given as the image of pairs (u,z) with $u = 0$. **

<u>Lemma 2</u>: Let ξ be the dual of the canonical bundle over $P(K^m)$. Then $\tilde{\Omega}_n^{SG}((X/A) \wedge T(\xi \oplus \xi \oplus \overline{\xi})) \cong \mathcal{W}_{n-3k}(K,m)(X,A)$.

<u>Proof</u>: Since $\xi \oplus \overline{\xi}$ is naturally oriented one has $f : BSG_s \times P(K^m) \longrightarrow BSG_{s+2} \times P(K^m)$ with $f^*(\mathcal{U}) = \mathcal{U} \oplus \xi \oplus \overline{\xi}$, $f^*(\xi) = \xi$ and by stability of BG_t, this is a homotopy equivalence up to dimension ks.

(Note: Being an induced fibration, stability follows for BSG_t. The maps $BSG_s \longrightarrow BSG_s \times P(K^m) \xrightarrow{f} BSG_{s+2} \times P(K^m) \longrightarrow BSG_{s+2}$ and $P(K^m) \longrightarrow BSG_s \times P(K^m) \xrightarrow{f} BSG_{s+2} \times P(K^m) \longrightarrow P(K^m)$ then give homotopy isomorphisms in low dimensions.) Thus in the limit

$$\begin{aligned}
\tilde{\Omega}_n^{SG}((X/A) \wedge T(\xi \oplus \xi \oplus \bar{\xi})) &= \lim_{s\to\infty} \pi_{n+ks}(TBSG_s \wedge T(\xi \oplus \xi \oplus \bar{\xi}) \wedge (X/A)), \\
&= \lim_{s\to\infty} \pi_{n+ks}(TBSG_{s+2} \wedge T(\xi) \wedge (X/A)), \\
&= \tilde{\Omega}_{n-2k}^{SG}((X/A) \wedge P(K^{m+1})), \\
&= \mathcal{W}_{n-3k}(K,m)(X,A). \quad **
\end{aligned}$$

Now consider the cofibration

$$P(K^3) \longrightarrow P(K^{n+3}) \longrightarrow P(K^{n+3})/P(K^3),$$

smash with (X/A) and apply $\tilde{\Omega}_m^{SG}(\quad)$ to obtain the exact sequence

$$\begin{array}{ccccccc}
\tilde{\Omega}_m^{SG}((X/A) \wedge P(K^3)) & \longrightarrow & \tilde{\Omega}_m^{SG}((X/A) \wedge P(K^{n+3})) & \longrightarrow & \tilde{\Omega}_m^{SG}((X/A) \wedge T(\xi \oplus \xi \oplus \bar{\xi})) & \longrightarrow & \cdots \\
\| & & \| & & \| & & \\
\mathcal{W}_{m-k}(K,2)(X,A) & \longrightarrow & \mathcal{W}_{m-k}(K,n+2)(X,A) & \longrightarrow & \mathcal{W}_{m-3k}(K,n)(X,A) & \longrightarrow & \cdots
\end{array}$$

and letting n go to ∞, this gives a long exact sequence

$$\mathcal{W}_{m-k}(K,2)(X,A) \longrightarrow \Omega_{m-k}^G(X,A) \longrightarrow \Omega_{m-3k}^G(X,A) \longrightarrow \mathcal{W}_{m-k-1}(K,2)(X,A) \longrightarrow \cdots$$

since for n large $\mathcal{W}_m(K,n)(X,A) = \Omega_m^G(X,A)$. This sequence splits up to give the Atiyah sequence, but the splitting requires one of the previous arguments.

Remark: If $\mathbb{F}$ is any spectrum and E_p is the two cell complex $S^1 \cup_p e^2$, one may form a new spectrum $E_p \wedge \mathbb{F}$, where $(E_p \wedge \mathbb{F})_s = E_p \wedge F_{s-1}$. This gives a homology theory

$$H_*(X,A;E_p \wedge \mathbb{F}) = \lim_{s\to\infty} \pi_{*+s}((X/A) \wedge E_p \wedge F_{s-1}),$$

$$= H_{*+1}((X/A) \wedge E_p;\underset{\sim}{\mathbb{F}}).$$

From the cofibration $S^1 \longrightarrow E_p \longrightarrow S^2$ one has an exact sequence

$$\begin{array}{c} H_*(X,A;\mathbb{F}) \xrightarrow{p} H_*(X,A;\underset{\sim}{\mathbb{F}}) \longrightarrow H_*(X,A;E_p \wedge \mathbb{F}) \\ \uparrow\underline{\qquad\qquad\qquad\delta\qquad\qquad\qquad}\downarrow \end{array}$$

where δ has degree -1. This is one way to introduce Z_p coefficients into spectral homology theory. (Another possibility is to use homotopy with Z_p coefficients rather than ordinary homotopy). With the given definition, $E_2 = P(R^3)$ and thus one may think of $\mathcal{W}_*(R,2)$ homology as oriented bordism with Z_2 coefficients. This was pointed out to me by D. Sullivan, and seems to explain the usefulness of $\mathcal{W}_*(R,2)$ theory, which at first glance appears extremely artifical.

<u>Remark</u>: From the Atiyah bordism approach, one has an exact sequence

$$\begin{array}{ccc} \mathcal{W}_*(K,2) & \xrightarrow{\quad i \quad} & \mathcal{W}_*(K,n+2) \\ & {}_{j}\nwarrow \quad \swarrow_{\Delta} & \\ & \mathcal{W}_*(K,n). & \end{array}$$

Then

$$\mathcal{W}_*(K,2) \xrightarrow{i} \mathcal{W}_*(K,n+2) \xrightarrow{i'} \Omega_*^G \xrightarrow{\Phi} \mathcal{W}_*(K,2)$$

is the identity, so this sequence splits. Thus $\mathcal{W}_*(K,n+2) \cong \mathcal{W}_*(K,2) \oplus \mathcal{W}_*(K,n)$ and these groups are then known inductively.

Chapter IX

Oriented Cobordism

With the exception of the unoriented cobordism problem, the most interesting manifold theoretic cobordism problem is the classification problem for "oriented" manifolds, where "oriented" is taken in the classical sense.

There are many equivalent descriptions of an "orientation" of a manifold, which may be given by:

a) A trivialization of the determinant bundle of the tangent (or normal) bundle;

b) A reduction of the structural group of the tangent (or normal) bundle to the special orthogonal group;

c) An integral cohomology orientation of the tangent (or normal) bundle in the sense of Dold; or

d) A fundamental integral homology class giving an orientation in the sense of Whitehead.

In addition to the desire to classify "oriented" manifolds because of their classical interest, definition (d) indicates a relation between "oriented" bordism and integral homology and full exploration of this relationship is desirable for geometric understanding of integral homology.

The analysis of "oriented" cobordism is a very complicated problem, the major outline of its solution having been:

1) Reduction to a homotopy problem and rational structure by Thom [127];

2) Calculation of odd primary and mod torsion structure by Milnor [81], Averbuh [21], and Novikov [93];

3) Calculation of 2 primary structure by Wall [130]; and

4) Analysis of oriented bordism by Conner and Floyd [36].

Using either definition (a) or (b) one has a classifying space BSO_n for oriented n plane bundles and from the Pontrjagin-Thom theorem, the "oriented" cobordism ring Ω_*^{SO} is given by the stable homotopy ring

$$\Omega_n^{SO} = \lim_{s\to\infty} \pi_{n+s}(TBSO_s,\infty).$$

From the analysis of the cohomology structure of BSO_n, one obtains:

<u>Theorem</u>: The groups Ω_n^{SO} are finitely generated and $\Omega_*^{SO} \otimes Q$ is the rational polynomial ring on the cobordism classes of the complex projective spaces $\mathbb{C}P(2i)$.

<u>Proof</u>: As noted in Chapter V, there are unique orientation classes $U_r \in H^r(TBSO_r;Z)$ which combine to define a Z cohomology orientation $U : \underset{\sim}{TBSO} \longrightarrow \underset{\sim}{K}(Z)$. By the Thom isomorphism theorem $\tilde{H}^n(\underset{\sim}{TBSO};Z) = \lim_{r\to\infty} H^{n+r}(TBSO_r,\infty,Z)$ is finitely generated, being isomorphic to $H^n(BSO_r;Z)$ for sufficiently large r by stability of these groups. Since $TBSO_r$ is $(r-1)$-connected, Serre's theorem shows that $\Omega_n^{SO} \longrightarrow \tilde{H}_n(\underset{\sim}{TBSO};Z)$ is an isomorphism modulo the class of finite groups. Thus Ω_n^{SO} is finitely generated.

Since $H^*(BSO;Q) = Q[\mathcal{P}_i]$, Ω_n^{SO} has rank equal to the number of partitions of $(n/4)$ if n is a multiple of 4, and is a finite group otherwise. In fact, $\Omega_*^{SO} \otimes Q \longrightarrow H_*(BSO;Q)$ is a ring isomorphism. From the diagonal formula $\Delta\mathcal{P}_i = \sum_{j+k=i} \mathcal{P}_j \otimes \mathcal{P}_k$, $\Omega_*^{SO} \otimes Q$ is a polynomial ring on generators x_{4i} of dimension $4i$, characterized by $s_{(i)}(\mathcal{P})[x_{4i}] \neq 0$ where $s_{(i)}(\mathcal{P})$ is the primitive class of dimension $4i$.

For the manifold $\mathbb{C}P(2i)$, $\tau \otimes 1 = (2i+1)\xi$ giving $c(\mathbb{C}P(2i)) = (1+\bar{\alpha})^{2i+1}$ and $c(\tau \otimes \mathbb{C}) = (1+\bar{\alpha})^{2i+1}(1-\bar{\alpha})^{2i+1} = (1-\bar{\alpha}^2)^{2i+1}$ so $\mathcal{P}(\mathbb{C}P(2i)) = (1+\bar{\alpha}^2)^{2i+1}$. In general, if $\mathcal{P}$ is written so that $\mathcal{P}_j$ is the j-th elementary symmetric function in classes β_t^2, $\dim \beta_t = 2$, the primitive class $s_{(i)}(\mathcal{P})$ is $\Sigma \beta_t^{2i}$. Thus

$$s_{(i)}(\mathcal{P})(\nu)[\mathbb{C}P(2i)] = -s_{(i)}(\mathcal{P})(\tau)[\mathbb{C}P(2i)], \quad \text{(primitivity)}$$

$$= -(2i+1)\bar{\alpha}^{2i}[\mathbb{C}P(2i)],$$

$$= -(2i+1)$$

for by uniqueness of the Thom space orientation, the SO induced fundamental class must agree with the fundamental class arising from the complex structure. Thus $\mathbb{C}P(2i)$ represents an acceptable generator of dimension $4i$. **

<u>Proposition</u>: The homomorphism

$$\gamma : \Omega_*^U \xrightarrow{S_*} \Omega_*^{SO} \xrightarrow{\pi} \Omega_*^{SO}/\text{Torsion}$$

with S_* induced by the forgetful functor, and π the quotient map, has kernel the ideal generated by the classes of dimension not a multiple of four and has cokernel a finite group of odd order (in each dimension).

<u>Proof</u>: Since Pontrjagin numbers are integer valued invariants, they annihilate the torsion subgroup and the homomorphism

$$\mathcal{P}' : \Omega_{4n}^{SO} \longrightarrow Z_2^{|\pi(n)|} : [M] \longrightarrow (\mathcal{P}_\omega(\tau)[M])$$

obtained by reducing the Pontrjagin numbers mod 2 factors through π. Then since $s_{(i)}(\mathcal{P}(\tau))[\mathbb{C}P(2i)]$ is odd, the map

$$\Omega_{4n}^U \xrightarrow{S_*} \Omega_{4n}^{SO} \xrightarrow{\mathcal{P}'} Z_2^{|\pi(n)|}$$

is epic.

Since $\Omega_{4n}^{SO}/\text{Torsion}$ has rank $|\pi(n)|$, this means $(\text{im}\gamma_{4n})$ has rank $|\pi(n)|$ and is a subgroup of odd index. Since $\Omega_{4n+2}^{SO}/\text{Torsion} = 0$, the kernel of γ

contains the ideal generated by classes of dimension not congruent to zero mod 4, and by rank this must be the precise kernel. **

In order to examine the odd primary situation, one has:

Proposition: The map $f : BSp \longrightarrow BSO$, obtained by considering a quaternionic vector bundle simply as an oriented vector bundle, is a homotopy equivalence mod the Serre class of 2 primary finite groups.

Proof: One has $H^*(BSp;Z) = Z[\mathcal{P}_i^s]$, where $\mathcal{P}_i^s$ is the i-th symplectic Pontrjagin class of the universal bundle $\mathcal{U}$, and considering $\mathcal{U}$ as a complex bundle, $\mathcal{P}_i^s(\mathcal{U}) = c_{2i}(\mathcal{U})$. Thus $\mathcal{P}_i(\mathcal{U}) = (-1)^i c_{2i}(\mathcal{U} \otimes C)$ and $c(\mathcal{U} \otimes C) = c(\tilde{\mathcal{U}})^2$, so

$$\mathcal{P}_i(\mathcal{U}) = (-1)^i 2\mathcal{P}_i^s(\mathcal{U}) + \text{decomposables}.$$

Thus the homomorphism

$$f^* : H^*(BSO;Z_p) = Z_p[\mathcal{P}_i] \longrightarrow H^*(BSp;Z_p) = Z_p[\mathcal{P}_i^s]$$

is an isomorphism for all odd primes p. **

Corollary: The forgetful homomorphism

$$F_* : \Omega_*^{Sp} \longrightarrow \Omega_*^{SO}$$

is an isomorphism modulo the Serre class of 2 primary finite groups.

Proof: The map $Tf : TBSp_n \longrightarrow TBSO_{4n}$ induces isomorphisms on Z_p cohomology for p any odd prime in dimensions less than 8n, using the knowledge of f^* and the Thom isomorphism. Thus by the generalized Whitehead theorem, $(Tf)_\#$ is an isomorphism on homotopy modulo 2 primary finite groups, and the Pontrjagin-Thom theorem completes the result. **

One then has the composition of homomorphisms defined by forgetful functors:

$$\Omega_*^{Sp} \xrightarrow{T_*} \Omega_*^{U} \xrightarrow{S_*} \Omega_*^{SO}$$

with $S_*T_* = F_*$, from which:

Theorem: All torsion in Ω_*^{SO} is 2 primary.

Proof: Since F_* is an isomorphism mod 2 primary torsion, while Ω_*^U is torsion free, neither Ω_*^{Sp} nor Ω_*^{SO} can have odd torsion. **

Theorem: The homomorphism

$$\gamma : \Omega_*^U \xrightarrow{S_*} \Omega_*^{SO} \xrightarrow{\pi} \Omega_*^{SO}/\text{Torsion}$$

is epic.

Proof: Since π is an isomorphism mod 2 primary torsion, $\pi F_* = \gamma \cdot T_*$ has finite 2 primary cokernel in each dimension, so that the same holds for γ, but it was previously noted that γ has finite odd order cokernel in each dimension. Thus γ is epic. **

Theorem: Ω_*^{SO}/Torsion is a polynomial ring over Z on classes x_i of dimension $4i$, and the classes x_i are characterized by

$$s_{(i)}(\mathcal{P}(\tau))[x_i] = \begin{cases} \pm 1 & 2i+1 \neq p^s \text{ for any prime } p \text{ and } s \in Z, \\ \pm p & 2i+1 = p^s \text{ for some prime } p \text{ and } s \in Z. \end{cases}$$

Proof: Ω_*^{SO}/Torsion is isomorphic to Ω_*^U mod the ideal generated by the classes having dimension not congruent to zero mod 4. Being a polynomial ring, generators are characterized by the s-number. The given values follow at once from the knowledge of the s-numbers of generators of complex cobordism, for if

M^n is a stably almost complex manifold with $c(M) = \Pi(1+x_j)$ formally, then $\mathcal{P}(M) = \Pi(1+x_j{}^2)$ and so $s_{(k)}(\mathcal{P}(\tau))(M) = \Sigma\, x_j{}^{2k} = s_{(2k)}(c(\tau))(M)$. **

To determine the two primary structure one uses the exact sequence

$$\Omega_*^{SO} \xrightarrow{t} \Omega_*^{SO} \xrightarrow{\rho} \mathcal{W}_*(R,2) \quad \text{with } \partial : \mathcal{W}_*(R,2) \to \Omega_*^{SO}$$

in which t becomes multiplication by 2 ($[S^0, \mathcal{O}]$ is two points, both positively oriented).

Let

$$\partial' = \rho\cdot\partial : \mathcal{W}_*(R,2) \longrightarrow \mathcal{W}_*(R,2)$$

be the composite homomorphism.

<u>Lemma</u>: $\partial' : \mathcal{W}_*(R,2) \longrightarrow \mathcal{W}_*(R,2)$ is a derivation with $(\partial')^2 = 0$, and choosing generators x_i, $i \neq 2^s-1$, for $\mathcal{N}_*$ such that

$$\mathcal{W}_*(R,2) = Z_2[x_j, (x_{2^s})^2 \mid j \neq 2^t,\ 2^t-1]$$

as in Chapter VIII, one has

$$\begin{cases} \partial' x_{2m-1} = 0, \\ \partial' x_{2m} = x_{2m-1}, \quad \text{and} \\ \partial'((x_{2^s})^2) = 0. \end{cases}$$

<u>Proof</u>: From Chapter VIII, one has $\partial'(ab) = \partial'a\cdot b + a\cdot\partial'b - [P(R^2)]\cdot\partial'a\cdot\partial'b$ but $[P(R^2)] = 0$ in $\mathcal{N}_*$, so ∂' is a derivation. Since $\partial'\partial' = \rho\partial\rho\partial$ and $\partial\rho = 0$, $(\partial')^2 = 0$. Examining the generators of $\mathcal{W}_*(R,2)$, the manifold N representing $(x_{2^s})^2$ is cobordant to $\rho[\mathbb{C}P(2^s)]$ and $\partial'\rho = 0$ so $\partial'[N] = 0$. For x_{2m}, one chose a representative

$$M \subset RP(1) \times RP(2^p) \times RP(2^{p+1}q)$$

dual to $\bar{\alpha}_1 + (2^p+1)\bar{\alpha}_2 + (2^{p+1}q+1)\bar{\alpha}_3$, so that $w_1(M) = \bar{\alpha}_1$. Then $\partial'[M]$ is represented by the submanifold dual to $w_1(M)$, i.e. the submanifold of $RP(1) \times RP(2^p) \times RP(2^{p+1}q)$ dual to

$$\bar{\alpha}_1(\bar{\alpha}_1 + (2^p+1)\bar{\alpha}_2 + (2^{p+1}q+1)\bar{\alpha}_3) = \bar{\alpha}_1((2^p+1)\bar{\alpha}_2 + (2^{p+1}q+1)\bar{\alpha}_3),$$

since $\bar{\alpha}_1^2 = 0$. Since the submanifold dual to $\bar{\alpha}_1$ is $RP(2^p) \times RP(2^{p+1}q)$, this is precisely the representative chosen for x_{2m-1}. Since $(\partial')^2 = 0$, this also gives $\partial' x_{2m-1} = 0$. **

Corollary: $\ker\partial'/\operatorname{im}\partial' = Z_2[(x_{2t})^2]$.

Proof: $\mathcal{W}_*(R,2)$ is the tensor product of algebras of the form:

a) $Z_2[x_{2m-1}, x_{2m}]$ m not a power of 2, with $\partial' x_{2m} = x_{2m-1}$, $\partial' x_{2m-1} = 0$; which has homology isomorphic to $Z_2[(x_{2m})^2]$; and

b) $Z_2[(x_{2^s})^2]$ with $\partial'(x_{2^s}{}^2) = 0$; isomorphic to its homology.

The Künneth theorem for the homology of a tensor product completes the computation. **

Proposition: $\rho\Omega_*^{SO} = \ker\partial = \ker\partial'$.

Proof: The composite $\ker\partial \hookrightarrow \ker\partial' \longrightarrow \ker\partial'/\operatorname{im}\partial'$ is epic, for if $a \in \ker\partial'/\operatorname{im}\partial'$, there is a class $b \in \mathcal{M}_*$ such that b^2 maps to a. Since b^2 is the class of a stably almost complex manifold, which is oriented, $b^2 \in \operatorname{im}\rho = \ker\partial$. Then if $\alpha \in \ker\partial'$, there is an $x \in \Omega_*^{SO}$ with $\alpha - \rho x$ mapping to zero in $\ker\partial'/\operatorname{im}\partial'$ or $\alpha - \rho x \in \operatorname{im}\partial' = \operatorname{im}\rho\cdot\partial \subset \operatorname{im}\rho$, so $\alpha \in \operatorname{im}\rho$. Thus $\operatorname{im}\rho = \ker\partial \subset \ker\partial' \subset \operatorname{im}\rho$. **

Theorem: All torsion in Ω_*^{SO} has order 2.

Proof: Suppose Ω_m^{SO} has some element of finite order which is not of order 2. There is then a class $x \in \Omega_m^{SO}$ with $2x \neq 0$, $4x = 0$. Then $t(2x) = 2(2x)$ is zero, so $2x = \partial y$ for some y. Since $\mathcal{W}_m(R,2)$ is a Z_2 vector space, $\partial' y = \rho\partial y = \rho(2x) = 2\rho(x) = 0$ and $y \in \ker\partial' = \ker\partial$. Thus $2x = \partial y = 0$, contradicting the choice of x. **

Corollary: Under the homomorphism $\rho : \Omega_*^{SO} \longrightarrow \mathcal{W}_*(R,2)$, the torsion subgroup maps isomorphically onto image ∂'.

Proof: If $x \in$ Torsion (Ω_*^{SO}), $2x = 0$ so $x = \partial y$ and $\rho x = \partial' y \in \operatorname{im}\partial'$. Thus $\rho(\text{Torsion } (\Omega_*^{SO})) \subset \operatorname{im}\partial'$. Conversely, $\partial' z = \rho\partial z$ so $\operatorname{im}\partial' \subset \rho \operatorname{im}\partial \subset \rho(\text{Torsion } (\Omega_*^{SO}))$. If $x \in$ Torsion (Ω_*^{SO}) and $\rho x = 0$, then $x = 2y$, but $y \in$ Torsion (Ω_*^{SO}) and hence $x = 2y = 0$. **

Corollary: The homomorphism $\Omega_*^{SO} \longrightarrow \ker\partial'/\operatorname{im}\partial'$ induces an isomorphism of $(\Omega_*^{SO}/\text{Torsion}) \otimes Z_2$ with the polynomial algebra $\ker\partial'/\operatorname{im}\partial'$.

Proof: The given homomorphism has kernel containing Torsion (Ω_*^{SO}) and defines a homomorphism $\Omega_*^{SO}/\text{Torsion} \longrightarrow \ker\partial'/\operatorname{im}\partial'$. This clearly sends $2(\Omega_*^{SO}/\text{Torsion})$ to zero, giving $(\Omega_*^{SO}/\text{Torsion}) \otimes Z_2 \longrightarrow Z_2[(x_{2t})^2]$. By the proposition this is epic and by the ranks must be an isomorphism. **

Theorem: Two oriented manifolds are cobordant if and only if they have the same Pontrjagin and Stiefel-Whitney numbers.

Proof: By Pontrjagin's theorem, cobordant manifolds have the same characteristic numbers. Conversely, suppose $x, x' \in \Omega_m^{SO}$ have the same Z and Z_2 cohomology characteristic numbers. Since all Z cohomology characteristic numbers of $y = x - x'$ are zero, y is a torsion class. Thus $2y = 0$ and $y = \partial z$ for some z. Since x and x' have the same Stiefel-Whitney numbers, all Z_2 cohomology characteristic numbers of ρy are zero, and thus $\partial' z = \rho\partial z = \rho y = 0$. This gives $z \in \ker\partial'$, so $y = \partial z = 0$ and $x = x'$. **

Turning to the two primary relationships among the characteristic numbers, one has:

Proposition: a) The homomorphism $\mathcal{P}'_\omega : \Omega_*^{SO} \longrightarrow Z_2$ which sends the class of M into the mod 2 reduction of the Pontrjagin number $\mathcal{P}_{i_1} \cdots \mathcal{P}_{i_r}(\tau)[M]$ $(\omega = (i_1,\ldots,i_r))$ coincides with evaluation of the Stiefel-Whitney number $w_{2\omega}{}^2 = (w_{2i_1} \cdots w_{2i_r})^2$.

b) The homomorphisms $\mathcal{P}'_\omega$ for $\omega \in \pi(m)$ form a base of $\mathrm{Hom}((\Omega_{4m}^{SO}/\mathrm{Torsion}) \otimes Z_2, Z_2)$.

c) There are no two primary relations among the Pontrjagin numbers of oriented manifolds.

Proof: a) If M is an oriented manifold, then reducing mod 2, $\mathcal{P}_i(\tau) \equiv c_{2i}(\tau \otimes \mathbb{C})$ and $c(\tau \otimes \mathbb{C}) \equiv w(\tau \oplus \tau) = w(\tau)^2$, so $\mathcal{P}_i(\tau) \equiv w_{2i}(\tau)^2$. Since the integral orientation reduces to the mod 2 orientation,

$$\begin{aligned}\mathcal{P}'_\omega([M]) &= \mathcal{P}_\omega(\tau)[M] \quad \mod 2\\ &= w_{2\omega}{}^2(\tau)[M].\end{aligned}$$

b) For any $\omega = (i_1,\ldots,i_r)$, let $\mathbb{C}P(2\omega) = \mathbb{C}P(2i_1) \times\ldots\times \mathbb{C}P(2i_r)$ and ordering partitions of m compatibly with refinement, the matrix

$$\| s_\omega(\mathcal{P}(\tau))[\mathbb{C}P(2\omega')] \|_{\omega,\omega' \in \pi(m)}$$

is triangular with odd diagonal entries. Thus

$$\mathcal{P}' = \prod_{\omega\in\pi(m)} \mathcal{P}'_\omega : \Omega_{4m}^{SO} \longrightarrow Z_2{}^{|\pi(m)|}$$

is epic, hence giving an isomorphism $(\Omega_{4m}^{SO}/\mathrm{Torsion}) \otimes Z_2 \cong Z_2{}^{|\pi(m)|}$.

c) If $\Sigma\, a_\omega \mathcal{P}_\omega$, $\omega \in \pi(m)$, $a_\omega \in Z$, always takes even values on 4m manifolds, then $\Sigma\, a_\omega \mathcal{P}'_\omega = 0$ so $a_\omega = 0$ in Z_2 for all ω.

Alternately phrased, $\Omega_*^{SO}/\text{Torsion} \longrightarrow \text{Hom}(Z[\mathcal{P}_i],Z)$ has image of odd index. **

Relations among the Stiefel-Whitney numbers are given by:

<u>Proposition</u>: The image of the forgetful homomorphism

$$F_* : \Omega_*^{SO} \longrightarrow \mathcal{N}_*$$

consists precisely of those classes for which all Stiefel-Whitney numbers with a factor w_1 vanish. Equivalently all relations among the Stiefel-Whitney numbers of oriented manifolds follow from the Wu formulae and the vanishing of w_1; explicitly, if $\phi : H^m(BO;Z_2) \longrightarrow Z_2$ is a homomorphism, then $\phi(x) = x(\tau)[M]$ for some oriented m-manifold if and only if $\phi(Sq^i\alpha + v_i\alpha) = 0$ for all i and all $\alpha \in H^{m-i}(BO;Z_2)$ and $\phi(w_1\beta) = 0$ for all $\beta \in H^{m-1}(BO;Z_2)$.

<u>Proof</u>: If M is an oriented manifold, $w_1(M) = w_1(\det\tau_M)$ is zero, since $\det\tau_M$ is trivial, and thus all Stiefel-Whitney numbers of M with w_1 as a factor must vanish. If $z \in \mathcal{N}_*$ has all numbers divisible by w_1 vanishing ($\phi(x) = x(z)$ for some such z), then $z \in \mathcal{W}_*(R,2)$ for all numbers with w_1^2 as a factor vanish. If N represents z and $K \subset N$ is dual to w_1, then

$$w(K) = w(N)/(1+w_1(N))$$

so $w_i(K) = w_i(N) + w_{i-1}(N)w_1(N)$ and $w_\omega(K)[K] = \{w_\omega(N) + w_1u_\omega\}w_1(N)[N]$ $= w_1w_\omega(N)[N]$. Since Stiefel-Whitney numbers determine class in $\mathcal{W}_*(R,2)$, $\partial'z = [K]$ is zero and thus $z = \rho([M])$ for some oriented M. **

Having described the two primary relations, one wishes to know the odd primary relations among the Pontrjagin numbers. From the study of complex cobordism, one knows that K-theory is a reasonable place to get relations. Unfortunately, oriented manifolds are not K-theory orientable. To circumvent this problem one has first a general construction.

Let K be one of the fields $\mathbb{R}$ or $\mathbb{C}$, and let K' be $\mathbb{C}$ or $\mathbb{H}$ respectively.

Let M^n be a closed, n-dimensional manifold with tangent bundle t, τ being a K vector bundle over M^n so that $\tau \cong t + u \cdot 1$ ($u \in Z$, $n+u \equiv 0 \bmod k$) as real vector bundles, defining a K structure on M.

Then the total space $E\tau$ is a differentiable manifold and if $\pi : E\tau \longrightarrow M$ is the projection, then the tangent bundle of $E\tau$ is $\pi^*(t) \oplus \pi^*(\tau)$, $\pi^*(\tau)$ being the bundle along the fibers and $\pi^*(t)$ its orthogonal complement. Thus the stable tangent bundle of $E\tau$ admits a K' vector bundle structure as $\pi^*(\tau) \oplus \pi^*(\bar{\tau}) = \pi^*(\tau) \otimes_{\mathbb{R}} \mathbb{C}$.

Letting $D\tau$ and $S\tau$ denote the disc and sphere bundles of τ, $D\tau$ has an induced stable K' structure as manifold with boundary, its boundary being $S\tau$.

Now suppose one is given a ring spectrum $\underset{\sim}{A}$ for which K' vector bundles are naturally oriented and suppose $U \in \tilde{H}^{n+u}(T\bar{\tau};\underset{\sim}{A}) = H^{n+u}(D\tau,S\tau;\underset{\sim}{A})$ is a fixed class. Then for $x \in H^j(M;\underset{\sim}{A})$ one may define a <u>number</u>

$$x_U[M] = \{\pi^*(x)\cdot U\}[D\tau,S\tau] \in H^{j-n}(pt;\underset{\sim}{A}).$$

<u>Notes</u>: 1) These numbers may not be meaningful invariants of M unless the class U is stable [i.e. replacing τ by $\tau \oplus 1$ gives $T(\overline{\tau \oplus 1}) = \Sigma^k T(\bar{\tau})$ and $U(\overline{\tau \oplus 1})$ should be the suspension of $U(\bar{\tau})$] or is defined for the tangent bundle t directly so that one need not make a choice of τ.

2) It is not at all clear that such numbers are cobordism invariants; indeed they won't be in the general case.

If M is imbedded in R^{n+kq} with K normal bundle ν such that $\nu \oplus \tau$ is trivial, the induced map on tangent spaces gives an imbedding $Et \hookrightarrow R^{n+kq} \times R^{n+kq}$ and hence imbeddings

$$E\tau \hookrightarrow R^{n+kq} \times R^{n+kq} \times R^{u},$$

$$D\tau \hookrightarrow R^{n+kq} \times D^{n+kq+u},$$

$$S\tau \hookrightarrow R^{n+kq} \times S^{n+kq+u-1},$$

and the normal bundle may be taken isomorphic to $\pi^*(\nu \oplus \bar{\nu})$.

<u>Note</u>: One has

$$\begin{array}{ccccc} M \times M & \longrightarrow & R^{n+kq} \times R^{n+kq} & \longrightarrow & R^{n+kq} \times R^{n+kq} \\ \Delta \uparrow & & \uparrow & (x,y) \longrightarrow (x,y-x) & \uparrow \\ M & \longrightarrow & \Delta & \longrightarrow & R^{n+kq} \times 0 \end{array}$$

corresponding to identification of $D(t)$ with a neighborhood of the diagonal in $M \times M$ or $D\tau$ with a neighborhood of $\Delta(M) \times 0$ in $M \times M \times R^{u}$.

One has the collapse map

$$c : S^{2n+2kq+u} \longrightarrow T(\pi^*(\nu \oplus \bar{\nu}))/T(\pi^*(\nu \oplus \bar{\nu})|_{S\tau}) = X$$

and thinking of M imbedded in $D\tau$ as the zero section, with normal bundle $\bar{\tau} \oplus \nu \oplus \bar{\nu}$ (its K structure) one has a collapse d of X onto $T(\bar{\tau} \oplus \nu \oplus \bar{\nu}) = S^{n+kq+u}T(\nu)$. Using the given imbedding of $D\tau$, the composite

$$S^{2n+2kq+u} \xrightarrow{c} X \xrightarrow{d} S^{n+kq+u}T(\nu)$$

is the $n+kq+u$-fold suspension of the map defined by the imbedding of M in R^{n+kq}.

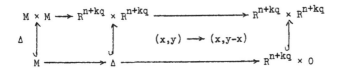

Note: d can be thought of as a homeomorphism since $X = T(\bar{\tau} \oplus \nu \oplus \bar{\nu})$.

Suppose $\underset{\sim}{A}$ is a ring spectrum for which the bundles $\tau, \bar{\tau}, \nu, \bar{\nu}$ all have a natural, stable, multiplicative orientation, and U is the orientation class $U_{\bar{\tau}}$. Then one has the map

$$\beta : X \longrightarrow (M/\emptyset) \wedge T(\bar{\tau}) \wedge T(\nu) \wedge T(\bar{\nu}) = Y$$

generalizing the usual $\emptyset$ to a 4 fold diagonal map. The cohomology class $x \otimes U_{\bar{\tau}} \otimes U_{\nu} \otimes U_{\bar{\nu}} = y$ may be pulled back to X in several different ways.

First one has

$$X = T(\bar{\tau} \oplus \nu \oplus \bar{\nu}) \longrightarrow T(\bar{\tau} \oplus \bar{\nu}) \wedge T\nu \longrightarrow (M/\emptyset) \wedge T(\bar{\tau} \oplus \bar{\nu}) \wedge T\nu \longrightarrow Y$$

and $T(\bar{\tau} \oplus \bar{\nu})$ is just the suspension of $(M/\emptyset)$ with $U_{\bar{\tau}} \otimes U_{\bar{\nu}}$ pulling back to the suspension class. Thus

$$c^*\beta^*(y)[S] = x[M].$$

Further one has

$$X \longrightarrow T\bar{\tau} \wedge X \xrightarrow{\emptyset \wedge T\pi^*} (M/\emptyset) \wedge T\bar{\tau} \wedge T(\nu \oplus \bar{\nu}) \longrightarrow Y$$

in which $x \otimes U_{\bar{\tau}}$ pulls back to $x \otimes U$ and then to $\pi^*x \cdot U$ while $U_{\nu} \otimes U_{\bar{\nu}}$ pulls back to the orientation class of X as a bundle over $D\tau$. Thus

$$c^*\beta^*(y)[S] = \{\pi^*(x) \cdot U\}[D\tau, S\tau].$$

Thus $x_U[M]$ coincides with $x[M]$ when everything makes sense, and thus x_U numbers give a natural generalization of ordinary characteristic numbers.

In order to make use of the above process of constructing "numbers" one must find a class $U \in H^{n+u}(D\tau, S\tau; \underset{\sim}{A})$. The reasonable theory to use is bundle theoretic cohomology, following Conner and Floyd [41] or Palais [97], p. 44.

Let K be $\mathbb{R}$ or $\mathbb{C}$ and let V be a K inner product space of dimension n, with inner product $<\ ,\ >$ which is K linear in the first factor and conjugate linear in the second factor (the conjugate of α being denoted by $\bar{\alpha}$).

Recall that the exterior algebra over K of V, $\Lambda(V) = \Sigma_0^n \Lambda^k(V)$ has an inner product given by

i) $\Lambda^j(V) \perp \Lambda^k(V)$ if $j \neq k$

ii) If $X = x_1 \wedge \cdots \wedge x_k$, $Y = y_1 \wedge \cdots \wedge y_k$, then $< X,Y > = \det|< x_i,y_j >|$.

If $e_1,\ldots,e_n$ is an orthonormal base of V, then $\{e_{i_1} \wedge \cdots \wedge e_{i_r} | i_1 < \cdots < i_r\}$ forms an orthonormal base of $\Lambda^r(V)$.

There is a canonical antiautomorphism of $\Lambda(V)$, $\alpha : \Lambda(V) \longrightarrow \Lambda(V)$, defined by

$$\alpha(v_1 \wedge \cdots \wedge v_k) = v_k \wedge \cdots \wedge v_1 = (-1)^{k(k-1)/2} v_1 \wedge \cdots \wedge v_k$$

which is K linear and preserves inner products.

Finally, recall that an orientation of V is a unit vector $\sigma \in \Lambda^n(V) = \det(V)$.

Being given an orientation σ of V, one may define a function $\tau : \Lambda^k(V) \longrightarrow \Lambda^{n-k}(V)$ as follows: $\tau X \in \Lambda^{n-k}(V)$ is the unique vector such that for all $Y \in \Lambda^{n-k}(V)$, one has

$$< \tau X, Y > = < \sigma, X \wedge Y > .$$

Lemma 1: τ is conjugate linear.

Proof:

$$< \tau(aX),Y > = < \sigma,aX \wedge Y >,$$

$$= < \sigma,X \wedge Y > \bar{a},$$

$$= < \tau X,Y > \bar{a},$$

$$= < \bar{a}\tau X,Y >$$

for all Y. **

Fix an orthonormal base $e_1,\ldots,e_n$ of V with $\sigma = e_1 \wedge\ldots\wedge e_n$. A monomial X is an element $\pm e_{r_1} \wedge\ldots\wedge e_{r_k}$ with $r_1 <\ldots< r_k$. Then for X and Y monomials:

$$< X,Y > = \begin{cases} 1 & \text{if } X = Y, \\ -1 & \text{if } X = -Y, \\ 0 & \text{otherwise.} \end{cases}$$

Further, for any monomial X there is a unique monomial $\tilde{X}$ with $X \wedge \tilde{X} = \sigma$. For X a monomial, τX is exactly $\tilde{X}$, for if Y is a monomial, $< \tau X,Y > = < \sigma,X \wedge Y >$ is given by $\{1,-1,0\}$ as Y is $\{\tilde{X},-\tilde{X}, \text{other}\}$. Then $\tau X \wedge X = (-1)^{k(n-k)}X \wedge \tau X = (-1)^{k(n-k)}\sigma = (-1)^{k(n-k)}\tau X \wedge \tau^2 X$ if $X \in \Lambda^k(V)$ is a monomial, so one has

Lemma 2: If $X \in \Lambda^k(V)$, then $\tau^2 X = (-1)^{k(n-k)}X$. **

Note: $\tau = *$ as defined in Palais [97]. If $a,b \in \Lambda^k(V)$, $< a,b > = \det(a \wedge *b) = < a \wedge *b,\sigma >$ defines $*$. Then $< b,a > = < \overline{a,b} > = < \overline{a \wedge *b,\sigma} > = < \sigma,a \wedge *b > = (-1)^{k(n-k)} < \sigma,*b \wedge a >$ so $\tau *b = (-1)^{k(n-k)}b$ or $\tau b = (-1)^{k(n-k)}\tau^2 *b = *b$ for $\tau^2 = (-1)^{k(n-k)}$ on $\Lambda^{n-k}(V)$.

Definition: Let $\mu : \Lambda(V) \longrightarrow \Lambda(V) : X \longrightarrow \tau\alpha X$.

Definition: If $K = \mathbb{R}$ and $\dim_{\mathbb{R}} V = 2n'$, let $\theta : \Lambda(V) \otimes_{\mathbb{R}} \mathbb{C} \longrightarrow \Lambda(V) \otimes_{\mathbb{R}} \mathbb{C}$ by letting $\theta(x) = i^{k(k-1)+n'} \cdot (\tau \otimes 1)(x)$ if $x \in \Lambda^k(V) \otimes_{\mathbb{R}} \mathbb{C}$.

Lemma 3: μ is conjugate linear and $\mu^2 = (-1)^{n(n-1)/2}$, while θ is complex linear and $\theta^2 = 1$.

Proof: If $X \in \Lambda^k(V)$, $\alpha X = (-1)^{k(k-1)/2} X$, $\alpha\tau X = (-1)^{(n-k)(n-k-1)/2} \tau X$ so $\mu^2 X = (-1)^r X$ where

$$\begin{aligned} r &= k(k-1)/2 + (n-k)(n-k-1)/2 + k(n-k), \\ &= k(n-1)/2 + (n-k)(n-1)/2, \\ &= n(n-1)/2. \end{aligned}$$

Letting $n = 2n'$, $x \in \Lambda^k(V)$,

$$\begin{aligned} \theta^2 x &= i^{(n-k)(n-k-1)+n'} \cdot i^{k(k-1)+n'} (-1)^{k(n-k)} x, \\ &= i^{(n-k)(n-k-1)+k(k-1)+n+2k(n-k)} x, \\ &= i^{n(n-1)+n-2k(n-k)+2k(n-k)} x, \quad \text{(by the above identity)} \\ &= i^{n^2} x, \\ &= i^{4n'^2} x, \\ &= x. \quad ** \end{aligned}$$

Lemma 4: If $SG(V)$ denotes the inner product preserving K linear transformations of V which fix σ, then for $g \in SG(V)$, $g\tau = \tau g$, $g\mu = \mu g$, and $g\theta = \theta g$.

Proof:

$$< g\tau X, gY > = < \tau X, Y > = < \sigma, X \wedge Y > = < g\sigma, gX \wedge gY >,$$

$$= < \sigma, gX \wedge gY > = < \tau gX, gY >$$

for all gY, so $g\tau X = \tau gX$ for all X. Since clearly $g\alpha = \alpha g$, $\mu g = g\mu$, while $ig = gi$ gives $g\theta = \theta g$. **

If V and W are K inner product spaces of dimensions n and m, the inclusions $V \hookrightarrow V \oplus W$, $W \hookrightarrow V \oplus W$ induce homomorphisms $\Lambda(V) \longrightarrow \Lambda(V \oplus W)$, $\Lambda(W) \longrightarrow \Lambda(V \oplus W)$ and then

$$\beta : \Lambda(V) \otimes_K \Lambda(W) \longrightarrow \Lambda(V \oplus W) \otimes \Lambda(V \oplus W) \xrightarrow{\wedge} \Lambda(V \oplus W)$$

is an isomorphism of graded algebras. $V \oplus W$ may be oriented by means of $\sigma = \beta(\sigma_V \otimes \sigma_W)$.

Lemma 5: If $X \in \Lambda^r(V)$, $Y \in \Lambda^s(W)$ then

$$\tau\beta(X \otimes Y) = (-1)^{s(n-r)}\beta(\tau_V X \otimes \tau_W Y),$$

$$\mu\beta(X \otimes Y) = (-1)^{ns}\beta(\mu_V X \otimes \mu_W Y),$$

$$\theta\beta(X \otimes Y) = \beta(\theta_V X \otimes \theta_W Y).$$

In particular, if $\dim_K V \equiv 0 \pmod 2$, then μ may be identified with $\mu_V \otimes \mu_W$.

Proof: Fix a base $e_1, \ldots, e_n$ for V; $e_{n+1}, \ldots, e_{n+m}$ for W with $\sigma_V = e_1 \wedge \ldots \wedge e_n$; $\sigma_W = e_{n+1} \wedge \ldots \wedge e_{n+m}$. Let X, Y be monomials, so that

$$\sigma = \beta((X \wedge \tau_V X) \otimes (Y \wedge \tau_W Y)),$$

$$= (-1)^{s(n-r)}\beta((X \otimes Y)) \wedge \beta(\tau_V X \otimes \tau_W Y)$$

and everything being monomials, $\tau\beta(X \otimes Y) = (-1)^{s(n-r)}\beta(\tau_V X \otimes \tau_W Y)$. Now

$\alpha\beta(X \otimes Y) = (-1)^{rs}\beta(\alpha(X) \otimes \alpha(Y))$ for $(-1)^{(r+s)(r+s-1)/2} =$ $(-1)^{rs}(-1)^{r(r-1)/2}(-1)^{s(s-1)/2}$ since

$$(r+s)(r+s-1) = r(r+s-1) + s(r+s-1),$$

$$= r(r-1) + rs + s(s-1) + sr,$$

which gives the formula for μ. To get the formula for θ, one has

$$\theta\beta(X \otimes Y) = i^{(r+s)(r+s-1)+(n'+m')}\tau\beta(X \otimes Y),$$

$$= i^{r(r-1)+s(s-1)+2rs+n'+m'}(-1)^{s(2n'-r)}\beta(\tau_V X \otimes \tau_W Y),$$

$$= (-1)^{rs}(-1)^{2n's-rs}\beta(i^{r(r-1)+n'}\tau_V X \otimes i^{s(s-1)+m'}\tau_W Y),$$

$$= \beta(\theta X \otimes \theta Y). \quad **$$

Now returning to V, one has for each $v \in V$ a map

$$F_v : \Lambda(V) \longrightarrow \Lambda(V) : x \longrightarrow v \wedge x$$

and its adjoint $(F_v)^* : \Lambda(V) \longrightarrow \Lambda(V)$ defined by

$$< X, F_v Y > = < F_v^* X, Y > \quad \text{for all} \quad X, Y \in \Lambda(V).$$

<u>Definition</u>: $\phi_v : \Lambda(V) \longrightarrow \Lambda(V)$ for $v \in V$ is the linear transformation $F_v + (F_v)^*$.

<u>Lemma 6</u>: For vector spaces V, W with $v \in V$, $w \in W$,

$$\phi_{v+w}(\beta(X \otimes Y)) = \beta(\phi_v X \otimes Y + (-1)^{\dim X} X \otimes \phi_w Y).$$

<u>Proof</u>:

$$F_{v+w}(\beta(X \otimes Y)) = \beta(v \otimes 1 + 1 \otimes w) \wedge \beta(X \otimes Y),$$

$$= \beta((v \wedge X) \otimes Y + (-1)^{\dim X} X \otimes (w \wedge Y)),$$

$$= \beta(F_v(X) \otimes Y + (-1)^{\dim X} X \otimes F_w(Y))$$

giving $F_{v+w} \circ \beta = \beta \circ [F_v \otimes 1 + \mathrm{sgn} \circ (1 \otimes F_w)]$ where $\mathrm{sgn} : \Lambda(V) \otimes \Lambda(W) \longrightarrow \Lambda(V) \otimes \Lambda(W)$ maps $\Lambda^k(V) \otimes \Lambda(W)$ by $(-1)^k$.

Then

$$\begin{aligned}
< F^*_{v+w}(\beta(X \otimes Y)), \beta(U \otimes V) > &= < \beta(X \otimes Y), F_{v+w}\beta(U \otimes V) >, \\
&= < \beta(X \otimes Y), \beta[F_v U \otimes V + (-1)^{\dim U} U \otimes F_w V] >, \\
&= < X, F_v U > . < Y, V > + (-1)^{\dim U} < X, U > . < Y, F_w V >, \\
&= < F^*_v X, U > . < Y, V > + (-1)^{\dim X} < X, U > . < F^*_w Y, V >, \\
&= < \beta[F^*_v X \otimes Y + (-1)^{\dim X} X \otimes F^*_w Y], \beta(U \otimes V) >
\end{aligned}$$

and thus $F^*_{v+w} \circ \beta = \beta \circ [F^*_v \otimes 1 + \mathrm{sgn} \circ (1 \otimes F^*_w)]$.

(<u>Note</u>: If $< X, U > \neq 0$, then $\dim X = \dim U$).

Thus $\phi_{v+w} \circ \beta = \beta \circ [\phi_v \otimes 1 + \mathrm{sgn} \circ (1 \otimes \phi_w)]$. **

<u>Corollary</u>: For each $v \in V$, $(\phi_v)^2 = \|v\|^2 \cdot 1_{\Lambda(V)}$.

<u>Proof</u>: If this holds for both V and W, then

$$\begin{aligned}
(\phi_{v+w})^2 \beta(X \otimes Y) &= \phi_{v+w}\beta[\phi_v X \otimes Y + (-1)^{\dim X} X \otimes \phi_w Y], \\
&= \beta[\phi_v^2 X \otimes Y + (-1)^{\dim X+1} \phi_v X \otimes \phi_w Y] \\
&\quad + (-1)^{\dim X} \beta[\phi_v X \otimes \phi_w Y + (-1)^{\dim X} X \otimes \phi_w^2 Y],
\end{aligned}$$

(<u>Note</u>: $\phi_v X$ has components of dimension $\dim X+1$ and $\dim X-1$, giving the same sign.)

$$\begin{aligned}
&= \beta[\|v\|^2 X \otimes Y + (-1)^{\dim X+1} \phi_v X \otimes \phi_w Y + \\
&\qquad (-1)^{\dim X} \phi_v X \otimes \phi_w Y + \|w\|^2 X \otimes Y], \\
&= (\|v\|^2 + \|w\|^2) \beta(X \otimes Y), \\
&= \|v+w\|^2 \beta(X \otimes Y)
\end{aligned}$$

and the result also holds for $V \oplus W$.

Thus, one need only check this result when $\dim V = 1$. Thus $\Lambda(V)$ has a base 1, σ, and $v = k\sigma$, so

$$F_v(1) = k\sigma, \qquad F_v(\sigma) = 0,$$

$$< F_v^*(1),\sigma > = < 1,F_v(\sigma) > = 0,$$

$$< F_v^*(\sigma),1 > = < \sigma,F_v(1) > = < \sigma,k\sigma > = \bar{k},$$

giving

$$F_v^*(1) = 0, \qquad F_v^*(\sigma) = \bar{k}.$$

Thus $\phi_v(1) = k\sigma$, $\phi_v(\sigma) = \bar{k}$, and so

$$\phi_v^2(1) = \phi_v(k\sigma) = k\bar{k} = \|k\|^2 \cdot 1,$$

$$\phi_v^2(\sigma) = \phi_v(\bar{k}) = \bar{k}k\sigma = \|k\|^2 \cdot \sigma. \quad **$$

Lemma 7: If $g : V \longrightarrow V$ is an inner product preserving linear transformation, then $g \circ \phi_v = \phi_{gv} \circ g$.

Proof: Clearly $gF_v(X) = g(v \wedge X) = gv \wedge gX = F_{gv}g(X)$ so $gF_v = F_{gv}g$ or $F_v g^{-1} = g^{-1}F_{gv}$, and then

$$< gF_v^*X,Y > = < F_v^*X,g^{-1}Y > = < X,F_v g^{-1}Y > = < X,g^{-1}F_{gv}Y >,$$

$$= < gX,gg^{-1}F_{gv}Y > = < gX,F_{gv}Y > = < F_{gv}^* gX,Y >. \quad **$$

Lemma 8: For $v \in V$, $\phi_v \circ \mu = \mu \circ \phi_v$ and $\phi_v \circ \theta = \theta \circ \phi_v$.

Proof: If $X \in \Lambda^k(V)$,

$$< \tau F_v X, Y > = < \sigma, v \wedge X \wedge Y >,$$
$$= (-1)^k < \sigma, X \wedge v \wedge Y >,$$
$$= (-1)^k < \sigma, X \wedge F_v Y >,$$
$$= (-1)^k < \tau X, F_v Y >,$$
$$= (-1)^k < F_v^* \tau X, Y >$$

so $\tau F_v X = (-1)^k F_v^* \tau X$. Then

$$\tau\alpha F_v X = \tau(-1)^{(k+1)k/2} F_v X,$$
$$= (-1)^{k(k+1)/2}(-1)^k F_v^* \tau X,$$
$$= (-1)^{k(k-1)/2} F_v^* \tau X,$$
$$= F_v^* \tau\alpha X$$

so $\mu F_v = F_v^* \mu$. Then $\mu^2 = (-1)^{n(n-1)/2}$ so

$$F_v \mu = \mu^2 F_v \mu\mu^2 = \mu(\mu F_v)\mu^3 = \mu F_v^* \mu\mu^3 = \mu F_v^*.$$

Thus $\phi_v \mu = \mu \phi_v$.

Also

$$\theta F_v X = i^{(k+1)k+n'} \tau F_v X,$$
$$= i^{(k+1)k+n'} i^{2k} F_v^* \tau X,$$
$$= i^{4k} i^{k(k-1)+n'} F_v^* \tau X,$$
$$= F_v^* \theta X$$

and $\theta^2 = 1$, so

$$F_v \theta = \theta^2 F_v \theta = \theta F_v^* \theta^2 = \theta F_v^*$$

giving

$$\phi_v \theta = \theta \phi_v. \quad **$$

Now returning to the geometric situation of interest, let ξ be an oriented $n = 2n'$ plane bundle over a space B, with Riemannian metric $\langle\ ,\ \rangle$ and $\sigma : B \longrightarrow S(\Lambda^n(\xi))$ a section of the sphere bundle of $\Lambda^n(\xi)$ defining the orientation. Let $\Lambda^{ev}(\xi) = \bigoplus_{k\ \text{even}} \Lambda^k(\xi)$, $\Lambda^{od}(\xi) = \bigoplus_{k\ \text{odd}} \Lambda^k(\xi)$ and with $\pi : D\xi \longrightarrow B$ the projection, let

$$\phi : \pi^*(\Lambda^{ev}(\xi) \otimes \mathbb{C}) \longrightarrow \pi^*(\Lambda^{od}(\xi) \otimes \mathbb{C})$$

be the bundle map defined at $e \in D\xi$ by sending (e,X) into $(e,\phi_e X)$ [Lemma 7 says this makes sense]. The restriction of ϕ to $S\xi$ is an isomorphism, for $\phi_e^2 = 1$ by the Corollary to Lemma 6. Thus the triple $(\pi^*\Lambda^{ev}(\xi) \otimes \mathbb{C},\ \pi^*\Lambda^{od}(\xi) \otimes \mathbb{C}, \phi)$ defines an element $\pi^*\Lambda^{ev}(\xi) \otimes \mathbb{C} - \pi^*\Lambda^{od}(\xi) \otimes \mathbb{C}$ in $K(D\xi)$ with an isomorphism to zero in $K(S\xi)$, and therefore an element of $K(D\xi,S\xi)$.

In addition, one may define the operator θ in $\pi^*(\Lambda(\xi) \otimes \mathbb{C})$ [Lemma 4 says this makes sense] as a bundle map of square 1. Since $\theta : \Lambda^k(\xi) \otimes \mathbb{C} \longrightarrow \Lambda^{n-k}(\xi) \otimes \mathbb{C}$, and n is even, θ takes $\pi^*(\Lambda^{ev}(\xi) \otimes \mathbb{C})$ and $\pi^*(\Lambda^{od}(\xi) \otimes \mathbb{C})$ into themselves and by Lemma 8, the map θ commutes with ϕ.

Let $\Lambda_+^a(\xi), \Lambda_-^a(\xi)$ denote the subbundles of $\pi^*(\Lambda^a(\xi) \otimes \mathbb{C})$, $a = ev$ or od, on which θ is $+1$ and -1 respectively, and let

$$\Delta(\xi) = (\Lambda_+^{ev}(\xi), \Lambda_+^{od}(\xi), \phi) - (\Lambda_-^{ev}(\xi), \Lambda_-^{od}(\xi), \phi)$$

in $K(D\xi,S\xi)$. Applying periodicity, one then has:

<u>Definition</u>: $U(\xi) = p^{-n'}\Delta(\xi) \in K^{2n'}(D\xi,S\xi)$.

<u>Assertion 1</u>: $U(\xi)$ is multiplicative.

Proof: It really suffices to verify this just for vector spaces. One has $\beta : \Lambda(\xi) \otimes \Lambda(\eta) \longrightarrow \Lambda(\xi \oplus \eta)$ giving isomorphisms

$$\Lambda^{ev}(\xi \oplus \eta) \cong \Lambda^{ev}(\xi) \otimes \Lambda^{ev}(\eta) \oplus \Lambda^{od}(\xi) \otimes \Lambda^{od}(\eta),$$

$$\Lambda^{od}(\xi \oplus \eta) \cong \Lambda^{od}(\xi) \otimes \Lambda^{ev}(\eta) \oplus \Lambda^{ev}(\xi) \otimes \Lambda^{od}(\eta),$$

while by Lemma 6, this decomposition is compatible with the map ϕ. By Lemma 5, one also has

$$\Lambda_+(\xi \oplus \eta) \cong \Lambda_+(\xi) \otimes \Lambda_+(\eta) \oplus \Lambda_-(\xi) \otimes \Lambda_-(\eta),$$

$$\Lambda_-(\xi \oplus \eta) \cong \Lambda_-(\xi) \otimes \Lambda_+(\eta) \oplus \Lambda_+(\xi) \otimes \Lambda_-(\eta),$$

being also compatible with the even-odd decomposition. Adding everything up with signs gives $\Delta(\xi \oplus \eta) = \Delta(\xi)\cdot\Delta(\eta)$, while periodicity is also multiplicative. **

Assertion 2: If ξ is a complex line bundle over B, considered as an oriented bundle, then $\Delta(\xi) = \pi^*\xi - \pi^*\bar{\xi}$ with $\pi^*\xi$ and $\pi^*\bar{\xi}$ identified over $S\xi$ by the standard trivializations over $S\xi$.

Proof: Let V be a 1 dimensional complex vector space with inner product [,] and $v \in V$ a unit vector. Then as a real vector space $\langle\ ,\ \rangle = \mathrm{Re}[\ ,\ ]$ is an inner product and V has an oriented base $\{v, iv\}$.

Thus $\Lambda(V)$ has a base $\{1, v, iv, \sigma\}$, $\sigma = v \wedge iv$, and being monomials, one gets immediately

$$\tau 1 = \sigma, \qquad \tau v = iv, \qquad \tau iv = -v, \qquad \tau\sigma = 1.$$

This gives

$$\theta(1 \otimes 1) = \sigma \otimes i, \qquad \theta(v \otimes 1) = iv \otimes i,$$

$$\theta(iv \otimes 1) = -v \otimes i, \qquad \theta(\sigma \otimes 1) = -1 \otimes i.$$

Thus $\Lambda_+(V)$ has a base $\{(1 \otimes 1 + \sigma \otimes i), (v \otimes 1 + iv \otimes i)\}$ and $\Lambda_-(V)$ has a base $\{(1 \otimes 1 - \sigma \otimes i), (v \otimes 1 - iv \otimes i)\}$. Then the maps

$$V \longrightarrow \Lambda_+^{od}(V) : x \longrightarrow x \otimes 1 + ix \otimes i,$$

$$V \longrightarrow \Lambda_-^{od}(V) : x \longrightarrow x \otimes 1 - ix \otimes i$$

are respectively conjugate linear and linear.

Thus, if ξ is a complex line bundle

$$\Lambda_+^{ev}(\xi) = 1, \qquad \Lambda_+^{od}(\xi) = \pi^*\bar{\xi},$$

$$\Lambda_-^{ev}(\xi) = 1, \qquad \Lambda_-^{od}(\xi) = \pi^*\xi.$$

To determine the map ϕ_x, let $x = (\alpha + \beta i)v$, so that:

$$F_x(1) = x, \quad F_x(v) = x \wedge v = -\beta\sigma, \quad F_x(iv) = x \wedge iv = \alpha\sigma, \quad F_x(\sigma) = 0,$$

and

$$< F_x^* v, 1 > = < v, x > = \alpha = < \alpha \cdot 1, 1 >,$$

$$< F_x^* iv, 1 > = < iv, x > = \beta = < \beta \cdot 1, 1 >,$$

$$< F_x^* \sigma, v > = < \sigma, x \wedge v > = -\beta = < -\beta v + \alpha iv, v >,$$

$$< F_x^* \sigma, iv > = < \sigma, x \wedge iv > = \alpha = < -\beta v + \alpha iv, iv >,$$

so

$$F_x^*(1) = 0, \quad F_x^*(v) = \alpha, \quad F_x^*(iv) = \beta, \quad F_x^*(\sigma) = (-\beta + \alpha i)v = ix.$$

Thus

$$\phi_x(1 \otimes 1 + \sigma \otimes i) = x \otimes 1 + ix \otimes i,$$

$$\phi_x(1 \otimes 1 + \sigma \otimes i) = x \otimes 1 - ix \otimes i$$

and if $x \in S(V)$, $\phi_x^2 = 1$, so ϕ induces the standard trivialization by sending $ze \in \pi^*(\xi)_e$ into z (or $\bar{z}$ for the conjugate bundle). **

Corollary: If ξ is the conjugate of the canonical bundle over $CP(n-1)$, then $U(\xi) = p^{-1}(\xi - \bar{\xi}) \in \tilde{K}(\mathbb{C}P(n))$.

Proof: Identifying $\mathbb{C}P(n)$ with $T(\xi)$, the bundle ξ over $\mathbb{C}P(n)$ is precisely the pullback of the bundle ξ over $CP(n-1)$ with the standard trivialization. **

Corollary: Let ξ be an oriented $2n$ plane bundle over B with Pontrjagin class $\mathcal{P}(\xi) = \prod_{i=1}^{n} (1 + x_i^2)$ and let $\Phi^H : \tilde{H}^*(T\xi;Q) \longrightarrow H^*(B;Q)$ be the Thom isomorphism defined by the orientation class of ξ. Then

$$\Phi^H ch(U(\xi)) = \prod_{i=1}^{n} ((e^{x_i} - e^{-x_i})/x_i).$$

Note: $(e^x - e^{-x})/x = (\Sigma\, x^j/j! - \Sigma(-x)^j/j!)/x = (2\sum_{j\text{ odd}} x^j/j!)/x$ is a power series in x^2. Thus $\Phi^H ch(U(\xi))$ is a rational power series in the Pontrjagin classes of ξ.

Proof: Since the Thom homomorphism and U are multiplicative, it suffices to prove this when ξ is the conjugate of the canonical bundle over $\mathbb{C}P(n-1)$. If $\bar{\alpha} = c_1(\xi) \in H^2(\mathbb{C}P(n);Z)$, $U(\xi) = p^{-1}(\xi - \bar{\xi}) \in \tilde{K}(\mathbb{C}P(n))$, so $chU(\xi) = e^{\bar{\alpha}} - e^{-\bar{\alpha}}$, while the orientation class is $\bar{\alpha}$, so $\Phi^H chU(\xi) = (e^{\bar{\alpha}} - e^{-\bar{\alpha}})/\bar{\alpha}$. **

From this one has the integrality theorem:

Theorem: Let M^{2n} be an oriented manifold and let $x \in K^j(M)$. Then

$$\{ch(x)\cdot\delta(\tau)\}[M] \in Z$$

where $\delta(\tau) \in H^*(M;Q)$ is given by the polynomial in the tangential Pontrjagin classes of M such that when $\mathcal{P}(\tau) = \prod_{i=1}^{n} (1+x_i^2)$ formally, then $\delta(\tau) = \prod_{i=1}^{n} (x_i/\tanh(x_i/2))$.

Proof: Since $\pi^*(x) \cup U(\tau) \in K^{j+2n}(D\tau, S\tau)$,

$$\{\pi^*(x) \cup U(\tau)\}[D\tau, S\tau] = ch[\pi^*x \cdot U(\tau)] \cdot \mathscr{J}(D\tau)[D\tau, S\tau] \cdot p(1)^{-n-j/2}$$

so $ch[\pi^*x \cdot U(\tau)] \cdot \mathscr{J}(D\tau)[D\tau, S\tau] \in Z$. Then $c(D\tau) = c(\pi^*\tau \otimes \mathbb{C}) = \pi^*c(\tau \otimes \mathbb{C})$ $= \pi^* \prod_{i=1}^{n} (1+x_i)(1-x_i)$, so $\mathscr{J}(D\tau) = \pi^*(\prod_{i=1}^{n} (x_i/e^{x_i}-1)(-x_i/e^{-x_i}-1))$ and

$$ch[\pi^*x \cdot U(\tau)] \cdot \mathscr{J}(D\tau) = (\phi^H)^{-1}(chx \cdot \prod_{i=1}^{n} (e^{x_i} - e^{-x_i}/x_i)(x_i/e^{x_i}-1)(-x_i/e^{-x_i}-1))$$

while

$$\frac{e^y - e^{-y}}{y} \cdot \frac{y}{(e^y - 1)} \cdot \frac{-y}{(e^{-y} - 1)} = y\left(\frac{e^y - e^{-y}}{(e^y - 1)(1 - e^{-y})}\right),$$

but

$$\frac{U - U^{-1}}{(U-1)(1-U^{-1})} = \frac{U^2 - 1}{(U-1)^2} = \frac{U+1}{U-1} = \frac{U^{1/2} + U^{-1/2}}{U^{1/2} - U^{-1/2}},$$

so

$$y\left(\frac{e^y - e^{-y}}{(e^y-1)(1-e^{-y})}\right) = y\left(\frac{e^{y/2} + e^{-y/2}}{e^{y/2} - e^{-y/2}}\right) = \frac{y}{\tanh(y/2)}.$$

Thus

$$(\phi^H)^{-1}(ch(x) \cdot \delta(\tau))[D\tau, S\tau] \in Z,$$

and since oriented bundles are naturally and multiplicatively oriented for integral cohomology, this is precisely $\{ch(x) \cdot \delta(\tau)\}[M]$. **

Note: Since

$$\begin{aligned} x/\tanh(x/2) &= x(e^{x/2} + e^{-x/2})/(e^{x/2} - e^{-x/2}), \\ &= x(2 + 2(x/2)^2/2! + \ldots)/(x + 2(x/2)^3/3! + \ldots), \\ &= 2 + \ldots \end{aligned}$$

the class $\delta(\tau) = \prod_1^n (x_i/\tanh(x_i/2))$ is far from stable. In order to eliminate the power of 2 which is causing the problem, one defines:

<u>Definition</u>: If ξ is a real vector bundle with Pontrjagin class $\mathcal{P}(\xi)$ expressed formally as $\prod_1^n (1 + x_i^2)$, $\dim x_i = 2$, then the Hirzebruch L class of ξ is given by the formal product

$$L(\xi) = \prod_1^n (x_i/\tanh(x_i)).$$

<u>Note</u>: $L(\xi)$ is a stable class and is closely related to δ. In particular, $2y/\tanh y$ differs from $y/\tanh(y/2)$ in that the component of y^k has been multiplied by 2^k. Thus

$$\begin{aligned} 2^n L(\tau)[M^{2n}] &= \{\prod_1^n (2x_i/\tanh x_i)\}_{2n}[M^{2n}], \\ &= 2^n\{\prod_1^n x_i/\tanh(x_i/2)\}_{2n}[M^{2n}], \\ &= 2^n\delta(\tau)[M^{2n}], \end{aligned}$$

giving

$$L(\tau)[M^{2n}] = \delta(\tau)[M^{2n}].$$

In order to find classes $x \in K(M)$, one may make use of the K theory Chern classes of $\tau \otimes \mathbb{C}$. Unfortunately, K theory is badly behaved for conjugation and one must make some modifications.

Let ξ be a real $2n$ plane bundle over a space B, and define K theory Pontrjagin classes $\pi^i(\xi) \in K(B)$ by

$$\begin{aligned} \Sigma\, s^i \pi^i(\xi) = \pi_s(\xi) &= \Sigma\, t^i p^i(\gamma_i(\xi \otimes \mathbb{C})), \\ &= \lambda_{t/(1-t)}(\xi \otimes \mathbb{C} - 2n_{\mathbb{C}}) \end{aligned}$$

where $s = t - t^2$. Note: $\pi^i(\xi)$ is a polynomial with integral coefficients in the Chern classes, hence belongs to $K(B)$.

Note: If η is a complex line bundle, then

$$\lambda_u(\eta \otimes \mathbb{C} - 2) = \lambda_u(\eta \oplus \bar{\eta} - 2),$$
$$= (1+u\eta)(1+u\bar{\eta})/(1+u)^2,$$

so

$$\lambda_{t/(1-t)}(\eta \otimes \mathbb{C} - 2) = \frac{(1+(t/1-t)\eta)}{(1/1-t)} \cdot \frac{(1+(t/1-t)\bar{\eta})}{(1/1-t)},$$
$$= (1 - t + t\eta)(1 - t + t\bar{\eta}),$$
$$= [1 + t(\eta-1)][1 + t(\bar{\eta}-1)],$$
$$= 1 + t(\eta + \bar{\eta} - 2) + t^2(2-\eta-\bar{\eta})$$

since $\eta \cdot \bar{\eta} = 1$.

If the Pontrjagin class of the bundle ξ is expressed formally as $\prod_1^n (1 + x_j^2)$, so that $c(\xi \otimes \mathbb{C}) = \prod_1^n (1 + x_j)(1 - x_j)$, then

$$\mathrm{ch}\, \Sigma\, t^i p^i(\gamma_i(\xi \otimes \mathbb{C})) = \prod_1^n (1 + t(e^{x_j} + e^{-x_j} - 2) + t^2(2 - e^{x_j} - e^{-x_j}))$$

so

$$\mathrm{ch}\, \pi_s(\xi) = \prod_1^n (1 + s(e^{x_j} + e^{-x_j} - 2))$$

and $\mathrm{ch}\pi^i(\xi)$ is the i-th elementary symmetric function in the variables $e^{x_j} + e^{-x_j} - 2$.

Definition: If ξ is a real vector bundle over B with $\mathcal{P}(\xi)$ expressed formally as $\Pi(1 + x_j^2)$, let $s_\omega(e_{\mathcal{P}})(\xi) \in H^*(B;Q)$ be given as the s_ω symmetric function in the variables $e^{x_j} + e^{-x_j} - 2$.

<u>Proposition</u>: Let M be an oriented manifold. Then for all $x \in K^*(M)$,

$$\{ch(x)L(\tau)\}[M] \in Z[1/2].$$

In particular,

$$\{s_\omega(e_\mathcal{O})L(\tau)\}[M] \in Z[1/2]$$

for all ω.

<u>Proof</u>: This will follow at once from $L(\tau) = ch(u)\delta(\tau)$ with $u \in K(M)[1/2]$. Let $(x/\tanh x) = v\cdot(x/\tanh(x/2))$, $a = e^x+e^{-x}-2$, and $b = e^x$, giving

$$\begin{aligned}
v &= \tanh(x/2)/\tanh(x), \\
&= \frac{(e^x-1)(1-e^{-x})}{e^x-e^{-x}} \Big/ \frac{e^x-e^{-x}}{e^x+e^{-x}}, \\
&= \frac{(b-1)(1-b^{-1})(b+b^{-1})}{(b-b^{-1})^2}\cdot\frac{b^2}{b^2}, \\
&= \frac{(b-1)^2(b^2+1)}{(b^2-1)^2} = \frac{b^2+1}{(b+1)^2}\cdot\frac{b^{-1}}{b^{-1}}, \\
&= (b+b^{-1})/(b+2+b^{-1}) = (a+2)/(a+4), \\
&= 1 - 2/(a+4), \\
&= 1 - 1/2(1/(1+a/4)), \\
&= 1 - 1/2[1 - a/4 + (a/4)^2 - \dots], \\
&= 1/2 + 1/2(a/4) - 1/2(a/4)^2 + 1/2(a/4)^3 - \dots, \\
&\in Z[1/2][e^x+e^{-x}-2].
\end{aligned}$$

Thus $L(\tau)/\delta(\tau) \in ch(K(M)[1/2])$, being a symmetric polynomial in the $e^{x_j}+e^{-x_j}-2$ with coefficients in $Z[1/2]$. **

In order to evaluate the expressions $s_\omega(e_{\mathcal{P}})L$, one may use:

Lemma: If ξ is a complex n-plane bundle then

$$s_\omega(e_{\mathcal{P}})(\xi) = s_{2\omega}(e)(\xi) + \Sigma\, a_\lambda s_\lambda(e)(\xi)$$

with $n(\lambda) > 2n(\omega)$, $a_\lambda \in Z$.

Proof:

$$e^x+e^{-x}-2 = (e^x-1)(1-e^{-x}),$$

$$= (e^x-1)^2 \cdot e^{-x},$$

$$= (e^x-1)^2 \cdot 1/(1+(e^x-1)),$$

$$= (e^x-1)^2 - (e^x-1)^3 + (e^x-1)^4 + \dots .$$

Thus the s_ω symmetric function in variables $e^x+e^{-x}-2$ given by

$$s_{2\omega}(e) + \Sigma\, a_\lambda s_\lambda(e)$$

where $a_\lambda \in Z$, $n(\lambda) > 2n(\omega)$, and $s_\mu(e)$ is the s_μ symmetric function in variables e^x-1. **

Lemma: If M is a stably almost complex manifold of real dimension $2n$ then

$$L(\tau) = \{1 + \Sigma\, b_\lambda s_\lambda(e)\} \cdot \mathcal{J}(M)$$

where $b_\lambda \in Z[1/2]$, $n(\lambda) > 0$.

Proof: If $c(M) = \Pi(1 + x_i)$, then

$$L(\tau) = \Pi((e^{x_i}-1)/\tanh(x_i)) \cdot \mathcal{J}(M)$$

and

$$\frac{e^x - 1}{\tanh x} = \frac{(e^x-1)(e^x+e^{-x})}{e^x-e^{-x}} \cdot \frac{e^x}{e^x},$$

$$= (u-1)(u^2+1)/(u^2-1), \qquad (u = e^x)$$

$$= (u^2+1)/(u+1),$$

$$= \frac{[(u-1)+1]^2+1}{(u-1)+2},$$

$$= (2+2a+a^2)/(2+a), \qquad (a = u-1 = e^x-1)$$

$$= \{1 + a + (a^2/2)\}\cdot\{1 - a/2 + (a/2)^2 -\ldots\},$$

so $L(\tau)/S(M) = 1 + \alpha$, where α is symmetric of positive degree in the $e^{x_i}-1$ with coefficients in $Z[1/2]$. **

In $H^*(BSO;Q)$ one writes the Pontrjagin class formally as $\Pi(1+x_j^2)$, $\dim x_j = 2$, and defines $s_\omega(\mathcal{P})$ and $s_\omega(e_{\mathcal{P}})$ as the s_ω symmetric functions of the variables x_j^2 and $e^{x_j}+e^{-x_j}-2$ respectively, and defines L as $\Pi(x_j/\tanh(x_j))$. Then $\Delta s_\omega(\mathcal{P}) = \sum_{\omega'\cup\omega''=\omega} s_{\omega'}(\mathcal{P}) \otimes s_{\omega''}(\mathcal{P})$, $\Delta s_\omega(e_{\mathcal{P}}) = \sum_{\omega'\cup\omega''=\omega} s_{\omega'}(e_{\mathcal{P}}) \otimes s_{\omega''}(e_{\mathcal{P}})$, and $\Delta L = L \otimes L$.

Let $\rho : H_*(BSO;Q) \longrightarrow Q[\beta_i] : z \longrightarrow \Sigma\{s_\omega(e_{\mathcal{P}})L\}[z]\cdot\beta_\omega$ and let $\rho' : H_*(BSO;Q) \longrightarrow Q[\alpha_i] : z \longrightarrow \Sigma\, s_\omega(\mathcal{P})[z]\cdot\alpha_\omega$. Then let

$$B_n = \{z \in H_n(BSO;Q) \mid \rho(z) \in Z[1/2][\beta_i],\ \rho'(z) \in Z[\alpha_i]\}$$

and

$$B_*^{\cdot} = \bigoplus_n B_n \subset H_*(BSO;Q).$$

For p an odd prime, let $\rho_p : B_* \longrightarrow Z_p[\beta_i]$ by letting $\rho_p(z)$ be $\rho(z)$ reduced mod p $(1/2 \in Z_p)$ and let $\rho_2' : B_* \longrightarrow Z_2[\alpha_i]$ by $\rho_2'(z) = \rho'(z)$ reduced mod 2.

One then has:

<u>Lemma</u>: There exist stably almost complex manifolds M^p_{2i} of dimension $4i$ for all primes p and integers i so that

a) For p odd, $\rho_p[\tau M^p_{2i}]$ has largest monomial

1) β_i if $2i+1 \neq p^s$ for any s,

2) $[\beta_{(p^{s-1}-1)/2}]^p$ if $2i+1 = p^s$ for some s.

b) $\rho_2'[\tau M^2_{2i}]$ has largest monomial α_i.

<u>Proof</u>: For the 2 primary case, one has $s_{(i)}(\mathcal{P})[\mathbb{C}P(2i)] = 2i+1$, so let $M^2_{2i} = \mathbb{C}P(2i)$. In the odd primary case, let M^p_{2i} be as given in Chapter VII. From the computations for complex manifolds, for M complex one has

$$(s_\omega(e_{\mathcal{P}})L)[M] = (s_{2\omega}(e)\mathcal{J})[M] + \Sigma\, a_\lambda(s_\lambda(e)\mathcal{J})[M]$$

with $a_\lambda \in Z[1/2]$, $n(\lambda) > 2n(\omega)$. For the manifolds M^p_{2i} the largest monomial is then known from Chapter VII. **

<u>Theorem</u>: a) $\Omega^{SO}_*/\text{Torsion}$ is a polynomial ring over Z on classes x_i of dimension $4i$, and the classes x_i are characterized by

$$s_{(i)}(\mathcal{P}(\tau))[x_i] = \begin{cases} \pm 1 & \text{if } 2i+1 \neq p^s \text{ for any prime } p \text{ and integer } s, \\ \pm p & \text{if } 2i+1 = p^s \text{ for some prime } p \text{ and integer } s. \end{cases}$$

b) The forgetful homomorphism

$$F_* : \Omega^U_* \longrightarrow \Omega^{SO}_*/\text{Torsion}$$

is epic.

c) All relations among the Pontrjagin numbers of oriented manifolds follow from the integrality of the Pontrjagin classes and the conditions $\{s_\omega(e_{\mathcal{P}})L\}[\tau M] \in Z[1/2]$ from K theory; i.e. $\tau\Omega^{SO}_* = B_*$.

Proof: One considers $\tau F_*\Omega_*^U \subset B_*$, with $c_i^p = \tau F_*[M_{2i}^p]$ as in Chapter VII, showing that B_* is polynomial with $\tau F_*\Omega_*^U = \tau\Omega_*^{SO} = B_*$. The condition on the characteristic numbers for generators in immediate since if $\dim M = 4i$

$$s_{(i)}(\mathcal{P})[M] = \{s_{(i)}(e_{\mathcal{P}})L\}[M],$$
$$= \{s_{(2i)}(e)\mathcal{S}\}[M],$$
$$= s_{(2i)}(c)[M],$$

and the characterizing numbers for generators of Ω_*^U are known. **

Note: The use of the Hirzebruch L class and relations arising as in the Atiyah-Singer index theorem was suggested to me by Hattori (private communication). It is also possible (as in Stong [117]) to use the $\hat{A}$ class defined by $\Pi(x_i/2)/\sinh(x_i/2)$. To see that this is equivalent to the above, one has the following argument (of Don Anderson)

$$(x/2)/\sinh(x/2) = u\,x/\tanh(x)$$

gives

$$u = (\tanh x/2\sinh(x/2)),$$
$$= (e^x-e^{-x})/(e^x+e^{-x})(e^{x/2}-e^{-x/2}),$$

and squaring this

$$u^2 = (e^{2x}+e^{-2x}-2)/(e^{2x}+e^{-2x}+2)(e^x+e^{-x}-2)$$

and letting $a = e^x+e^{-x}-2$, $e^x+e^{-x} = a+2$, so

$$u^2 = (a+2)^2-4/(a+2)^2\cdot a = (a^2+4a)/a\cdot(a+2)^2,$$
$$= (a+4)/(a+2)^2 = (1+a/4)\cdot(1 - a/2 + (a/2)^2 - \dots)^2.$$

Thus u^2 is a power series over $Z[1/2]$ in a, with leading term 1, and from the binomial theorem

$$\sqrt{1+v} = 1 + \sum_{k=0}^{\infty} 1/2(-1/2) \ldots (-(2k-1)/2)v^{k+1},$$

so u is a power series over $Z[1/2]$ in a with leading term 1. Thus

$$\hat{A} = ch(\xi)L$$

with $\xi \in K(BSO)[1/2]$ an invertible class.

Oriented Bordism

As previously noted, one of the main reasons for interest in oriented cobordism is the realizability of integral homology. The main study of these bordism groups was made by Conner and Floyd [36].

Theorem: For every CW pair (X,A), $\Omega_*^{SO}(X,A) \otimes Q$ is a free $\Omega_*^{SO} \otimes Q$ module isomorphic to $H_*(X,A;Q) \otimes_Q (\Omega_*^{SO} \otimes Q)$.

Proof: $\pi_*((X/A) \wedge \underset{\sim}{TBSO}) \longrightarrow \tilde{H}_*((X/A) \wedge \underset{\sim}{TBSO};Z)$ is an isomorphism modulo torsion. **

Lemma: There is a 2 primary homotopy equivalence

$$f : \underset{\sim}{TBSO} \longrightarrow \underset{\sim}{K}(\Omega_*^{SO}).$$

Proof: Let $\underset{\sim}{TBSO}$ be mapped into a product of spectra $\underset{\sim}{K}(Z,n(\omega))$ realizing the classes $\mathcal{P}_\omega U$, and spectra $K(Z_2,n_i)$ realizing classes dual to the torsion of Ω_*^{SO} to define a map f. The induced homotopy homomorphism is monic, with finite odd order cokernel in each dimension. **

Theorem: For any CW pair (X,A), there is an isomorphism mod the Serre class of finite groups of odd order

$$f_* : \Omega_*^{SO}(X,A) \longrightarrow H_*(X,A;\Omega_*^{SO}).$$

Proof: By the lemma, the induced homomorphism

$$f_* : \pi_*^S((X/A) \wedge \underset{\sim}{TBSO}) \longrightarrow \pi_*^S((X/A) \wedge \underset{\sim}{K}(\Omega_*^{SO}))$$

is an isomorphism mod odd torsion. **

Theorem: Let (X,A) be any CW pair. For each class $c \in H_n(X,A;Z)$ there is an integer k with $(2k+1)c$ represented by $g_*([M^n,\partial M^n])$ with $g : (M,\partial M) \longrightarrow (X,A)$ an oriented bordism element of (X,A).

Proof: The evaluation homomorphism $e_n : \Omega_n^{SO}(X,A) \longrightarrow H_n(X,A;Z)$ is induced by the composite $\underset{\sim}{TBSO} \longrightarrow \underset{\sim}{K}(\Omega_*^{SO}) \xrightarrow{\pi} \underset{\sim}{K}(\Omega_0^{SO}) = \underset{\sim}{K}(Z)$, π being the projection. By the previous theorem, coker e_n is finite of odd order. **

To determine the odd primary structure of $\Omega_*^{SO}(X,A)$, one has the homomorphisms

$$\Omega_*^{Sp}(X,A) \xrightarrow{T*} \Omega_*^{U}(X,A) \xrightarrow{S*} \Omega_*^{SO}(X,A)$$

with the composite being an isomorphism modulo 2 primary torsion.

Theorem: If (X,A) has no torsion in its integral homology then $\Omega_*^{SO}(X,A)$ is a free Ω_*^{SO} module isomorphic to $H_*(X,A;Z) \otimes \Omega_*^{SO}$. In particular, the evaluation homomorphism $e : \Omega_*^{SO}(X,A) \longrightarrow H_*(X,A;Z)$ is epic. If $\{x_i\}$ is a homogeneous base of $H_*(X,A;Z)$ and $f_i : (M_i,\partial M_i) \longrightarrow (X,A)$ is a map of an oriented manifold into (X,A) with $f_{i*}([M_i,\partial M_i]) = x_i$, then $\Omega_*^{SO}(X,A)$ is the free Ω_*^{SO} module on the classes of the (M_i,f_i).

<u>Proof</u>: $e' : \Omega_*^U(X,A) \longrightarrow H_*(X,A;Z)$ is epic, so e is epic. Choose a collection of maps (M_i, f_i) as above, defining a homomorphism

$$\mathfrak{m} : H_*(X,A;Z) \otimes \Omega_*^{SO} \longrightarrow \Omega_*^{SO}(X,A).$$

Considering the composite $f_* \circ \mathfrak{m} : H_*(X,A;Z) \otimes \Omega_*^{SO} \longrightarrow H_*(X,A;\Omega_*^{SO})$, one may write $H_*(X,A;\Omega_*^{SO})$ as $H_*(X,A;Z) \otimes \Omega_*^{SO}$ by the universal coefficient theorem (since $H_*(X,A;Z)$ is torsion free), with $f_* \circ \mathfrak{m}(x_i \otimes 1) = x_i \otimes 1$. In this form $f_* \cdot \mathfrak{m}$ is simply

$$1 \otimes f_{\#} : H_*(X,A;Z) \otimes \Omega_*^{SO} \longrightarrow H_*(X,A;Z) \otimes \Omega_*^{SO}$$

where $f_{\#}$ is the homotopy homomorphism induced by f. The construction of f shows that $f_{\#}$ and hence $f_* \cdot \mathfrak{m}$ and $\mathfrak{m}$ are monic with odd primary cokernel.

Let $g_i : (N_i, \partial N_i) \longrightarrow (X,A)$ be complex bordism elements with $g_{i*}([N_i, \partial N_i]) = x_i$ and use these to define homomorphisms giving a commutative diagram

$$\begin{array}{ccccc}
H_*(X,A;Z) \otimes \Omega_*^U & \xrightarrow{1 \otimes S_*} & H_*(X,A;Z) \otimes \Omega_*^{SO} & & \\
\mathfrak{n}' \downarrow \cong & & \downarrow \mathfrak{n} & & H_*(X,A;Z) \otimes \Omega_*^{SO}. \\
\Omega_*^U(X,A) & \xrightarrow{S_*} & \Omega_*^{SO}(X,A) & \xleftarrow{\mathfrak{m}} &
\end{array}$$

Suppose $\mathfrak{m}$ maps onto $\Omega_j^{SO}(X,A)$ for $j < n$. Then since S_* has 2 primary cokernel, there is an integer k with $2^k \alpha \in \operatorname{im} S_* \subset \operatorname{im} \mathfrak{n}$ for all $\alpha \in \Omega_n^{SO}(X,A)$. In particular, for all i with $\dim x_i = n$,

$$2^k([M_i, f_i]) = \Sigma\, [N_j, g_j] P_j$$

with $P_j = S_* Q_j \in \Omega_*^{SO}$. Applying e gives at once

$$2^k([M_i, f_i]) = 2^k([N_i, g_i]) + \Sigma\, [N_j, g_j] P_j'$$

with $P_j' \in \Omega_*^{SO}$ having positive dimension. By the inductive assumption, this gives $2^k[N_i,g_i] \in \operatorname{im}\mathfrak{m}$. In particular, for any $\alpha \in \Omega_n^{SO}(X,A)$

$$2^k\alpha = \Sigma\ [N_j,g_j]R_j$$

with $R_j \in S_*\Omega_*^U$ and $\dim N_j \leq n$. Thus $2^k(2^k\alpha) \in \operatorname{im}\mathfrak{m}$. Since $\mathfrak{m}$ has odd primary cokernel, this gives $\alpha \in \operatorname{im}\mathfrak{m}$. Thus $\mathfrak{m}$ is epic by induction. **

Theorem: Let (X,A) be a finite CW pair such that all torsion of $H_*(X,A;Z)$ has order 2. Then two classes in $\Omega_*^{SO}(X,A)$ are the same if and only if they have the same Z and Z_2 cohomology characteristic numbers.

Proof: One has $\Omega_*^{SO}(X,A) \cong H_*(X,A;\Omega_*^{SO})$ since neither group has odd torsion (the first by being a direct summand of $\Omega_*^U(X,A)$ except for the prime 2, the second by the universal coefficient theorem) and thus all torsion in $\Omega_*^{SO}(X,A)$ has order 2. If all Z cohomology characteristic numbers of α vanish, then α is a torsion class so $2\alpha = 0$. If also all Z_2 characteristic numbers vanish, then α maps to zero in $\mathcal{W}_*(R,2)(X,A)$ and thus $\alpha = 2\beta$. Since β is also a torsion class, $0 = 2\beta = \alpha$. **

Making use of the arguments for $\Omega_*^U(X,A)$ one also obtains:

Theorem: For any finite CW pair (X,A) having no 2 primary torsion, there are no 2 primary relations among the integral characteristic numbers for $\Omega_*^{SO}(X,A)$. If (X,A) has no p-primary torsion for an odd prime p, then all p primary relations among the integral characteristic numbers for $\Omega_*^{SO}(X,A)$ are given by

$$\{f^*ch(x)\cdot s_\omega(e_{\mathcal{P}}(\tau))L(\tau)\}[M,\partial M] \in Z[1/2]$$

(where $f : (M,\partial M) \longrightarrow (X,A)$) for all ω and all $x \in K^*(X,A)$. **

Relation to Framed Cobordism

Proposition: A framed manifold of positive dimension bounds an oriented manifold; i.e. the homomorphism $F_n : \Omega_n^{fr} \longrightarrow \Omega_n^{SO}$ induced by the forgetful functor is the zero homomorphism if $n > 0$. Further, $F_0 : \Omega_0^{fr} \longrightarrow \Omega_0^{SO} = Z$ is an isomorphism.

Proof: Oriented cobordism class is determined by Z and Z_2 cohomology characteristic numbers which must vanish on positive dimensional framed manifolds. Note: One may also prove this by noting that F_* factors through complex cobordism. **

Forming the relative cobordism theory $\Omega_*(F) = \lim_{r\to\infty} \pi_{*+r}(TBSO_r, S^r, \infty)$, the resulting exact sequence will split up to give short exact sequences

$$0 \longrightarrow \Omega_n^{SO} \longrightarrow \Omega_n(F) \longrightarrow \Omega_{n-1}^{fr} \longrightarrow 0$$

for $n-1 > 0$, and

$$\begin{array}{ccccccccccc} 0 & \longrightarrow & \Omega_1^{SO} & \longrightarrow & \Omega_1(F) & \longrightarrow & \Omega_0^{fr} & \longrightarrow & \Omega_0^{SO} & \longrightarrow & 0. \\ & & \| & & \| & & \| & & \| & & \\ & & 0 & & 0 & & Z & & Z & & \end{array}$$

The main questions are then the nature of the extension in these sequences and the invariants of framed cobordism obtainable by characteristic numbers.

First examining the torsion subgroup of Ω_n^{SO}, one notes that since Stiefel-Whitney numbers detect the torsion, this subgroup may be split. In particular, the torsion subgroup may be analyzed by mapping this sequence into the relative sequence for framed and unoriented cobordism. The only invariant of framed cobordism arising from Stiefel-Whitney numbers is the 2 primary Hopf invariant obtained by evaluating the top dimensional

Stiefel-Whitney class. (Note: Since $w_1(\nu) = 0$ for an oriented manifold with framed boundary, this invariant will be non-zero only for $n = 2, 4,$ or 8.)

To see that no additional 2 primary information is obtainable, consider

$$S^r \hookrightarrow TBSO_r \xrightarrow{f_r} \Pi K(Z,*+r) \times \Pi K(Z_2,*+r)$$

where f_r is the unstable 2 primary homotopy equivalence similar to that of the previous section. The only 2 primary information obtainable from oriented cobordism is then the same as that obtained from the map $S^r \longrightarrow K(Z,r)$ realizing the fundamental class. In particular, since the Z_2 cohomology map induced by $K(Z,r) \longrightarrow K(Z_2,r)$ is epic, corresponding to the map $\underset{\sim}{TBSO} \longrightarrow \underset{\sim}{TBO}$, no new information can be obtained.

Turning to the torsion free structure, consider an oriented manifold V^n with framed boundary. Corresponding to an imbedding $V^n \hookrightarrow H^{n+r}$ one has the normal map $\tilde{\nu} : (V,\partial V) \longrightarrow (BSO_r,*)$ by interpreting the framing of the boundary as a deformation of the normal map of ∂V to a point. One may then form the Pontrjagin numbers of $(V,\partial V)$ which will be integers. Since the only 2 primary relations among the integral characteristic numbers of oriented manifolds follow from integrality of the Pontrjagin numbers, this shows that the relative sequence splits insofar as the prime 2 is concerned and that no 2 primary information about framed cobordism is obtainable from integral cohomology characteristic classes.

Turning to the characteristic numbers $s_\omega(e_{\mathcal{P}})L(\tau)[V,\partial V]$, one has the class $U(\tau) \in K^*(D\tau,S\tau)$ and for any $x \in K^*(V,\partial V)$, one has $\pi^*(x)\cdot U(\tau) \in K^*(D\tau,\partial D\tau)$. Note: $\partial D\tau \neq S\tau$. In particular, $ch(x)\cdot\delta(\tau)[V,\partial V]$ is integral for all $x \in K^*(V,\partial V)$. Since $\tau \otimes \mathbf{C} - n_{\mathbf{C}} \in K(V,\partial V)$ one has $s_\omega(e_{\mathcal{P}})(\tau) \in chK(V,\partial V)$ if $n(\omega) > 0$, and $L(\tau) = ((1/2)^{n/2} + ch(\theta))\delta(\tau)$ with

$\theta \in K(V,\partial V)[1/2]$. Thus $s_\omega(e_{\mathcal{P}})L(\tau)[V,\partial V] \in Z[1/2]$ for all ω if and only if $L(\tau)[V,\partial V] \in Z[1/2]$.

Since a closed oriented manifold has $L(\tau)[M] = \mathcal{S}(\tau)[M] \in Z$, this gives:

Theorem: A necessary and sufficient condition that an oriented manifold with framed boundary have the same Pontrjagin numbers as a closed oriented manifold is that the L number be integral.

Note: If the class $L(\xi)_{4n}$ is expressed in the form of an integral polynomial in the Pontrjagin classes with relatively prime coefficients, divided by an integer $\mu(L_{4n})$, then $\mu(L_{4n})$ is odd. This is immediate from the lack of 2 primary relations.

Since the homomorphism $L' : \Omega_*(F) \longrightarrow Q$ sends Ω_*^{SO} into Z, one has induced a homomorphism $L'' : \Omega_*^{fr} \longrightarrow Q/Z$. One then has:

Theorem: The homomorphism $L'' : \Omega_*^{fr} \longrightarrow Q/Z$ coincides with the odd primary part of the Adams invariant $e_{\mathbb{C}}$; i.e. for $\alpha \in \Omega_*^{fr}$, $e_{\mathbb{C}}(\alpha) = (a/b) + (c/2^k)$ for some integers a,b,c, and k, with b odd, and $L''(\alpha) = (a/b)$. In particular, L'' and $e_{\mathbb{C}}$ coincide when reduced to $Q/Z[1/2]$.

Proof: If $\alpha = [M]$, choose a stably almost complex manifold V with $\partial V = M$. Then

$$\begin{aligned} L''(\alpha) &= L(\tau)[V,M], \\ &= \{1 + \Sigma\, b_\lambda s_\lambda(e)\}\mathcal{S}(V)[V,M], \\ &= e_{\mathbb{C}}(\alpha) + \Sigma\, b_\lambda s_\lambda(e)S(V)[V,M], \end{aligned}$$

where $b_\lambda \in Z[1/2]$, $n(\lambda) > 0$. Since each $s_\lambda(e)S(V)[V,M] \in Z$, $L''(\alpha) = e_{\mathbb{C}}(\alpha) + (d/2^m)$ for some integers d and m. Since the denominator of the L polynomial is odd, one also has $L''(\alpha) = (a'/b')$ for some integers a' and b' with b' odd. The result is immediate by combining these expressions. **

Note: This shows that the L" invariant gives less information than $e_{\mathbb{C}}$.

Relation to Unoriented Cobordism

Letting $G_* : \Omega_*^{SO} \to \mathfrak{N}_*$ be the homomorphism induced by the forgetful functor, the two primary analysis of Ω_*^{SO} has given fairly complete knowledge of G_*. In particular, the kernel of G_* is the ideal generated by 2, which is free abelian and so the relative group splits. The knowledge of $\mathcal{W}_*(R,2)$ gives essentially complete description of the cokernel of G_*, which is a Z_2 vector space and equal to the torsion subgroup of the relative group.

There are several approaches to finding a description of the relative groups $\Omega_n^{0,SO}$.

One approach is to link the exact sequences

$$\Omega_*^{SO} \xrightarrow{2} \Omega_*^{SO} \xrightarrow{\rho} \mathcal{W}_*(R,2) \quad (\partial : \mathcal{W}_*(R,2) \to \Omega_*^{SO})$$

$$0 \to \mathcal{W}_*(R,2) \xrightarrow{F_*} \mathfrak{N}_* \xrightarrow{d} \mathfrak{N}_* \to 0$$

to give a long exact sequence

$$\cdots \to \Omega_n^{SO} \xrightarrow{F_*\rho} \mathfrak{N}_n \xrightarrow{(\bar{\partial},d)} \Omega_{n-1}^{SO} \oplus \mathfrak{N}_{n-2} \xrightarrow{(2,0)} \Omega_{n-1}^{SO} \to \cdots,$$

where $\bar{\partial}$ is the homomorphism taking the submanifold dual to w_1. (Note: One has $\partial = \bar{\partial}\circ F_*$.) From this sequence, it is clear that

$$\Omega_n^{0,SO} \cong \Omega_{n-1}^{SO} \oplus \mathfrak{N}_{n-2}.$$

This exact sequence was first noticed by Dold [46].

A semigeometric argument may be given as follows: Let (V^n, M^{n-1}) be a manifold with oriented boundary and let $f : V \longrightarrow RP(N)$ be a map with $f^*\xi \cong \det\tau_V$, sending M to a point not in $RP(N-1)$. Make f transverse regular on $RP(N-1)$ keeping M fixed, to obtain a closed manifold $W^{n-1} = f^{-1}(RP(N-1))$ and map $f|_W : W \longrightarrow RP(\infty)$. Letting ν denote the normal bundle of W In V, $\det\nu \cong \nu \cong f^*\xi \cong \det\tau_V|_W$, so $\det\tau_W \cong \det\tau_V|_W \otimes \det\nu$ is trivialized. Thus one has an oriented bordism element of $RP(\infty)$. Identifying $D\nu$ with a tubular neighborhood of W in V, $W-(D\nu)^0$ is an oriented manifold with boundary $M - S\nu$ and may be used to give a cobordism of (V,M) to $(D\nu, S\nu)$ (the union of the three pieces bounds $V \times I$). Thus one has defined a homomorphism $\Omega_n^{O,SO} \longrightarrow \Omega_{n-1}^{SO}(RP(\infty))$.

The inverse homomorphism may be described as follows: Let $g : X \longrightarrow RP(\infty)$ be an oriented bordism element and to it assign the class of $(Dg^*\xi, Sg^*\xi)$ in $\Omega_n^{O,SO}$. Since g extends to $Dg^*\xi \longrightarrow D\xi \subset RP(\infty+1) = RP(\infty)$, automatically transverse recovering X, this is clearly an inverse to the above.

Thus $\Omega_n^{O,SO} \cong \Omega_{n-1}^{SO}(RP(\infty))$ and the isomorphism with $\Omega_{n-1}^{SO} \oplus \mathcal{N}_{n-2}$ is obtained by sending (X,g) to (X,Y) where $Y \subset X$ is the submanifold dual to $g^*\xi$. That this is an isomorphism may be seen by noting that $\Omega_{n-1}^{SO}(RP(\infty))$ is isomorphic to Ω_{n-1}^{SO} plus the reduced group $\tilde{\Omega}_{n-1}^{SO}(RP(\infty)) \cong \mathcal{N}_{n-2}$.

Another proof may be given by using the Atiyah bordism approach. One has the cofibration sequence $RP(1) \longrightarrow RP(N) \longrightarrow RP(N)/RP(1)$, and $RP(N)/RP(1)$ may be identified as the Thom space of 2ξ over $RP(N-2)$. Since 2ξ is an oriented bundle, there is a Thom isomorphism for oriented bordism (Note: If $f : X \longrightarrow BSO_n$ is a map, then $Tf : Tf^*\mathcal{U}_n \longrightarrow TBSO_n$ may be thought of as a cohomology class in TBSO theory, which defines the orientation.) and one has the exact sequence obtained by applying $\tilde{\Omega}_{n+1}^{SO}$ and letting N go to ∞

$$\begin{array}{ccccccc} \cdots \longrightarrow & \tilde{\Omega}^{SO}_{n+1}(RP(1)) & \longrightarrow & \tilde{\Omega}^{SO}_{n+1}(RP(\infty)) & \longrightarrow & \tilde{\Omega}^{SO}_{n+1}(T(2\xi)) & \longrightarrow \cdots \\ & \Vert\wr & & \Vert\wr & & \Vert\wr & \\ \cdots \longrightarrow & \Omega^{SO}_{n} & \longrightarrow & \mathcal{M}_n & \longrightarrow & \Omega^{SO}_{n-1}(RP(\infty)) & \longrightarrow \cdots . \end{array}$$

Since the map $TBSO \wedge RP(1) = \Sigma\, TBSO \longrightarrow TBSO \wedge RP(\infty) = \Sigma\, TBO$ is just the suspension of the inclusion $TBSO \longrightarrow TBO$ induced by G, the relative group is precisely the homotopy of the cofiber or the bordism of $T(2\xi)$ up to dimension shifts.

This situation has been generalized slightly by George Mitchell (Thesis; University of Virginia) who considers the bordism theory defined by maps $(V,\partial V) \longrightarrow (X,A)$ with ∂V oriented. This is denoted $\Omega^{O,SO}_*(X,A)$, and is given by the homotopy of the cofiber of the map

$$(A/\emptyset) \wedge TBSO \wedge RP(1) \longrightarrow (X/\emptyset) \wedge TBSO \wedge RP(\infty)$$

given by suspending

$$(A/\emptyset) \wedge TBSO \longrightarrow (X/\emptyset) \wedge TBO.$$

Up to dimension shift, this is the oriented bordism of the pair $((X/\emptyset) \wedge RP(\infty),\ (A/\emptyset) \wedge RP(1))$ or $(X \times RP(\infty),\ A \times RP(1) \cup X \times *)$.

Relation to Complex Cobordism

The homomorphism $S_* : \Omega^U_* \longrightarrow \Omega^{SO}_*$ has previously been examined in considerable detail. In particular, the kernel of S_* is free abelian so that the relative group $\Omega_*(S)$ splits as the direct sum of kernel S_* and cokernel S_*.

If one considers the composite

$$\Omega_*^U \xrightarrow{S*} \Omega_*^{SO} \xrightarrow{\pi\times\rho} (\Omega_*^{SO}/\text{Torsion}) \oplus \mathcal{N}_*,$$

writing $\Omega_*^U \cong Z[b_i]$ in the usual way, then since $\pi \times \rho$ is monic, the kernel of S_* is the intersection of the ideals kernel(ρS_*), generated by 2 and the b_{2^s-1}, and kernel(πX_*), generated by the b_{2i+1}. Thus kernel(S_*) is the ideal generated by the elements b_{2^s-1} and $2b_{2i+1}(2i+1 \neq 2^s-1)$.

Since $Z[b_{2i}]$ maps isomorphically to Ω_*^{SO}/Torsion, the subgroup $S_*(Z[b_{2i}])$ of Ω_*^{SO} forms a complementary summand for Torsion(Ω_*^{SO}). Thus, the torsion subgroup maps onto cokernel S_*, which is therefore a Z_2 vector space and forms the torsion subgroup of $\Omega_*(S)$. In particular, S_* maps onto $2\Omega_*^{SO}$ so that cokernel $S_* \cong (\Omega_*^{SO}/2\Omega_*^{SO})/\mathcal{N}_*^2$, where $\Omega_*^{SO}/2\Omega_*^{SO}$ is thought of as a subgroup of $\mathcal{N}_*$. Since cokernel S_* maps monomorphically into the torsion subgroup of $\Omega_*^{O,U}$, this subgroup is detected by Z_2 cohomology characteristic numbers, while the torsion free part is detected by integral cohomology characteristic numbers.

The Index

Let M^n be a closed oriented manifold of dimension $n = 4k$. By Poincaré duality and the universal coefficient theorem, the pairing

$$H^{2k}(M;\mathbb{R}) \otimes_{\mathbb{R}} H^{2k}(M;\mathbb{R}) \longrightarrow \mathbb{R} : x \otimes y \longrightarrow x \cup y[M]$$

where $\mathbb{R}$ denotes the reals, is a nondegenerate pairing. Since $\dim x = \dim y = 2k$ is even, one has $x \cup y[M] = y \cup x[M]$, and this pairing is symmetric. One may then choose a base for $H^{2k}(M;\mathbb{R})$ so that the matrix of the pairing is diagonal. One then defines the index of M, $I(M)$, to be the number of positive diagonal entries minus the number of negative diagonal entries. This function is extended to manifolds of dimension not divisible by 4 by letting $I(M) = 0$ in these cases.

Recalling that the only invariants of symmetric bilinear forms over the reals are the rank and the index, while an orientation preserving homotopy equivalence of closed manifolds must preserve the pairing, one has:

Theorem: The index of M is an invariant of the oriented homotopy type of M. **

Theorem: The index has the following properties:

a) $I(M + N) = I(M) + I(N)$, $I(-M) = -I(M)$;

b) $I(M \times N) = I(M)\cdot I(N)$;

c) If M bounds then $I(M) = 0$; and

d) $I(\mathbb{C}P(2k)) = 1$.

Thus, $I : \Omega_*^{SO} \longrightarrow Z$ is the unique ring homomorphism taking the value 1 on each $\mathbb{C}P(2k)$.

Proof: a) is clear for $H^{2k}(M + N;\mathbb{R})$ is the direct sum of $H^{2k}(M;\mathbb{R})$ and $H^{2k}(N;\mathbb{R})$, with the pairing being the 'sum' of the two pairings, while the pairing for M with orientation reversed is just the negative of that for M.

To prove b), let $P = M \times N$, with dimensions $p, m,$ and n respectively. If $p \not\equiv 0 \pmod 4$, then at least one of m and n is not zero mod 4 so that $I(P)$ and $I(M)\cdot I(N)$ are both zero. If $p = 4k$, then

$$H^{2k}(P;\mathbb{R}) \cong \sum_{s=0}^{2k} H^s(M;\mathbb{R}) \otimes_{\mathbb{R}} H^{2k-s}(N;\mathbb{R})$$

by the Künneth theorem. This vector space decomposes into the subspaces

$$H^s(M;\mathbb{R}) \otimes H^{2k-s}(N;\mathbb{R}) \oplus H^{m-s}(M;\mathbb{R}) \otimes H^{2k+s-m}(N;\mathbb{R})$$

for $s < m/2$, and the space

$$H^{m/2}(M;\mathbb{R}) \otimes H^{n/2}(N;\mathbb{R}),$$

with distinct summands being 'orthogonal' under the pairing and with the restriction of the pairing to each summand being nondegenerate.

If $s < m/2$, choose a base x_i for $H^s(M;\mathbb{R})$ and a base y_j for $H^{2k-s}(N;\mathbb{R})$ with dual bases x_p^* and y_q^* in $H^{m-s}(M;\mathbb{R})$ and $H^{2k+s-m}(N;\mathbb{R})$ respectively. Using the base $x_i \otimes y_j$, $x_p^* \otimes y_q^*$ for the s-subspace, the pairing sends all pairs of basis vectors to zero except for the pairs $(x_i \otimes y_j, x_i^* \otimes y_j^*)$ and $(x_i^* \otimes y_j^*, x_i \otimes y_j)$ on which the value is $(-1)^t$ where $t = (2k-s)(m-x)$. Thus the pairing matrix decomposes as a direct sum of 'orthogonal' 2 dimensional subspaces with matirx $(-1)^t \begin{pmatrix} 0 & 1 \\ 1 & 0 \end{pmatrix}$. Since the index of this 2×2 matrix is zero, each of the s-subspaces contributes zero to the index of P.

Thus, the index of P is precisely the same as the index of the pairing on $H^{m/2}(M;\mathbb{R}) \otimes H^{n/2}(N;\mathbb{R})$. If both m and n are congruent to zero mod 4, choosing bases for which the forms of M and N are both diagonal gives the basis of products in which the form on $H^{m/2}(M;\mathbb{R}) \otimes H^{n/2}(N;\mathbb{R})$ is diagonal. Looking at the diagonal entries gives immediately $I(P) = I(M)\cdot I(N)$. If both m and n are congruent to 2 mod 4, then the pairings $H^{m/2}(M;\mathbb{R}) \otimes H^{m/2}(M;\mathbb{R}) \longrightarrow \mathbb{R}$ and that of N are both skew-symmetric. Thus one may choose a base of $H^{m/2}(M;\mathbb{R})$ so that the pairing matrix is a direct sum of copies of $\begin{pmatrix} 0 & 1 \\ -1 & 0 \end{pmatrix}$, and similarly for N. Looking in the product of two such two dimensional subspaces, the pairing matrix is $\begin{pmatrix} 0 & J \\ J' & 0 \end{pmatrix}$ with $J = \begin{pmatrix} 0 & 1 \\ -1 & 0 \end{pmatrix}$ and $J' = \begin{pmatrix} 0 & -1 \\ 1 & 0 \end{pmatrix}$. This matrix has index 0, so $I(P) = 0 = I(M)\cdot I(N)$.

The proof of c) is due to Thom [126]. Suppose $M^n = \partial W^{n+1}$ with $n = 4k$ and M and W are oriented. By Lefschetz duality, one then has a commutative exact ladder

$$\begin{array}{ccccccccccc} \cdots & \longrightarrow & H^r(W) & \xrightarrow{f^*} & H^r(M) & \longrightarrow & H^{r+1}(W,M) & \longrightarrow & H^{r+1}(W) & \longrightarrow & \cdots \\ & & \downarrow \cong & & \downarrow \cong & & \downarrow \cong & & \downarrow \cong & & \\ \cdots & \longrightarrow & H_{n+1-r}(W,M) & \longrightarrow & H_{n-r}(M) & \xrightarrow{f^*} & H_{n-r}(W) & \longrightarrow & H_{n-r}(W,M) & \longrightarrow & \cdots \end{array}$$

where $f : M \longrightarrow W$ is the inclusion, all groups having real coefficients.

Let $A^r = \text{Image}((f^*)^r)$ and $K_{n-r} = \text{kernel}((f_*)_{n-r})$.

By exactness, one has $A^r \xrightarrow{\cong} K_{n-r}$.

If $a \in A^r$, $b \in A^{n-r}$, then $< a \cup b, [M] > = 0$. To see this, one has $< a \cup b, [M] > = < f^*(\alpha \cup \beta), \partial[W,M] > = < \delta f^*(\alpha \cup \beta), [W,M] > = < 0, [W,M] > = 0$.

Since the coefficients are a field, one has by the universal coefficient theorem $H^i(M) \cong H_i(M)$ and f^* is the dual map of f_*: i.e. the diagram

$$\begin{array}{ccc} H_{n-p}(W) & \xleftarrow{f^*} & H_{n-p}(M) \\ \| \wr & & \| \wr \\ H^{n-p}(W) & \xrightarrow{f^*} & H^{n-p}(M) \end{array}$$

commutes, giving $H_{n-p}(M)/K_{n-p} \cong$ dual of A^{n-p}. Thus A^p is precisely the annihilator of A^{n-p}.

With $\dim M = 4k$, this gives $H^{2k}(M) = A^{2k} \oplus B^{2k}$ with A and B dually paired and with dual bases a_i, b_j such that $a_i b_j = \delta_{ij}$, $a_i a_j = b_i b_j = 0$. Ordering the basis as $a_1, b_1, a_2, b_2, \ldots$, the matrix of the pairing consists of 2×2 blocks $\begin{pmatrix} 0 & 1 \\ 1 & 0 \end{pmatrix}$ along the diagonal, with zeros elsewhere. One may then compute the index, which is zero, giving $I(M) = 0$.

For part d), $H^{2k}(\mathbb{C}P(2k);\mathbb{R})$ has a base given by α^k, where $\alpha \in H^2(\mathbb{C}P(2k);\mathbb{R})$ is the first Chern class of the canonical bundle. Under the pairing, $\alpha^k \otimes \alpha^k$ is sent to $\alpha^{2k}[\mathbb{C}P(2k)] = (-1)^{2k} = 1$. Thus the pairing matrix is (1), and the index is 1.

Finally, properties a)-d) indicate that the index defines a ring homomorphism $I : \Omega_*^{SO} \longrightarrow Z$ sending each $\mathbb{C}P(2k)$ to 1. Since any ring homomorphism to Z or Q must annihilate the torsion subgroup, while the $\mathbb{C}P(2k)$ *generate* $\Omega_*^{SO} \otimes Q$ *as a ring (over* Q*), such homomorphisms are* completely determined by their values on the $\mathbb{C}P(2k)$, proving uniqueness. **

Since the index defines a homomorphism of Ω_*^{SO} into Q, there must be an expression for the index of an oriented n manifold as a rational linear combination of the Pontrjagin numbers. The precise expression for the index is the Hirzebruch index theorem [55]:

Theorem: The index homomorphism $I : \Omega_*^{SO} \longrightarrow Z$ is given by the evaluation of the L class; i.e. for any closed oriented manifold, $I(M) = L(\tau)[M]$.

Proof: Let $L' : \Omega_*^{SO} \longrightarrow Q : [M] \longrightarrow L(\tau)[M]$, be the homomorphism defined by the L class evaluation. From the diagonal formula $\Delta L = L \otimes L$, it is immediate that L' is a ring homomorphism. In order to show that $I = L'$, it then suffices to show that $L(\tau)[\mathbb{C}P(2k)] = 1$ for each k. For $\mathbb{C}P(2k)$, one has $\mathcal{P}(\tau) = (1 + \bar{\alpha}^2)^{2k+1}$, where $\bar{\alpha} \in H^2(\mathbb{C}P(2k);Z)$ is the first Chern class of ξ, and hence

$$
\begin{aligned}
L(\tau)[\mathbb{C}P(2k)] &= (\bar{\alpha}/\tanh\bar{\alpha})^{2k+1}[\mathbb{C}P(2k)], \\
&= \text{coefficient of } \bar{\alpha}^{2k} \text{ in } (\bar{\alpha}/\tanh\bar{\alpha})^{2k+1}, \\
&= (1/2\pi i) \oint dz/(\tanh z)^{2k+1}, \\
&= (1/2\pi i) \oint du/u^{2k+1}(1-u^2), \qquad (u = \tanh z) \\
&= (1/2\pi i) \oint (1/u^{2k+1})(1 + u^2 + \dots)du, \\
&= 1. \quad **
\end{aligned}
$$

It is convenient to know the form of the power series $x/\tanh x$. Since one has

$$x + \frac{x}{\tanh x} = \frac{(-2x)}{(e^{-2x}-1)},$$

the knowledge of the power series for $y/(e^y-1)$ gives

$$x/\tanh x = 1 + \frac{1}{3}x^2 - \frac{1}{45}x^4 + \ldots + (-1)^{k-1}\frac{2^{2k}}{(2k)!}B_k x^{2k} + \ldots,$$

where B_k is the k-th Bernoulli number.

Odd Primary Data

It is frequently convenient to know something of the p primary structure of BSO and the Z_p cohomology characteristic number structure of oriented cobordism, which the chosen approach to cobordism has made unnecessary. It is possible to approach oriented cobordism in this fashion also. First, consider the case p odd.

Proposition: For p an odd prime, neither $H^*(\underset{\sim}{BSO};Z)$ nor $\tilde{H}^*(\underset{\sim}{TBSO};Z)$ have p-primary torsion. The Bockstein operator Q_0 is trivial in $\tilde{H}^*(\underset{\sim}{TBSO};Z_p)$, making this an $\mathcal{A}_p/(Q_0)$ module, and as such it is a free module.

Proof: Since $H^*(BSO;Z_p) \cong Z_p[\mathcal{P}_i]$ is nonzero only in dimensions congruent to zero mod 4, the universal coefficient theorem shows that there is no p-primary torsion in the integral groups. By the Thom isomorphism theorem, the same is true of $\tilde{H}^*(\underset{\sim}{TBSO};Z)$. Since the groups $\tilde{H}^*(\underset{\sim}{TBSO};Z_p)$ are nonzero in dimensions congruent to zero mod 4 only, while $\dim Q_0 = 1$, Q_0 must act trivially. Using the map $BU \longrightarrow BSO$, one has induced a homomorphism $\tilde{H}^*(\underset{\sim}{TBSO};Z_p) \longrightarrow \tilde{H}^*(\underset{\sim}{TBU};Z_p)$ sending the Thom class of $\underset{\sim}{TBSO}$ to that of $\underset{\sim}{TBU}$, and the homomorphism $\mathcal{A}_p/(Q_0) \longrightarrow \tilde{H}^*(\underset{\sim}{TBSO};Z_p)$ induced by action on the Thom class is monic, since the composite homomorphism to $\tilde{H}^*(\underset{\sim}{TBU};Z_p)$ is monic. By the Milnor-Moore theorem, $\tilde{H}^*(\underset{\sim}{TBSO};Z_p)$ is a free $\mathcal{A}_p/(Q_0)$ module. **

Using the mod p Steenrod algebra one may then duplicate for oriented manifolds almost all of the constructions made for the prime 2 in unoriented theory.

If M^n is an oriented manifold, then Poincaré duality and the universal coefficient theorem imply that $H^i(M;Z_p) \otimes H^{n-i}(M;Z_p) \longrightarrow Z_p : a \otimes b \longrightarrow a \cup b[M]$ is a dual pairing. Thus there are unique Wu classes $v_i \in H^{2i(p-1)}(M;Z_p)$ such that

$$\mathcal{P}^i a[M] = v_i \cup a[M]$$

for all $a \in H^{n-2i(p-1)}(M;Z_p)$. Letting $v = 1 + v_1 + \ldots \in H^*(M;Z_p)$, one defines a class $Q = 1 + Q_1 + \ldots \in H^*(M;Z_p)$ where $\dim Q_i = 2i(p-1)$ by $Q = \mathcal{P}v$.

<u>Theorem</u>: If M^n is a closed oriented manifold, then the class Q_i is the mod p reduction of the class $s_{(\underbrace{(p-1)/2,\ldots,(p-1)/2}_{i})}(\mathcal{P}(\tau))$; i.e. if the tangential Pontrjagin class of M is expressed formally as $\Pi(1 + x_j^2)$, then the class Q is given by $\Pi(1 + x_j^{p-1})$.

<u>Proof</u>: Duplicating the proof for the relation between Wu class and tangential Stiefel-Whitney classes, it suffices to consider the effect of applying $\mathcal{P}^i$ to the Thom class in $\tilde{H}^*(TBSO_{2k};Z_p)$. Using the splitting principle, U may be written as a product $x_1 \ldots x_k$ of 2 dimensional classes, so that $\mathcal{P}^i(x_1 \ldots x_k)$ is the sum of all monomials $x_1 \ldots x_{j_1}^p \ldots x_{j_i}^p \ldots x_k$. This is the i-th elementary symmetric function in the variables x_j^{p-1} times the class $x_1 \ldots x_k$. **

Note: Writing the tangential Pontrjagin class of M as $\Pi(1 + x_j^2)$, the Wu class v is the mod p reduction of

$$\Pi\{1 + (x_j - x_j^p + x_j^{p^2} + \ldots + (-1)^k x_j^{p^k} + \ldots)^{p-1}\}.$$

To see this, one has $\mathcal{P}y = 1 + x^{p-1}$ giving $y = 1 + (\mathcal{P}^{-1}x)^{p-1}$, and if $\dim x = 2$, then $\mathcal{P}^{-1}x = x - x^p + x^{p^2} + \ldots + (-1)^k x^{p^k} + \ldots$.

One then has the mod p analogue of the Dold theorem:

Theorem: All relations among the mod p reductions of the Pontrjagin numbers of closed oriented n manifolds are given by the Wu relations; i.e. if $\phi : H^n(BSO; Z_p) \longrightarrow Z_p$ is a homomorphism, there is a closed oriented n dimensional manifold with $\phi(a) = (\tau^*(a))[M]$ for all a if and only if $\phi(\mathcal{P}b - vb) = 0$ for all $b \in H^*(BSO; Z_p)$.

Proof: From the free module structure of $\tilde{H}^*(\underset{\sim}{TBSO}; Z_p)$ as an $\mathcal{A}_p/(Q_0)$ module, and the knowledge of the homotopy of spectra of this type, it is immediate that the image of $\pi_*(\underset{\sim}{TBSO})$ in $\tilde{H}_*(\underset{\sim}{TBSO}; Z_p)$ consists precisely of those classes annihilating $\bar{\mathcal{A}}_p\tilde{H}^*(\underset{\sim}{TBSO}; Z_p)$. Using the proof given for Dold's theorem, the result is immediate. **

Note: This result and the analogue for complex manifolds (which is proved in exactly the same way using the fact that $\tilde{H}^*(\underset{\sim}{TBU}; Z_p)$ is a free $\mathcal{A}_p/(Q_0)$ module) were first proven by Atiyah and Hirzebruch [19]. Since all p-primary relations among the integral characteristic numbers follow from K theory, these Z_p relations should be derived from the K theory. The derivation, which follows, is due to Atiyah and Hirzebruch.

Theorem: For each ω, let $\theta_\omega \in H^*(BSO; \mathbb{R})$ be the class obtained from $s_\omega(e_{\mathcal{P}})L$ by multiplying the component of dimension $2i + 4n(\omega)$ by q^i where $q = p^{1/(p-1)}$. Then each component of θ_ω is expressible as a power of q

times a rational polynomial in the Pontrjagin classes with denominator relatively prime to p, so that θ_ω has a meaningful mod p reduction, $\rho_p(\theta_\omega)$. In fact $\rho_p(\theta_\omega) = \mathcal{P}^{-1}s_\omega(\mathcal{P})\cdot v$. Thus

$$\{\mathcal{P}^{-1}s_\omega(\mathcal{P})\cdot v - s_\omega(\mathcal{P})\}[M^n] = q^{\{n/2-2n(\omega)\}}\{s_\omega(e_{\mathcal{P}})L - s_\omega(\mathcal{P})\}[M^n]$$

which is zero mod p, and hence the K-theoretic relations imply the Wu relations: $\{\mathcal{P}\tau^*(b) - v\cdot\tau^*(b)\}[M] = 0$ for all $b \in H^*(BSO;Z_p)$.

Proof: It suffices to apply the splitting principle and write each class as a symmetric function.

The power of p in $k!$ is at most $(k-1)/(p-1)$ and equality holds if and only if k is a power of p. By the Wilson theorem, $(p^j)!/p^e \equiv (-1)^e \equiv (-1)^j$ mod p, where $e = (p^j-1)/(p-1)$, so

$$\begin{aligned}(e^{qx} - 1)/q &\equiv \sum_{j=0}^{\infty} (-1)^j x^{p^j} \mod p \\ &\equiv \mathcal{P}^{-1}(x)\end{aligned}$$

or

$$\begin{aligned}\frac{(e^{qx}+e^{-qx}-2)}{q^2} &= \frac{(e^{qx}-1)(1-e^{-qx})}{q\qquad q}, \\ &\equiv \mathcal{P}^{-1}(x)\cdot\mathcal{P}^{-1}(x), \\ &\equiv \mathcal{P}^{-1}(x^2)\end{aligned}$$

and letting ϕ_ω be otained from $s_\omega(e_{\mathcal{P}})$ by multiplying the term of degree $2i + 4n(\omega)$ by q^i,

$$\phi_\omega = s_\omega\left(\frac{e^{qx_j}+e^{-qx_j}-2}{q^2}\right) \equiv s_\omega(\mathcal{P}^{-1}(x_j^2)) = \mathcal{P}^{-1}s_\omega(\mathcal{P}).$$

Also

$$\frac{qx}{\tanh qx} = \frac{2qx}{(e^{2qx}-1)} - qx = \{1/[\sum_{j=0}^{\infty} (-1)^j (2x)^{p^j-1}]\} - qx$$

but $2^{p^j-1} \equiv 1\ (p)$ and $qx \equiv 0\ (p)$, while letting $y = \sum_{j=0}^{\infty} (-1)^j (x)^{p^j-1}$, $(xy)^p = (\sum_{j=0}^{\infty} (-1)^j x^{p^j})^p = \sum_{j=0}^{\infty} (-1)^j x^{p^{j+1}} = -xy + x$, since $(a+b)^p = a^p+b^p \bmod p$. Thus $x = xy + (xy)^p$ or $1/y = 1 + (xy)^{p-1}$. This gives

$$\frac{qx}{\tanh qx} \equiv 1 + (\sum_{j=0}^{\infty} (-1)^j x^{p^j})^{p-1} \mod p,$$

and letting L^* be obtained from L by multiplying the component of dimension $2i$ by q^i, L^* has mod p reduction equal to v.

Then

$$\rho_p(\theta_\omega) = \rho_p(\phi_\omega)\cdot\rho_p(L^*) = \mathcal{P}^{-1} s_\omega(\mathcal{P})\cdot v.$$

This gives

$$\{\mathcal{P}^{-1} s_\omega(\mathcal{P})\cdot v - s_\omega(\mathcal{P})\}[M] = q^{\{n/2-2n(\omega)\}}\{s_\omega(e_{\mathcal{P}})L - s_\omega(\mathcal{P})\}[M]$$

reduced mod p, and since $s_\omega(e_{\mathcal{P}})L[M]$ and $s_\omega(\mathcal{P})[M] \in Z[1/2]$, this is zero mod p.

Then for any b, $\mathcal{P}b = \sum \lambda_\omega s_\omega(\mathcal{P})$, $\lambda_\omega \in Z_p$ so $b\cdot v - \mathcal{P}b = \mathcal{P}^{-1}(\mathcal{P}b)\cdot v - (\mathcal{P}b) = \sum \lambda_\omega\{\mathcal{P}^{-1}s_\omega(\mathcal{P})\cdot v - s_\omega(\mathcal{P})\}$, giving the Wu relation. **

<u>Note</u>: One may use the same techniques in the complex situation. In fact, if the component of dimension $2i + 2n(\omega)$ in $s_\omega(e)\mathcal{J}$ is multiplied by q^i, the resulting class reduces mod p to give $\mathcal{P}^{-1}s_\omega(c)\cdot v$, and the Wu relations all follow from the K-theory relations. This also works for $p = 2$, since the terms involving 2^{p^j-1} never appear and since $s_\omega(e)\mathcal{J}[M] \in Z$.

To complete the p-primary study, note that just as in the Z_2 cohomology situation, one has:

<u>Theorem</u>: *If p is an odd prime, then for any framed manifold* M^n *there is an oriented manifold* V^{n+1} *with* $\partial V = M$ *and the mod p Hopf invariant of the homotopy class represented by M is given by*

$$s_{\underbrace{(\frac{p-1}{2},\ldots,\frac{p-1}{2})}_{i}}(\mathcal{P}(\nu))[V,M]$$

where $2i(p-1) = n+1$. This is the only invariant of framed cobordims defined by Z_p cohomology characteristic numbers of this type.

<u>Note</u>: From the work of Liulevicius [72] on the decomposability of the operations $\mathcal{P}^i$, it follows that the mod p Hopf invariant corresponding to $\mathcal{P}^i$ can be nonzero only for $i = 1$. For $n = 2p - 3$, one has

$$H_p([M]) = Q_1(\nu)[V,M] = v_1(\nu)[V,M] = -v_1(\tau)[V,M]$$

$$= -pL_{\frac{p-1}{2}}(\tau)[V,M] = -pe_{\mathbb{C}}([M]) \mod p.$$

Thus, if the Adams invariant $e_{\mathbb{C}}([M])$ is written as $(a/p) + (b/c)$ with $a,b,c \in Z$ and c relatively prime to p, then $H_p([M])$ is the class of $-a$ mod p. Thus the Adams invariant determines the mod p Hopf invariant in a precise fashion.

<u>Two Primary Data</u>

To complete the study of oriented cobordism and oriented vector bundles it seems desirable to have a basic knowledge of the 2 primary structure of BSO, which has not been necessary in the approach to cobordism taken here.

__Theorem__: The cohomology ring $H^*(BSO(n);Z_2)$ is the polynomial algebra over Z_2 on the Stiefel-Whitney classes $w_i(\tilde{\gamma}^n)$ for $1 < i \leq n$.

__Proof__: Let $f_n : BSO(n) \longrightarrow BO(n)$ be the map classifying $\tilde{\gamma}^n$. Then $f_n^* : H^*(BO(n);Z_2) = Z_2[w_i | 1 \leq i \leq n] \longrightarrow H^*(BSO(n);Z_2)$, sending w_i to $w_i(\tilde{\gamma}^n)$. Since $w_1(\tilde{\gamma}^n) = w_1(\det\tilde{\gamma}^n) = 0$ because $\det\tilde{\gamma}^n$ is trivial, this induces

$$f_n^* : P_n = Z_2[w_i | 1 < i \leq n] \longrightarrow H^*(BSO(n);Z_2).$$

To see that this is monic, let $g_n : BO(n-1) \longrightarrow BSO(n)$ be a map classifying the bundle $\gamma^{n-1} \oplus \det\gamma^{n-1}$, which is orientable. Then $g_n^* f_n^*(w_i)$ is given by $w_i + w_1 w_{i-1}$ if $i < n$, and $w_1 w_{n-1}$ if $i = n$. Since these elements are algebraically independent in $Z_2[w_i | 1 \leq i \leq n-1]$, f_n^* must be monic (on P_n).

To see that this is epic, use induction on n. For $n = 1$, $BSO(1)$ is a point while for $n = 2$, $BSO(2) = BU(1) = \mathbb{C}P(\infty)$ whose cohomology is the polynomial algebra generated by $w_2(\tilde{\gamma}^2) = c_1(\tilde{\gamma}^2)$. Assuming that f_{n-1}^* is epic, one has the diagram of the pair $(D\tilde{\gamma}^n, S\tilde{\gamma}^n)$

$$\begin{array}{ccccc} BSO(n-1) & \longrightarrow & BSO(n) & \longrightarrow & TBSO(n) \\ \downarrow & & \downarrow & & \downarrow \\ BO(n-1) & \longrightarrow & BO(n) & \longrightarrow & TBO(n) \end{array}$$

giving a commutative diagram

$$\begin{array}{ccccccccc} 0 & \longleftarrow & H^i(BSO(n-1)) & \xleftarrow{r} & H^i(BSO(n)) & \longleftarrow & H^{i-n}(BSO(n)) & \longleftarrow & 0 \\ & & \uparrow f_{n-1}^i & & \uparrow f_n^i & & \uparrow f_n^{i-n} & & \\ 0 & \longleftarrow & H^i(BO(n-1)) & \xleftarrow{s} & H^i(BO(n)) & \longleftarrow & H^{i-n}(BO(n)) & \longleftarrow & 0 \end{array}$$

in which the cohomology of the Thom space is replaced by that of the base by means of the Thom isomorphism. (Note: This sequence splits up since s is epic, and r *is epic since* $f^i_{n-1}s$ *is epic.) Using induction on* i, f^{i-n}_n is epic, and f^i_{n-1} is epic, so f^i_n is epic. Hence f^*_n is epic. **

<u>Note</u>: This justifies the fact that the only Z_2 cohomology characteristic numbers of oriented cobordism were Stiefel-Whitney numbers.

<u>Lemma</u>: In $H^*(BO;Z_2)$, one has $Sq^1 w_i = w_1 w_i + (i+1)w_{i+1}$.

<u>Proof</u>: Apply the splitting principle to write $w_i = \sum x_1 \ldots x_i$. Then $Sq^1 w_i = \sum x_1^2 x_2 \ldots x_i$. On the other hand $w_1 \cdot w_i$ is the sum of the monomials $x_1 \ldots x_j^2 \ldots x_i$ and monomials $x_1 \ldots x_{i+1}$, the latter occurring once for each subscript which came from the w_1 factor. Thus

$w_1 w_i = s_{(2,1,\ldots,1)} + (i+1)w_{i+1}$. **

Now consider the operation $Sq^1 : H^*(BO;Z_2) \longrightarrow H^*(BO;Z_2)$. One has $Sq^1(a\cdot b) = Sq^1 a\cdot b + a\cdot Sq^1 b$ and (by the Adem relations) $Sq^1 Sq^1 = 0$. Since Sq^1 is a derivation of square zero, one may form the homology with respect to Sq^1.

<u>Lemma</u>: The homology groups with respect to Sq^1 are given by

$$H(H^*(BO;Z_2),Sq^1) = Z_2[{w_{2j}}^2],$$

$$H(H^*(BSO(n);Z_2),Sq^1) = Z_2[{w_{2j}}^2 \mid 2j < n] \text{ if } n \text{ is odd,}$$

$$= Z_2[{w_{2j}}^2, w_n \mid 2j < n] \text{ if } n \text{ is even.}$$

<u>Proof</u>: Since $Sq^1 w_{2i} = w_{2i+1} + w_1 w_{2i}$, one may write $H^*(BO;Z_2)$ as the polynomial algebra on w_1, w_{2i}, and $Sq^1 w_{2i}$. Thus $H^*(BO;Z_2)$ is the tensor product of polynomial algebras of the forms

$$Z_2[w_{2i}, Sq^1 w_{2i}] \quad \text{and} \quad Z_2[w_1]$$

on which Sq^1 act. Applying the Künneth theorem, the homology of $H^*(BO)$ is the tensor product of the homology groups of the factors, being $Z_2[w_{2i}{}^2]$ and Z_2 respectively.

For the groups $H^*(BSO(n);Z_2)$ the given computation still applies. One has $H^*(BSO(2k-1);Z_2) = Z_2[w_{2i}, Sq^1 w_{2i} | i < k]$, and $H^*(BSO(2k);Z_2) = Z_2[w_{2i}, Sq^1 w_{2i}, w_{2k} | i < k]$, with $Sq^1 w_{2k} = 0$ in the latter. **

Corollary: All torsion in $H^*(BSO(n);Z)$ has order 2.

Proof: It has previously been noted that all torsion is 2 primary. If some torsion class in $H_*(BSO(n);Z)$ has order 2^k, $k > 1$, then the homology of $H^*(BSO(n);Z_2)$ with respect to Sq^1 must be nonzero in two consecutive dimensions. **

Turning attention to the Thom spectrum $\underset{\sim}{TBSO}$, one has:

Lemma: The homomorphism $\nu : \mathcal{A}_2 \longrightarrow H^*(\underset{\sim}{TBSO};Z_2) : a \longrightarrow a(U)$ has kernel precisely $\mathcal{A}_2 Sq^1$.

Proof: Using the pair $(D\tilde{\gamma}^n, S\tilde{\gamma}^n)$ one has the exact sequence

$$0 \longleftarrow H^*(BSO(n-1);Z_2) \longleftarrow H^*(BSO(n);Z_2) \longleftarrow \tilde{H}^*(TBSO(n);Z_2) \longleftarrow 0$$

under which the Thom class is sent to w_n, and the cohomology of the Thom space is identified with the multiples of w_n. Since $Sq^1 w_n = w_1 w_n = 0$, Sq^1 annihilates the Thom class and the kernel of ν contains $\mathcal{A}_2 Sq^1$.

Letting $g_n : BO(n-1) \longrightarrow BSO(n)$ classify $\gamma^{n-1} \oplus \det\gamma^{n-1}$ one has $g_n^*(w_n) = w_1 w_{n-1} = Sq^1 w_{n-1}$. Considering $\tilde{H}^*(TBO(n-1);Z_2)$ as the multiples of w_{n-1} in $H^*(BO(n-1);Z_2)$, one knows that within the stable range the homomorphism $\mathcal{A}_2 \longrightarrow \tilde{H}^*(\underset{\sim}{TBO};Z_2) : a \longrightarrow a(U)$ is monic. Thus the kernel of ν is contained in the kernel of $\mathcal{A}_2 \longrightarrow \tilde{H}^*(\underset{\sim}{TBO}:Z_2) : a \longrightarrow aSq^1(U)$, and since the kernel of $\mathcal{A}_2 \longrightarrow \mathcal{A}_2 : a \longrightarrow aSq^1$ is $\mathcal{A}_2 Sq^1$, the kernel of ν is $\mathcal{A}_2 Sq^1$. **

Corollary: The homomorphism $\mathcal{A}_2 \longrightarrow \tilde{H}^*(\underset{\sim}{TBR}^{(2)};Z_2)$ obtained by evaluation on the Thom class is monic, and thus $\tilde{H}^*(\underset{\sim}{TBR}^{(2)};Z_2)$ is a free $\mathcal{A}_2$ module.

Proof: Since $\underset{\sim}{TBR}^{(2)} = \underset{\sim}{TBSO} \wedge RP(2)$, one may consider $\tilde{H}^*(\underset{\sim}{TBR}^{(2)};Z_2)$ as $\tilde{H}^*(\underset{\sim}{TBSO} \wedge RP(2);Z_2) \cong \tilde{H}^*(\underset{\sim}{TBSO};Z_2) \otimes \tilde{H}^*(RP(2);Z_2)$, with the Thom class being $U \otimes x$, $x \in H^1(RP(2);Z_2)$. One may take as a base of $\mathcal{A}_2/\mathcal{A}_2 Sq^1$ the admissible sequences Sq^I, $I = (i_1,\ldots,i_r)$ with $i_r > 1$, and $\mathcal{A}_2$ has a base $\{Sq^I, Sq^I Sq^1\}$. Then

$$Sq^I(U \otimes x) = (Sq^I U) \otimes x + \text{terms divisible by } x^2,$$

$$Sq^I Sq^1(U \otimes x) = Sq^I(U \otimes x^2) = (Sq^I U) \otimes x^2.$$

Since the $Sq^I U$ are linearly independent over Z_2, these are also linearly independent, so evaluation on $U \otimes x$ is monic.

As previously noted S^1 is a group, making $BR^{(2)}$ into an H-space. This makes $\tilde{H}^*(\underset{\sim}{TBR}^{(2)};Z_2)$ a coalgebra with counit the Thom class, and by the Theorem of Milnor-Moore, $\tilde{H}^*(\underset{\sim}{TBR}^{(2)};Z_2)$ is a free $\mathcal{A}_2$ module. **

Note: From this and the fact that the cohomology of BO maps onto that of $BR^{(2)}$ one may conclude that $\mathcal{W}_*(R,2)$ maps monomorphically into $\mathcal{N}_*$, and in fact draw out all of the structure of $\mathcal{W}_*(R,2)$.

One can also obtain from this the result of Wall:

Theorem: As a module over the Steenrod algebra, $\tilde{H}^*(\underset{\sim}{TBSO};Z_2)$ is a direct sum of copies of $\mathcal{A}_2$ and $\mathcal{A}_2/\mathcal{A}_2 Sq^1$. Further, there is a map of $\underset{\sim}{TBSO}$ into a product of spectra of the types $\underset{\sim}{K}(Z)$ and $\underset{\sim}{K}(Z_2)$ which is a 2 primary homotopy equivalence.

Proof: Let $T = \tilde{H}^*(\underset{\sim}{TBSO};Z_2)$. From the cofibration $S^1 \longrightarrow RP(2) \longrightarrow S^2$, one obtains by smashing with $\underset{\sim}{TBSO}$ and taking Z_2 cohomology, an exact sequence

$$0 \longleftarrow T \longleftarrow T\{x,x^2\} \longleftarrow T \longleftarrow 0$$

$$t \otimes x^2 \longleftarrow t$$

$$t \longleftarrow t \otimes x$$

where $U = T\{x,x^2\}$ denotes the free T module on x and x^2, which represents $\tilde{H}^*(\underset{\sim}{TBSO} \wedge RP(2);Z_2) \cong \tilde{H}^*(\underset{\sim}{TBR}^{(2)};Z_2)$ and is a free $\mathcal{A}_2$ module.

Let $\pi : T \longrightarrow T/\bar{\mathcal{A}}_2 T$ be the projection. Let K be a subspace of kernel(Sq^1) in T mapping isomorphically onto $\pi(\text{kernel}(Sq^1))$ and let $L \subset T$ be a subspace mapping isomorphically onto a complementary summand for $\pi(\text{kernel}(Sq^1))$.'

The natural homomorphism of $\mathcal{A}_2$ modules $\mathcal{A}_2 \otimes (L \oplus K) \longrightarrow T$ is epic (as in the Milnor-Moore theorem), and since Sq^1 annihilates K, induces a homomorphism $f : \mathcal{A}_2 \otimes L \oplus \mathcal{A}_2/\mathcal{A}_2 Sq^1 \otimes K \longrightarrow T$.

For $a \in \mathcal{A}_2$ and $t \in T$, one has

$$a(t \otimes x^2) = (at) \otimes x^2,$$

$$a(t \otimes x) = (at) \otimes x + (a't) \otimes x^2,$$

where $\Delta a = a \otimes 1 + a' \otimes Sq^1 + \ldots$. Letting $F : \mathcal{A}_2 \otimes (L \otimes x^2 \oplus L \otimes x \oplus K \otimes x) \longrightarrow U$, this gives

$$ak \otimes x = a(k \otimes x) + a'Sq^1(k \otimes x), \qquad ak \otimes x^2 = aSq^1(k \otimes x),$$

$$a\ell \otimes x = a(\ell \otimes x) + a'(\ell \otimes x^2), \qquad a\ell \otimes x^2 = a(\ell \otimes x^2),$$

and since f is epic, F is epic.

Since the composite $L \otimes x + K \otimes x \xrightarrow{F} U \longrightarrow U/\bar{\mathcal{A}}_2 U \longrightarrow T/\bar{\mathcal{A}}_2 T$ is an isomorphism, $L \otimes x + K \otimes x$ forms part of a base for U as $\mathcal{A}_2$ module. Since F is epic, one may find a subspace $L' \subset L$ so that $L' \otimes x^2 + L \otimes x + K \otimes x$ is a base for U. In particular, $f : \mathcal{A}_2 \otimes L' \oplus \mathcal{A}_2/\mathcal{A}_2 Sq^1 \otimes K \longrightarrow T$ is monic (with image T') since the composite into U maps isomorphically onto $\mathcal{A}_2 \otimes L' \otimes x^2 \oplus \mathcal{A}_2 Sq^1(K \otimes x)$.

Choose a complementary summand M for L' in L, $L = M \oplus L'$, and suppose $M^j = 0$ for $j < i$, and $m \in M^i$, with $m \neq 0$. Let $U' = U/\mathcal{A}_2(L' \otimes x^2 + L' \otimes x + K \otimes x)$ and consider the composite $T \longrightarrow U'$. Since the $T \otimes x^2$ components of all classes in $\mathcal{A}_2(L' \otimes x^2 + L' \otimes x + K \otimes x)$ belong to $T' \otimes x^2$, the map $T/T' \longrightarrow U'$ is monic. Since $m \neq 0$ in T/T' because it maps into the complement of $\pi T'$, $m \otimes x^2 \in U' \cong \mathcal{A}_2 \otimes M \otimes x$ is nonzero. Thus the natural map $U' \longrightarrow T/T'$ must have nonzero kernel in dimension $i + 2$. Thus, there are elements $m' \in M^i$ and $m'' \in M^{i+1}$ which are not both zero so that $Sq^1(m' \otimes x) + m'' \otimes x \in (\mathcal{A}_2 \otimes M \otimes x)^{i+2} = Sq^1(M^i \otimes x) \oplus M^{i+1} \otimes x$ which maps isomorphically onto $(U')^{i+2}$ is sent to zero under the map into T/T'. Thus

$$
\begin{aligned}
Sq^1 m' + m'' &= \text{image of } (Sq^1(m' \otimes x) + m'' \otimes x) \\
&= \sum a_i \ell_i' + \sum b_j k_j
\end{aligned}
$$

with $\ell_i' \in L'$, $k_j \in K$, and $a_i, b_j \in \mathcal{A}_2$. Applying π to this with the independence of M, L', and K shows that $m'' = 0$, $a_i, b_j \in \bar{\mathcal{A}}_2$. Since $Sq^1(Sq^1 m') = 0$, $Sq^1(\sum a_i \ell_i' + \sum b_j k_j) = 0$, and since $\mathrm{kernel}(Sq^1)/\mathrm{image}(Sq^1) = K$ in $\mathcal{A}_2 \otimes L' + \mathcal{A}_2/\mathcal{A}_2 Sq^1 \otimes K$, one has

$$
Sq^1 m' = Sq^1(\sum \tilde{a}_i \ell_i') + Sq^1(\sum \tilde{b}_j k_j).
$$

Letting ℓ' be the sum of the terms $\tilde{a}_i\ell_i'$ for which $\tilde{a}_i \notin \bar{\mathcal{a}}_2$, this gives

$$Sq^1\ell = Sq^1(\sum \hat{a}_i\ell_i' + \sum \tilde{b}_j k_j) \in Sq^1(\bar{\mathcal{a}}_2 T)$$

where $\ell = m' + \ell'$, $\sum \hat{a}_i\ell_i'$ is the sum of terms with $\tilde{a}_i \in \bar{\mathcal{a}}_2$, while Sq^1 annihilates terms with $\tilde{b}_j \notin \bar{\mathcal{a}}_2$. This gives $\ell + \sum c_i t_i \in \text{kernel}(Sq^1)$ with $c_i \in \bar{\mathcal{a}}_2$, so that $\pi\ell$ belongs to the image of K, contradicting the choice of L unless $\ell = 0$. Since $L = M \oplus L'$, this gives $m' = \ell' = 0$, so that both m' and m'' are zero. This contradicts the choice of m' and m'' and thus $M = 0$. Thus $T = T'$ and f is an isomorphism.

To complete the proof, one has $f : \mathcal{a}_2 \otimes L \oplus \mathcal{a}_2/\mathcal{a}_2 Sq^1 \otimes K \xrightarrow{\cong} T$. Under the obvious map one has an isomorphism $\text{kernel}(Sq^1)/\text{image}(Sq^1) \to \text{kernel}(Sq^1)/\bar{\mathcal{a}}_2 T$, both being isomorphic to K. Since $Sq^1U = 0$, one has $H(\tilde{H}^*(\underset{\sim}{TBSO};Z_2),Sq^1)$ isomorphic to $H(H^*(BSO;Z_2),Z_2)$ by the Thom isomorphism, for the Thom isomorphism sends both $\text{kernel}(Sq^1)$ and $\text{image}(Sq^1)$ into the corresponding groups. Thus, one may choose K to be the span of the classes $w_{2\omega}{}^2U$, which are the reductions of integral classes, the $\mathcal{P}_\omega U$. This gives a map of $\underset{\sim}{TBSO}$ into a product of $\underset{\sim}{K}(Z)$ spectra realizing the summand $\mathcal{a}_2/\mathcal{a}_2 Sq^1 \otimes K$ in the cohomology. One may also map $\underset{\sim}{TBSO}$ into a product of $\underset{\sim}{K}(Z_2)$ spectra to realize the summand $\mathcal{a}_2 \otimes L$. The product map sends $\underset{\sim}{TBSO}$ into a product of $\underset{\sim}{K}(Z)$ and $\underset{\sim}{K}(Z_2)$ spectra, inducing an isomorphism of Z_2 cohomology, and thus giving a two primary homotopy equivalence. **

Note: This result has previously been proved by geometric arguments. This proof, using only cohomological methods, is essentially that of Wall [132].

Chapter X

Special Unitary Cobordism

Having already built up the machinery to study special unitary cobordism, the 'oriented' analogue of complex cobordism, one may obtain much of the structure in fairly easy fashion. The one new feature which arises is the use of KO-theory characteristic numbers.

Since SU cobordism is the (B,f) theory in which $B_{2r} = B_{2r+1} = BSU_r$, one has the determination theorem:

$$\Omega_n^{SU} \cong \lim_{r\to\infty} \pi_{n+2r}(TBSU_r,\infty).$$

The primary requisite for the study of these groups is then a knowledge of the structure of BSU. In cohomology, this is provided by:

<u>Lemma</u>: The cohomology ring $H^*(BSU_n;Z)$ is the integral polynomial ring on the Chern classes $c_i(\tilde{\gamma}^n)$, $1 < i \leq n$, where $\tilde{\gamma}^n$ is the universal oriented complex n-plane bundle over BSU_n.

<u>Proof</u>: Let $f_n : BSU_n \longrightarrow BU_n$ be the map classifying $\tilde{\gamma}^n$. Then $f_n^* : H^*(BU_n;Z) = Z[c_i | 1 \leq i \leq n] \longrightarrow H^*(BSU_n;Z)$ sends c_i to $c_i(\tilde{\gamma}^n)$. Since $c_1(\tilde{\gamma}^n) = c_1(\det\tilde{\gamma}^n) = 0$ because $\det\tilde{\gamma}^n$ is trivial, this induces

$$f_n^* : P_n = Z[c_i | 1 < i \leq n] \longrightarrow H^*(BSU_n;Z).$$

To see that this is monic, let $g_n : BU_{n-1} \longrightarrow BSU_n$ be a map classifying the orientable bundle $\gamma^{n-1} \oplus \det\gamma^{n-1}$, so that $g_n^* f_n^*(c_i)$ is $c_i - c_1 c_{i-1}$ if $i < n$ and $c_1 c_{n-1}$ if $i = n$. Thus the elements $g_n^* f_n^*(c_i)$, $1 < i \leq n$, are algebraically independent in $Z[c_i | 1 \leq i \leq n-1]$ and so f_n^* is monic on P_n.

To see that this is epic, one may use induction on n exactly as in the study of $H^*(BSO_n;Z_2)$. To begin the induction, BSU_1 is a point and $BSU_2 = BSp_1 = \mathbb{H}P(\infty)$ for which the result is known. **

Since the spectrum $\underset{\sim}{TBU}$ is oriented for integral cohomology, the same holds for the spectrum $\underset{\sim}{TBSU}$ and one has:

<u>Proposition</u>: *The groups* Ω_n^{SU} *are finitely generated and* $\Omega_*^{SU} \otimes Q$ *is a rational polynomial algebra on classes* x_{2i} *of dimension* $2i$, $i > 1$.

<u>Proof</u>: The standard methods give everything here, proving that $\Omega_*^{SU} \otimes Q \cong H_*(BSU;Q)$. To see that this is a polynomial ring, consider the submanifold M^{2n} of $\mathbb{C}P(n+1)$ dual to $\det\tau$. Then $c(M^{2n}) = (1+\alpha)^{n+2}/(1+(n+2)\alpha)$ so $s_n(c(\tau))[M^{2n}] = (n+2)[(n+2)\alpha^n - \{(n+2)\alpha\}^n]\alpha[\mathbb{C}P(n+1)] = (n+2)^2 - (n+2)^{n+1}$ which is nonzero if $n \neq 1$. Thus $\Omega_*^{SU} \otimes Q$ maps onto a polynomial subalgebra of $\Omega_*^U \otimes Q$. **

<u>Corollary</u>: *The kernel of the forgetful homomorphism* $F_* : \Omega_*^{SU} \longrightarrow \Omega_*^U$ *is precisely the torsion subgroup.*

For the odd primary structure one has as always that $\tilde{H}^*(\underset{\sim}{TBSU};Z_p)$ is a connected coalgebra over Z_p with counit $U \in \tilde{H}^0(\underset{\sim}{TBSU};Z_p)$. Since the cohomology is all even dimensional, this is an $\mathcal{A}_p/(Q_0)$ module and one has:

<u>Lemma</u>: *The homomorphism* $\nu : \mathcal{A}_p/(Q_0) \longrightarrow \tilde{H}^*(\underset{\sim}{TBSU};Z_p) : a \longrightarrow a(U)$ *is monic if* p *is an odd prime.*

<u>Proof</u>: From the map $g_n : BU_{n-1} \longrightarrow BSU_n$, one has $\tilde{H}^*(\underset{\sim}{TBSU}_n;Z_p)$ identified with $g_n^*(H^*(BSU_n;Z_p))\cdot c_{n-1}c_1 \subset H^*(BU_{n-1};Z_p)$. In stable dimensions $\tilde{H}^*(\underset{\sim}{TBU}_{n-1};Z_p) \cong H^*(BU_{n-1};Z_p)\cdot c_{n-1} \subset H^*(BU_{n-1};Z_p)$ is a free $\mathcal{A}_p/(Q_0)$ module. Since $c_1(\mathbb{C}P(1)) = 2 \not\equiv 0(p)$, one generator of this module may be taken to be c_1c_{n-1}, and thus ν is monic. **

Remarks: 1) An alternate proof is obtainable by letting $h : BSp \longrightarrow BSU$ classify the universal bundle, since an Sp bundle is SU, and $(Th)^*\nu$ is monic, so ν is also.

2) For $p = 2$, doubling gives an isomorphism of $\tilde{H}^*(\underset{\sim}{TBSO};Z_2)$ and $\tilde{H}^*(\underset{\sim}{TBSU};Z_2)$. Thus $\tilde{H}^*(\underset{\sim}{TBSU};Z_2)$ is a direct sum of copies of $\mathcal{A}_2/(Sq^1)$ and $(\mathcal{A}_2/(Sq^1))/(\mathcal{A}_2/(Sq^1))\cdot Sq^2 \tilde{=} \mathcal{A}_2/(Sq^1) + \mathcal{A}_2 Sq^2$. This is not a particularly useful description.

Corollary: All torsion in Ω_*^{SU} is 2 primary.

To complete the calculation of the odd primary structure, let $\Omega_*^{TSU} \subset \Omega_*^U$ be the set of cobordism classes for which all Chern numbers divisible by c_1 are zero. It is clear, since c_1 is zero for SU manifolds, that $F_*\Omega_*^{SU} \subset \Omega_*^{TSU}$. One also has:

Lemma 1: (Conner and Floyd [39], (11.5)) $2\Omega_*^{TSU} \subset F_*\Omega_*^{SU}$.

Proof: If M^n has all Chern numbers with c_1 as a factor zero, let $N^n \subset M^n \times \mathbb{C}P(1)$ be the submanifold dual to c_1 (or $\det\tau$), so N^n has an SU structure. One has

$$c(N) = \frac{c(M)\cdot(1+\bar{\alpha})^2}{1+c_1(M)+2\bar{\alpha}}$$

giving for characteristic numbers

$$c_\omega[N] = (c_\omega(M) + \bar{\alpha}u_\omega + (c_1(M)+2\bar{\alpha})v_\omega)(c_1(M)+2\bar{\alpha})[M \times \mathbb{C}P(1)],$$

$$= c_\omega(M)[M]\cdot 2\bar{\alpha}[CP(1)],$$

$$= 2c_\omega[M]$$

where u_ω, v_ω are polynomials in the $c_i(M)$ and $\bar{\alpha}$ (Note: These additional terms all vanish since $\bar{\alpha}^2 = 0$ and Chern numbers of M with a factor c_1 are zero). Thus $2[M] = [N] \in F_*\Omega_*^{SU}$. **

Insofar as odd primary structure is concerned, this identifies Ω_*^{SU} with Ω_*^{TSU}. For example, all odd primary relations among the Chern numbers of SU manifolds follow from the vanishing of c_1 and the K-theory relations for stably almost complex manifolds. The multiplicative structure follows from:

Proposition: Let p be an odd prime. There exist SU manifolds $M_i^p \in \Omega_{2i}^{SU}$, $i \geq 2$, such that $\rho_p(\tau(M_i^p))$, the mod p reduction of $\rho(\tau(M_i^p)) = \sum s_\omega(e)\mathscr{S}[M]\cdot\alpha_\omega$, has largest monomial

(1) α_i if $i \neq p^s$, p^s-1 for any s,

(2) $\alpha^p_{p^{s-1}}$ if $i = p^s$ for some s, and

(3) $\alpha^p_{p^{s-1}-1}$ if $i = p^s-1$ for some s.

Proof: For any almost complex M, $\partial M \subset M$ denotes the submanifold dual to c_1, which admits an SU structure.

For part (1), one has $s_i(e)\mathscr{S}[M] = s_i(c)[M]$ if $\dim M = 2i$, and it suffices to find an $\omega \in \pi(i+1)$ for which $s_i(c)[\partial\mathbb{C}P(\omega)]$ is nonzero mod p. One choice of such ω is:

(a) i, $i+1 \not\equiv 0$ (p) : $\omega = (1,1,i-1)$;

(b) $i+1 = p^r(pu+v)$, $r > 0$, $0 < v < p$, $i \neq p^s-1$:

(1) $u > 0 : \omega = (p^r v, p^{r+1}u)$,

(2) $u = 0 : \omega = (p^r, p^r(v-1))$; and

(c) $i = p^r(pu+v)$, $r > 0$, $0 < v < p$, $i \neq p^s$:

(1) $u > 0 : \omega = (1, p^r v, p^{r+1}u)$,

(2) $u = 0 : \omega = (1, p^r, p^r(v-1))$.

For part (2), $i = p^s$ and let $M_i^p = \partial(\mathbb{C}P(1) \times \mathbb{C}P(p^{s-1}) \times \ldots \times \mathbb{C}P(p^{s-1}))$ with p copies of $\mathbb{C}P(p^{s-1})$. The total Chern class of M_i^p is $(1+x)^2 \cdot \prod_1^p (1+x_j)^{p^{s-1}+1}/(1+2x+(p^{s-1}+1)\sum x_j)$ and M_i^p is dual to $2x+(p^{s-1}+1)\sum x_j$. Working mod p, the terms $\sum x_j$ give p equal terms in characteristic numbers, so for mod p numbers this is the same as if $c(M_i^p) = \prod_1^p (1+x_j)^{p^{s-1}+1}$ and M_i^p were dual to $2x$. Thus the mod p Chern numbers of M_i^p are the same as those of $2\mathbb{C}P(p^{s-1})^p$. For $\omega \in \pi(i)$, $s_\omega(e)\mathcal{S}[M_i^p] = s_\omega(c)[M_i^p] \equiv 2s_\omega(c)[\mathbb{C}P(p^{s-1})^p] = 2s_\omega(e)\mathcal{S}[\mathbb{C}P(p^{s-1})^p]$, and since $\rho_p[\mathbb{C}P(p^{s-1})]$ has largest monomial $\alpha_{p^{s-1}}$, the result follows.

For part (3), $i = p^s-1$, $s \geq 1$ and let $M = M_i^p = \partial(CP(p^{s-1}) \times \ldots \times CP(p^{s-1}))$ with p copies of $\mathbb{C}P(p^{s-1})$, with $H \subset \mathbb{C}P(p^{s-1}) \times \ldots \times \mathbb{C}P(p^{s-1})$ dual to $\xi_1 \otimes \ldots \otimes \xi_p$ as considered in Chapter VII. (M is dual to $(\xi_1 \otimes \ldots \otimes \xi_p)^{p^{s-1}+1}$). M has total Chern class $\prod_1^p (1+x_j)^{p^{s-1}+1}/(1+(p^{s-1}+1)\sum x_j)$ and is dual to $(p^{s-1}+1)\sum x_j$, while H has total Chern class $\prod_1^p (1+x_j)^{p^{s-1}+1}/(1+\sum x_j)$ and is dual to $\sum x_j$. Thus for $s \geq 2$, $c_\omega(M) = c_\omega(H) + p^{s-1}v_\omega$ where v_ω is symmetric in the x_j, i.e. $v_\omega = a \sum x_1^{p^{s-1}} \ldots x_j^{p^{s-1}-1} \ldots x_p^{p^{s-1}}$, and multiplying by $(p^{s-1}+1)\sum x_j$ and evaluating on the product of $\mathbb{C}P$'s gives $c_\omega[M] = (p^{s-1}+1)c_\omega[H] \bmod p^2$. For $\mu \in \pi(k)$, $k \geq p^s-p+1$, $\dim V = p^s-1$, $s_\mu(e)\mathcal{S}[V] = \sum (a_\omega/b_\omega)c_\omega[V]$, $a_\omega, b_\omega \in Z$, $b_\omega \not\equiv 0\ (p^2)$, so $s_\mu(e)\mathcal{S}[M] = (p^{s-1}+1)s_\mu(e)\mathcal{S}[H] \bmod p$. Thus M has the same largest monomial as H, which is as given. For $s = 1$, $c_\omega[M] \equiv 0 \bmod p$ by symmetry in the x's, so $s_\mu(e)\mathcal{S}[M] = \sum (a_\omega/b_\omega)c_\omega[M] \equiv 0 \bmod p$ if $\mu \in \pi(k)$, $k > 0$ (since then $b_\omega \not\equiv 0\ (p)$). Thus one needs only $\mathcal{S}[M] \not\equiv 0\ (p)$. However,

$$\mathscr{S}[M] = \prod_{j=1}^{p} (x_j/(e^{x_j}-1))^2 \cdot (e^{2\Sigma x_j}-1)[CP(1)^p],$$

$$= \left(\frac{1}{2\pi i}\oint \frac{e^{2z}dz}{(e^z-1)^2}\right)^p - \left(\frac{1}{2\pi i}\oint \frac{dz}{(e^z-1)^2}\right)^p,$$

$$= \left(\frac{1}{2\pi i}\oint \frac{(u+1)^2du}{u^2(u+1)}\right)^p - \left(\frac{1}{2\pi i}\oint \frac{du}{u^2(u+1)}\right)^p, \qquad (u = e^z-1),$$

$$= 1 - (-1)^p,$$

$$= 2,$$

$$\neq 0\ (p). \quad **$$

Corollary: $\Omega_*^{SU} \otimes Z[1/2]$ is a polynomial ring over $Z[1/2]$ on classes x_{2i}, $i > 1$.

Notes: 1) The manifolds M_i^p given in the proposition provide mod p generators of Ω_*^{SU}.

2) The odd primary structure of Ω_*^{SU} was first calculated by Novikov [93] using the Adams spectral sequence method.

The calculation of the 2 primary structure was done by Conner and Floyd [39], whose methods are used here.

One has exact sequences

$$\begin{array}{ccc} \Omega_*^{SU} & \xrightarrow{t} & \Omega_*^{SU} \\ & {}_{\partial}\nwarrow \quad \swarrow_{\rho} & \\ & \mathcal{W}_*(\mathbb{C},2) & \end{array}$$

and

$$0 \longrightarrow \mathcal{W}_*(\mathbb{C},2) \xrightarrow{F_*} \Omega_*^U \xrightarrow{d} \Omega_*^U \longrightarrow 0.$$

Since $\mathcal{W}_{2j-1}(\mathbb{C},2) \subset \Omega^U_{2j-1} = 0$, this gives an exact sequence

$$0 \longrightarrow \Omega^{SU}_{2j-1} \xrightarrow{t} \Omega^{SU}_{2j} \xrightarrow{\rho} \mathcal{W}_{2j}(\mathbb{C},2) \xrightarrow{\partial} \Omega^{SU}_{2j-2} \xrightarrow{t} \Omega^{SU}_{2j-1} \longrightarrow 0.$$

Lemma 2: $\Omega^{SU}_0 \cong Z$, $\Omega^{SU}_1 \cong Z_2$, and $\Omega^{SU}_2 \cong Z_2$. If $\theta \in \Omega^{SU}_1$ is the nonzero class, then θ^2 is the nonzero class in Ω^{SU}_2.

Proof: One has $0 \longrightarrow \mathcal{W}_0(\mathbb{C},2) \xrightarrow{F_*} \Omega^U_0 = Z \xrightarrow{d} \Omega^U_{-4} = 0 \longrightarrow 0$ and $0 \longrightarrow \mathcal{W}_2(\mathbb{C},2) \xrightarrow{F^*} \Omega^U_2 = Z \xrightarrow{d} \Omega^U_{-2} = 0 \longrightarrow 0$. Then $\Omega^{SU}_0 \xrightarrow{\rho} \mathcal{W}_0(\mathbb{C},2) \xrightarrow{F_*} \Omega^U_0 = Z$ are isomorphisms. Since $\mathcal{W}_2(\mathbb{C},2) \cong Z$ generated by $\mathbb{C}P(1)$, with $\partial\mathbb{C}P(1) = 2$, the homomorphism $Z = \mathcal{W}_2(\mathbb{C},2) \xrightarrow{\partial} \Omega^{SU}_0 = Z$ has cokernel $Z_2 \cong \Omega^{SU}_1$ with $\theta = t(1)$ the nonzero element. Since ∂ is monic on $\mathcal{W}_2(\mathbb{C},2)$, this also gives an isomorphism $t : \Omega^{SU}_1 \longrightarrow \Omega^{SU}_2$. Since the homomorphism t is multiplication by $\theta = t(1)$, this gives $\Omega^{SU}_2 \cong Z_2$ with nonzero element θ^2. **

Proposition: All torsion in Ω^{SU}_* has order 2.

Proof: Since $t : \Omega^{SU}_{2j-2} \longrightarrow \Omega^{SU}_{2j-1}$ is epic and given by multiplication by θ, Ω^{SU}_{2j-1} consists of elements of order 2. The torsion subgroup of Ω^{SU}_{2j} is the kernel of the composite $\Omega^{SU}_{2j} \xrightarrow{\rho} \mathcal{W}_{2j}(\mathbb{C},2) \xrightarrow{F_*} \Omega^U_{2j}$, but F_* is monic, so $\text{Torsion}(\Omega^{SU}_{2j}) = \text{kernel}(\rho) = \text{image}(t)$, which consists of elements of order 2. **

Lemma 3: $\Omega^{SU}_3 = 0$.

Proof: (Due to Lashof and Rothenberg). One has the forgetful homomorphism $S_* : \Omega^{fr}_* \longrightarrow \Omega^{SU}_*$ induced by the inclusion $j : \underset{\sim}{S} \longrightarrow \underset{\sim}{TBSU}$ with $j^* : \tilde{H}^*(\underset{\sim}{TBSU};Z) \longrightarrow \tilde{H}^*(\underset{\sim}{S};Z)$ an isomorphism in dimensions less than 4 and epic in dimension 4, so the homotopy map S_* is an isomorphism in dimensions less than 3 and epic in dimension 3. Let $\alpha \in \Omega^{fr}_1$ with $S_*(\alpha) = \theta$. Since $2\alpha = 0$, $2\alpha^3 = 0$, and since $\Omega^{fr}_3 = Z_{24}$, $\alpha^3 = 2\beta$. Thus $\theta^3 = S_*(\alpha^3) = 2S_*(\beta)$ but all torsion in Ω^{SU}_* has order 2, so $2S_*(\beta) = 0$. Finally $t : \Omega^{SU}_2 \longrightarrow \Omega^{SU}_3$ is epic, so $\theta^3 = 0$ implies $t\theta^2 = 0$ and $\Omega^{SU}_3 = 0$. **

<u>Note</u>: An alternate proof may be given as follows. From $0 \longrightarrow \mathcal{W}_4(\mathbb{C},2) \longrightarrow \Omega_4^U = Z \oplus Z \xrightarrow{d} \Omega_0^U = Z \longrightarrow 0$, $\mathcal{W}_4(\mathbb{C},2) \cong Z$ and a generator is represented by $9\mathbb{C}P(1)^2 - 8\mathbb{C}P(2)$ with $c_1^2 = 0$, $c_2 = 12$, and with $\mathcal{S}$ number 1. For a 4 dimensional SU manifold, the $\mathcal{S}$ number must be even (KO theory characteristic number argument to be given later), so that $\rho : \Omega_4^{SU} \longrightarrow \mathcal{W}_4(\mathbb{C},2)$ is not epic. Thus $\partial : \mathcal{W}_4(\mathbb{C},2) \longrightarrow \Omega_2^{SU} \cong Z_2$ is epic and $t : \Omega_2^{SU} \longrightarrow \Omega_3^{SU}$ is both epic and zero, giving $\Omega_3^{SU} = 0$. **

Considering the exact triangle

$$\begin{array}{ccc} \Omega_*^{SU} & \xrightarrow{t} & \Omega_*^{SU} \\ {\scriptstyle\partial}\nwarrow & & \swarrow{\scriptstyle\rho} \\ & \mathcal{W}_*(\mathbb{C},2) & \end{array}$$

as an exact couple, one has a derived couple

$$\begin{array}{ccc} \text{Image } t & \xrightarrow{t} & \text{Image } t \\ {\scriptstyle\partial'}\nwarrow & & \swarrow{\scriptstyle\rho'} \\ & H(\mathcal{W}) & \end{array}$$

where $H(\mathcal{W})$ is the homology of $\mathcal{W}_*(\mathbb{C},2)$ with respect to the differential $\rho\partial : \mathcal{W}_*(\mathbb{C},2) \longrightarrow \mathcal{W}_*(\mathbb{C},2)$. This gives an exact sequence

$$\cdots \longrightarrow t(\Omega_{2j-1}^{SU}) \xrightarrow{t} t(\Omega_{2j}^{SU}) \xrightarrow{\rho'} H_{2j}(\mathcal{W}) \xrightarrow{\partial'} t(\Omega_{2j-3}^{SU}) \xrightarrow{t} t(\Omega_{2j-2}^{SU}) \longrightarrow \cdots .$$

Now $\Omega_{2j-1}^{SU} = t(\Omega_{2j-2}^{SU})$ since t is epic so $t : t(\Omega_{2j-1}^{SU}) \longrightarrow t(\Omega_{2j}^{SU})$ has image $t^3(\Omega_{2j-2}^{SU})$, but $t^3 = 0$ since $\theta^3 = 0$. This sequence then splits up as

$$\begin{array}{ccccccccc} 0 & \longrightarrow & t(\Omega_{2j}^{SU}) & \longrightarrow & H_{2j}(\mathcal{W}) & \longrightarrow & t(\Omega_{2j-3}^{SU}) & \longrightarrow & 0 . \\ & & \| & & & & \wr\| & & \\ & & \Omega_{2j+1}^{SU} & & & & \Omega_{2j-3}^{SU} & & \end{array}$$

Since $\text{kernel}(\rho\partial) = \Omega_*^{TSU}$, while Lemma 1 proved that $2\Omega_*^{TSU} \subset \text{image}(\rho\partial)$, $H_{2j}(\mathcal{W})$ is a Z_2 vector space, and the above sequence splits to give

Lemma 4: $H_{2j}(\mathcal{W}) \cong \Omega_{2j+1}^{SU} \oplus \Omega_{2j-3}^{SU}$. **

Since the torsion structure of Ω_*^{SU} is entirely determined by the groups Ω_{2j+1}^{SU}, knowing $H_*(\mathcal{W})$ would give the torsion structure.

To begin this computation, one has an exact sequence

$$0 \longrightarrow \mathcal{W}_*(\mathbb{C},2) \xrightarrow{2} \mathcal{W}_*(\mathbb{C},2) \longrightarrow \mathcal{W}_*(\mathbb{C},2) \otimes Z_2 \longrightarrow 0$$

giving a homology exact triangle

$$\begin{array}{ccc} H_*(\mathcal{W}) & \xrightarrow{\quad 2 \quad} & H_*(\mathcal{W}) \\ & \nwarrow \quad \swarrow & \\ & H_*(\mathcal{W} \otimes Z_2) & \end{array}$$

but every element in $H_*(\mathcal{W})$ has order 2, so one has

$$0 \longrightarrow H_{2k}(\mathcal{W}) \longrightarrow H_{2k}(\mathcal{W} \otimes Z_2) \longrightarrow H_{2k-2}(\mathcal{W}) \longrightarrow 0.$$

From Chapter VIII one has:

Assertion: $\mathcal{W}_*(\mathbb{C},2) \otimes Z_2$ is a polynomial algebra over Z_2 on classes z_{2n} for $n \neq 2$. The boundary homomorphism is given by $\partial z_2 = 0$, $\partial z_{4n} = z_{4n-2}$ if $n \geq 2$, and satisfies $\partial(ab) = (\partial a)b + a(\partial b) + z_2(\partial a)(\partial b)$.

Since $\partial z_2 = 0$, $\partial(z_2 a) = z_2(\partial a)$, and the ideal $W'' \subset \mathcal{W}_*(\mathbb{C},2) \otimes Z_2$ generated by z_2 is a subcomplex. From the short exact sequence

$$0 \longrightarrow W'' \longrightarrow \mathcal{W}_*(\mathbb{C},2) \otimes Z_2 \longrightarrow W' = \mathcal{W}_*(\mathbb{C},2) \otimes Z_2/W'' \longrightarrow 0$$

one has an exact sequence

$$\begin{array}{ccc} H_*(W'') & \longrightarrow & H_*(\mathcal{W} \otimes Z_2) \\ & \nwarrow \quad \swarrow & \\ & H_*(W'). & \end{array}$$

Since $W' \cong Z_2[z_{2n} | n \geq 2]$ with $\partial z_{4n} = z_{4n-2}$ and $\partial(xy) = x(\partial y) + (\partial x)y$, $H_*(W') \cong Z_2[z_{4n}^2]$.

From the product formula for ∂ in $\mathcal{W}_*(\mathbb{C},2) \otimes Z_2$, $H_*(\mathcal{W} \otimes Z_2)$ is a ring, and the homomorphism to $H_*(W')$ is a ring homomorphism. Now

$$\partial[(z_{4n})^2 + z_2 z_{4n-2} z_{4n}] = (2z_{4n} z_{4n-2} + z_2 z_{4n-2}^2) + z_2 z_{4n-2}^2 = 0$$

so the classes $h_{8n} = z_{4n}^2 + z_2 z_{4n-2} z_{4n}$ in $H_*(\mathcal{W} \otimes Z_2)$ map onto the polynomial generators of $H_*(W')$. This splits the exact sequence to give

$$0 \longrightarrow H_*(W'') \longrightarrow H_*(\mathcal{W} \otimes Z_2) \longrightarrow H_*(W') \longrightarrow 0.$$

From the formula $\partial(z_2 x) = z_2(\partial x)$ previously noted, one has $H_*(W'') = z_2 \cdot H_*(\mathcal{W} \otimes Z_2)$, giving:

Lemma 5: $H_*(\mathcal{W} \otimes Z_2)$ is a polynomial algebra over Z_2 with generators h_2 (represented by z_2) and h_{8n}, $n \geq 2$, (represented by $(z_{4n})^2 + z_2 z_{4n-2} z_{4n}$).

Returning now to $\mathcal{W}_*(\mathbb{C},2)$, one notes that the generators z_{2n}, $n \neq 2$, for $\mathcal{W}_*(\mathbb{C},2) \otimes Z_2$ are represented by classes $z'_{2n} \in \mathcal{W}_*(\mathbb{C},2)$ with $\rho\partial z'_{4n} = z'_{4n-2}$ if $n \geq 2$, and $\rho\partial z'_2 = 2$. Using the product in Ω_*^U and the extension ∂' of $\rho\partial$ to Ω_*^U, one also has

$$\Phi(a \cdot b) = a \cdot b + 2[V^4] \cdot \partial' a \cdot \partial' b$$

and

$$\partial'(a \cdot b) = a(\partial' b) + (\partial' a)b - z'_2(\partial' a)(\partial' b)$$

for $a,b \in \mathcal{W}_*(\mathbb{C},2)$, $[V^4]$ being given by $z_2'^2 - \mathbb{C}P(2)$.

If M has all numbers divisible by c_1 zero $(\partial'[M] = 0)$ and X is any stably almost complex manifold, then the submanifold of $M \times X$ dual to

c_1 has the same Chern numbers as $M \times N$ where $N \subset X$ is dual to c_1. Thus $\partial'([M]\cdot[X]) = [M]\cdot\partial'([X])$. In particular, $H_*(\mathcal{W})$ is a ring and the homomorphism into $H_*(\mathcal{W} \otimes Z_2)$ is a ring homomorphism.

One then has:

Lemma 6: The homomorphism $H_*(\mathcal{W}) \longrightarrow H_*(\mathcal{W} \otimes Z_2)$ maps $H_*(\mathcal{W})$ isomorphically onto the subalgebra generated by $(h_2)^2$ and h_{8k}, $k \geq 2$. Thus $H_*(\mathcal{W})$ is a polynomial algebra on classes c_4 and c_{8k}, $k \geq 2$.

Proof: $\Phi(\mathbb{C}P(1)^2) = \mathbb{C}P(1)^2 + 8[V^4] = 9\mathbb{C}P(1)^2 - 8\mathbb{C}P(2) \in \mathcal{W}_*(\mathbb{C},2)$ has all Chern numbers divisible by c_1 zero ($c_1^2 = 0$, $c_2 = 12$), so is a cycle in $\mathcal{W}_*(\mathbb{C},2)$. This class represents the product z_2^2 in $\mathcal{W}_*(\mathbb{C},2) \otimes Z_2$ or $(h_2)^2$ in $H_*(\mathcal{W} \otimes Z_2)$.

For $n \geq 2$,

$$\begin{aligned}\partial'\Phi(z_{4n}'^2) &= \partial'(z_{4n}'^2 + 2[V^4](\partial' z_{4n}')^2), \\ &= (2z_{4n}'\cdot z_{4n-2}' - z_2'\cdot z_{4n-2}'^2) + 2\partial'[V^4]\cdot z_{4n-2}'^2, \\ &= 2z_{4n}'\cdot z_{4n-2}' - z_2' z_{4n-2}'^2\end{aligned}$$

for $\partial'[V^4]$ is a 2 dimensional class with $c_1(\partial'[V^4]) = 0$, so $\partial'[V^4] = 0$. Also

$$\begin{aligned}\partial'\Phi(z_2' z_{4n}') &= \partial'(z_2' z_{4n}' + 4[V^4]z_{4n-2}'), \\ &= z_2' z_{4n-2}' + 2z_{4n}' - 2z_2' z_{4n-2}' + 4\partial'[V^4]z_{4n-2}', \\ &= 2z_{4n}' - z_2' z_{4n-2}'.\end{aligned}$$

Thus

$$\Phi(z_{4n}'^2) - z_{4n-2}' \, \Phi\,(z_2' z_{4n}') \in \mathcal{W}_*(\mathbb{C},2)$$

is a cycle $(\partial' z'_{4n-2} = 0)$ and reduces to $(z_{4n})^2 + z_2 z_{4n-2} z_{4n} \in \mathcal{W}_*(\mathbb{C},2) \otimes Z_2$ so to $h_{8n} \in H_*(\mathcal{W} \otimes Z_2)$.

Thus $H_*(\mathcal{W})$ maps onto the asserted subring of $H_*(\mathcal{W} \otimes Z_2)$, while the exact sequence $0 \longrightarrow H_*(\mathcal{W}) \longrightarrow H_*(\mathcal{W} \otimes Z_2) \longrightarrow H_*(\mathcal{W}) \longrightarrow 0$ together with a dimension count make this an isomorphism. **

Returning to the isomorphism $H_{2n}(\mathcal{W}) \cong \Omega^{SU}_{2n+1} \oplus \Omega^{SU}_{2n-3}$, one has $H_{8k+2}(\mathcal{W}) \cong H_{8k+6}(\mathcal{W}) \cong 0$ so $\Omega^{SU}_{8k+3} \cong \Omega^{SU}_{8k+7} = 0$. Since $H_{8k+4}(\mathcal{W}) = H_{8k}(\mathcal{W})$ (via multiplication by c_4), $\Omega^{SU}_{8k+1} \oplus \Omega^{SU}_{8k-3} \cong \Omega^{SU}_{8k+5} \oplus \Omega^{SU}_{8k+1}$ or $\Omega^{SU}_{8k+5} \cong \Omega^{SU}_{8k-3}$, and induction on k, beginning with $\Omega^{SU}_{-3} = 0$ gives $\Omega^{SU}_{8k+5} = 0$. This then gives $\Omega^{SU}_{8k+1} \cong H_{8k}(\mathcal{W})$, and one has the result of Conner and Floyd [39] (18.3):

<u>Theorem</u>: The torsion of Ω^{SU}_* is given as follows: $\text{Torsion}(\Omega^{SU}_n) = 0$ unless $n = 8k+1$ or $8k+2$, in which case $\text{Torsion}(\Omega^{SU}_n)$ is a Z_2 vector space of rank the number of partitions of k.

<u>Proof</u>: Since $H_{8*}(\mathcal{W})$ is the Z_2 polynomial algebra on the c_{8k}, $k \geq 2$, and c_4^2, with $\Omega^{SU}_{8k+1} \cong H_{8k}(\mathcal{W})$, the odd groups Ω^{SU}_n satisfy the given conditions. Since $t : \Omega^{SU}_{2n-1} \longrightarrow \text{Torsion}(\Omega^{SU}_{2n})$ is an isomorphism, the torsion subgroup is known in the even dimensional case also. **

Returning to the exact sequence

$$0 \longrightarrow \Omega^{SU}_{2j-1} \xrightarrow{t} \Omega^{SU}_{2j} \xrightarrow{\rho} \mathcal{W}_{2j}(\mathbb{C},2) \xrightarrow{\partial} \Omega^{SU}_{2j-2} \xrightarrow{t} \Omega^{SU}_{2j-1} \longrightarrow 0$$

one has:

<u>Theorem</u>: *The homomorphism* $\rho : \Omega^{SU}_{2j} \longrightarrow \mathcal{W}_{2j}(\mathbb{C},2)$ has image equal to the group $Z(\mathcal{W}_{2j}(\mathbb{C},2),\rho\partial)$ of cycles if $2j \not\equiv 4 \ (8)$ and has image equal to the group $B(\mathcal{W}_{2j}(\mathbb{C},2),\rho\partial)$ of boundaries if $2j \equiv 4 \ (8)$.

Proof: If $2j \not\equiv 4\ (8)$, $2j-2 \neq 8k+2$ so Ω^{SU}_{2j-2} is torsion free and $\rho : \Omega^{SU}_{2j-2} \longrightarrow \mathcal{W}_{2j-2}(\mathbb{C},2)$ is monic. Thus $\text{kernel}(\partial : \mathcal{W}_{2j}(\mathbb{C},2) \longrightarrow \Omega^{SU}_{2j-2}) = \text{kernel}(\rho\partial : \mathcal{W}_{2j}(\mathbb{C},2) \longrightarrow \mathcal{W}_{2j-2}(\mathbb{C},2)) = Z(\mathcal{W}_{2j}(\mathbb{C},2),\rho\partial)$. If $2j = 8k+4$, one has $\partial : \mathcal{W}_{2j}(\mathbb{C},2) \longrightarrow \Omega^{SU}_{2j-2}$ epic, while $\rho : \Omega^{SU}_{2j-2} \longrightarrow \mathcal{W}_{2j-2}(\mathbb{C},2)$ has kernel isomorphic to $H_{2j-4}(\mathcal{W}) \cong H_{2j}(\mathcal{W})$. Thus $(\text{kernel}\rho\partial/\text{kernel}\partial)_{2j} \cong H_{2j}(\mathcal{W})$, and $\rho\Omega^{SU}_{2j} = B(\mathcal{W}_{2j}(\mathbb{C},2),\rho\partial)$. **

Corollary: Let $\partial' : \Omega^{U}_{*} \longrightarrow \Omega^{U}_{*}$ be the homomorphism sending M into the submanifold $N \subset M$ dual to $\det\tau_M$. The forgetful homomorphism $F_* : \Omega^{SU}_{*} \longrightarrow \Omega^{U}_{*}$ has image containing $\text{image}(\partial')$. There exist SU manifolds W^{8k}, $k \geq 1$, such that $(\text{image } F_*)/(\text{image } \partial') \cong Z_2[W^{8k}]$. Every torsion element of Ω^{SU}_{*} is uniquely expressible in the form $V^{8n}\cdot\theta$ or $V^{8n}\cdot\theta^2$ where V^{8n} is a polynomial in the W^{8k} with coefficients 0 or 1.

Proof: Dualizing $\det\tau_M$ gives a submanifold admitting an SU structure, so image $\partial' \subset$ image F_*. Then image $F_* = \rho\Omega^{SU}_{*} \subset \mathcal{W}_*(\mathbb{C},2)$ is described in the theorem as $Z(\mathcal{W})$ (or $B(\mathcal{W})$ if $\dim \equiv 4\ (8)$) while image $\partial' = B(\mathcal{W})$. Thus $(\text{image } F_*)/(\text{image } \partial')_n \cong H_n(\mathcal{W})$ if $n \not\equiv 4\ (8)$ and is zero if $n \equiv 4\ (8)$, which proves the polynomial structure. The torsion group Ω^{SU}_{8k+1} is $t\Omega^{SU}_{8k}$, and t annihilates image ∂, while $t^2\Omega^{SU}_{8k} = \text{Torsion}(\Omega^{SU}_{8k+2})$ giving the structure of the torsion. **

In order to examine the structure of Ω^{SU}_{*} more closely, it seems necessary to consider KO-theory characteristic numbers. Briefly, there is a multiplicative cohomology theory KO^* indexed by the integers (positive and negative) for which $KO^0(X)$ is the Grothendieck group of isomorphism classes of real vector bundles over X, and $KO^{-4}(X)$ is the Grothendieck group of isomorphism classes of quaternionic vector bundles over X (denoted $KSp(X)$).

This cohomology theory is periodic of period 8, the periodicity isomorphism $\bar{p} : KO^i(X) \longrightarrow KO^{i-8}(X)$ being given by multiplication by a generator $\bar{p}(1) \in KO^{-8}(pt) \cong \widetilde{KO}^0(S^8) \cong Z$.

In order to describe elements in $KO^*(X)$ geometrically, it is convenient to consider a complex vector bundle V over X together with an automorphism $J : V \longrightarrow V$ such that $Ji = -iJ$ (J is conjugate linear). Then (V,J) is

1) Symplectic if $J^2 = -1$,

2) Real if $J^2 = 1$.

(That this is justifiable follows from the fact that if $J^2 = 1$, then $V = (\frac{1+J}{2})V \oplus (\frac{1-J}{2})V$, decomposing as the $+1$ and -1 eigenspaces of J, and the summands are interchanged under multiplication by i. Thus V is isomorphic to the complexification of $(\frac{1+J}{2})V$.)

Being given two such pairs (V_1,J_1) and (V_2,J_2), the tensor product $(V_1 \otimes_{\mathbb{C}} V_2)$ admits a conjugate linear automorphism $J_1 \otimes J_2$ and $(V_1 \otimes_{\mathbb{C}} V_2, J_1 \otimes J_2)$ is:

1) Real if J_1 and J_2 are both either real or symplectic;

2) Symplectic if one is symplectic and the other is real.

This describes the product which relates $KO(X)$ and $KSp(X)$ in $KO^*(X)$. (<u>Note</u>: If (V_1,J_1) and (V_2,J_2) are symplectic, $J_1 \otimes J_2$ acts on $V_1 \otimes_{\mathbb{C}} V_2$, thought of as complex vector bundles, and the real eigenbundle is $V_1 \otimes_{\mathbb{H}} V_2$, where V_1 is made a right vector bundle over $\mathbb{H}$ by means of conjugation.)

<u>Lemma</u>: $KO^*(\mathbb{H}P(n))$ is a free $KO^*(pt)$ module with base $1,\tilde{\alpha},\ldots,\tilde{\alpha}^n$, where $\tilde{\alpha} \in \widetilde{KO}^4(\mathbb{H}P(n))$ is represented as $\bar{p}^{-1}(1-\lambda)$, λ being the canonical quaternionic line bundle.

<u>Proof</u>: From the change of fields section of Chapter V one has quotient maps $S^{4n+3} \longrightarrow \mathbb{C}P(2n+1) \xrightarrow{f} \mathbb{H}P(n)$ with $f^*(\lambda) = \lambda \oplus \bar{\lambda}$, and $\mathbb{H}P(n)$ has

proper integral cohomology, with $H^*(\mathbb{H}P(n);Z) = Z[\hat{\alpha}]/\hat{\alpha}^{n+1} = 0$ and $f^*(\hat{\alpha}) = -\alpha^2$ in $H^*(\mathbb{C}P(2n+1);Z)$. In particular, under the collapse $\mathbb{H}P(n) \xrightarrow{d} \mathbb{H}P(n)/\mathbb{H}P(n-1) = S^{4n}$, $d^*(\tilde{\iota}) = (-\hat{\alpha})^n$.

Assuming the result true for $\mathbb{H}P(n-1)$, one has the exact sequence

$$0 \longrightarrow \widetilde{KO}^*(S^{4n}) \xrightarrow{d^*} KO^*(\mathbb{H}P(n)) \xrightarrow{i^*} KO^*(\mathbb{H}P(n-1)) \longrightarrow 0$$

arising from the cofibration $\mathbb{H}P(n-1) \xrightarrow{i} \mathbb{H}P(n) \xrightarrow{d} S^{4n}$, i^* being epic by the assumption. Since $\tilde{\alpha}^n$ is trivial over the $4n-1$ skeleton, $i^*\tilde{\alpha}^n = 0$, and $\tilde{\alpha}^n = d^*(w)$ for some $w \in \widetilde{KO}^{4n}(S^{4n})$. To prove the result for $\mathbb{H}P(n)$ by induction, it suffices to show that w is a generator of $\widetilde{KO}^{4n}(S^{4n})$.

Considering λ as a complex bundle, one has $f^*ch(1_H-\lambda) = chf^*(1_H-\lambda) = ch(2-\lambda \oplus \bar{\lambda}) = 2-e^{\alpha}-e^{-\alpha} = -\alpha^2$ + higher terms, so $ch(\tilde{\alpha}) = \hat{\alpha}$ + higher terms. For n odd, $\tilde{\alpha}^n$ is represented as an Sp 'bundle', and thought of as complex, $ch\tilde{\alpha}^n = +\hat{\alpha}^n = d^*(-\dot{\iota})$, so $ch(w) = -\dot{\iota}$ and w is a generator. For n even, $\tilde{\alpha}^n$ is a real 'bundle' and $ch(\tilde{\alpha}^n \otimes \mathbb{C}) = \hat{\alpha}^n = d^*\dot{\iota}$, so $ch(w \otimes C) = \dot{\iota}$ and w is a generator. **

From this it is clear that $\tilde{\alpha}$ satisfies all the conditions to give $\mathbb{H}P(n)$ proper cohomology, with $w \in \widetilde{KO}^{4n}(S^{4n})$ being the standard orientation class $\tilde{\iota}$, chosen so that for n odd $ch(\tilde{\iota}) = \dot{\iota}$ and for n even $ch(\tilde{\iota} \otimes C) = \dot{\iota}$, where $\dot{\iota} \in \tilde{H}^{4n}(S^{4n};Z)$ is the standard generator.

Since the operation $(V,J) \longrightarrow V$ obtained by taking the underlying bundle maps the generator of $\widetilde{KSp}(S^4)$ onto that of $\tilde{K}(S^4)$, this operation is compatible with suspension and one may define a ring homomorphism $\psi : KO^*(X) \longrightarrow K^*(X)$ represented geometrically by $(V,J) \longrightarrow V$ in degrees congruent to zero mod 4.

It is now reasonable to describe $KO^*(pt)$. Briefly, $KO^*(pt)$ contains a subring consisting of Laurent series on the class $\bar{p}(1) \in KO^{-8}(pt)$ and is a module over this subring with base $1, a, b, z$, $a \in KO^{-1}(pt)$, $b \in KO^{-2}(pt)$, $z \in KO^{-4}(pt)$ with relations $2a = 2b = 0$, $a^2 = b$ and $z^2 = 4\bar{p}(1)$.

Note: $z \in KO^{-4}(pt)$ is represented by the trivial symplectic line bundle, or $\psi(z)$ by the 2 dimensional complex bundle. Thus $\psi(z)^2$ is represented by the image under periodicity of the trivial complex 4 plane bundle, which is $4\psi(\bar{p}(1))$.

Under $\psi : KO^*(pt) \longrightarrow K^*(pt)$ one has $\psi(a) = \psi(b) = 0$, $\psi(z) = 2p(1)^2$, $\psi(\bar{p}(1)) = p(1)^4$.

Being given a $U(n)$ bundle ξ over a space B, with complex inner product $< \; , \; >$, one has defined a bundle map

$$\phi : \pi^*(\Lambda^{ev}(\xi)) \longrightarrow \pi^*(\Lambda^{od}(\xi))$$

where π is the projection of $D(\xi)$ onto B (Lemma 7, Chapter IX makes this meaningful) and over $S(\xi)$, this is an isomorphism (Corollary to Lemma 6). Thus $\pi^*(\Lambda^{ev}(\xi)) - \pi^*(\Lambda^{od}(\xi)) \in K(D(\xi))$ is trivialized over $S(\xi)$ by ϕ, defining a class $d(\pi^*\Lambda^{ev}(\xi), \; \pi^*\Lambda^{od}(\xi), \phi) \in K(D(\xi), S(\xi))$. Let $\hat{U}(\xi) = p^{-n} d(\pi^*\Lambda^{ev}(\xi), \pi^*\Lambda^{od}(\xi), \phi) \in \tilde{K}^{2n}(T(\xi))$.

Proposition: $\hat{U}(\xi) = (-1)^n \tilde{U}(\xi)$, where $\tilde{U}(\xi)$ is the orientation defined by K-theory Chern classes.

Proof: As in Assertion 1, $\hat{U}(\xi)$ is multiplicative and so one may verify this for line bundles. If V is a 1 dimensional vector space, $\Lambda(V)$ has a base $\{1, v\}$, v a unit vector in V, and

$$F_x(1) = x, \qquad F_x(v) = 0,$$

$$F_x^*(v) = < v, x >, \qquad F_x^*(1) = 0,$$

so

$$\phi_x(1) = x, \qquad \phi_x(v) = < v, x >$$

and ϕ defines the standard trivialization over $S(\xi)$ by sending $ze \in \pi^*(\xi)_e$ into z. Thus $\hat{U}(\xi) = p^{-1}(1-\xi)$ for the conjugate of the canonical bundle over $\mathbb{C}P(n-1)$. Since $\tilde{U}(\xi) = p^{-1}(\xi-1)$, this gives the result. **

If ξ is an $SU(2n)$ bundle over B with $\sigma : B \longrightarrow S(\Lambda^{2n}(\xi))$ defining the orientation, then by Lemma 4 (Chapter IX) one has defined an operator $\mu : \Lambda(\xi) \longrightarrow \Lambda(\xi)$, anticommuting with i (Lemma 3). Since $\mu : \Lambda^k(\xi) \longrightarrow \Lambda^{2n-k}(\xi)$, μ preserves the even/odd decomposition, and by Lemma 8, μ commutes with ϕ. Since $\mu^2 = (-1)^{2n(2n-1)/2}$, one has:

<u>Proposition</u>: If ξ is an $SU(2n)$ bundle, the class

$$d(\pi^*\Lambda^{ev}(\xi), \pi^*\Lambda^{od}(\xi), \phi) \in \tilde{K}(T(\xi))$$

admits a conjugate linear operator μ. Since $\mu^2 = (-1)^n$, this defines a class

$$u(\xi) \in \widetilde{KO}^{4n}(T(\xi)).$$

<u>Proof</u>: If n is odd, $d = d(\pi^*\Lambda^{ev}(\xi), \pi^*\Lambda^{od}(\xi), \phi)$ is given an Sp structure, while for n even d has a real structure. Applying real periodicity this gives an element in $\widetilde{KO}^{4n}(T(\xi))$.

By Lemma 5, Chapter IX, μ is multiplicative for $SU(2n)$ bundles and so $u(\xi)$ is a multiplicative Thom class.

<u>Assertion</u>: $u(\xi)$ is an orientation.

<u>Proof</u>: Letting ξ be the trivial complex $2n$ plane bundle over a point, ξ is given an SU structure by the trivialization. Thus $u(\xi) \in \widetilde{KO}^{4n}(S^{4n})$ is defined. Applying ψ to $u(\xi)$ gives $\tilde{U}(\xi) \in \tilde{K}^{4n}(S^{4n})$ which is the standard orientation, so $u(\xi)$ is the generator of $\widetilde{KO}^{4n}(S^{4n})$. **

Note: This shows that SU bundles are KO* orientable. For ξ an $SU(2k-1)$ bundle, $u(\xi \oplus 1) \in \widetilde{KO}^{4k}(T(\xi \oplus 1)) = \widetilde{KO}^{4k}(\Sigma^2 T(\xi)) = \widetilde{KO}^{4k-2}(T(\xi))$ is an orientation. The difficulty in doing this case geometrically is that one has no nice way to describe KO^{4k+2}.

Having an orientation for SU bundles, it would be desireable to have characteristic classes also. For a symplectic bundle ξ over X one has KO*-characteristic classes, $\pi_i^s(\xi) \in KO^{4i}(X)$, defined by the general theory.

For ξ a complex bundle over X, $\xi \otimes_{\mathbb{R}} \mathbb{C} \cong \xi \oplus \bar{\xi}$ has a symplectic structure and thus one has KO* characteristic classes $\pi_i^s(\xi \otimes \mathbb{C}) \in KO^{4i}(X)$.

Lemma: $\psi(\pi_i^s(\xi \otimes \mathbb{C})) = (-1)^i \cdot p(1)^{-2i}\pi^i(\xi)$ where $\pi^i(\xi) \in K(X)$ is the K-theory Pontrjagin class defined by the underlying real bundle of ξ. (See Chapter IX).

Proof: Clearly $\psi(\pi_i^s(\xi \otimes \mathbb{C})) = p(1)^{-2i}\beta$, with $\beta \in K(X)$ and to evaluate β it suffices to apply the splitting principle. Then for $\xi = \lambda$, $i = 1$, over $\mathbb{C}P(n)$, one has $\beta = 2 - \lambda \oplus \bar{\lambda} = -\pi^1(\lambda)$, giving the result. **

It is also convenient to reexamine the classes $\pi^i(\xi)$ as follows: For any real vector bundle ξ over X, one defines

$$\lambda_t^{\mathbb{R}}(\xi) = \sum_{i=0}^{\infty} \Lambda_{\mathbb{R}}^i(\xi)t^i \in KO(X)[[t]],$$

so that $\lambda_t^{\mathbb{R}}(\xi \oplus \eta) = \lambda_t^{\mathbb{R}}(\xi)\cdot\lambda_t^{\mathbb{R}}(\eta)$, permitting extension to $KO(X)$. Thus

$$\lambda_t^{\mathbb{R}}(\xi - \dim\xi) = \frac{\sum \Lambda_{\mathbb{R}}^i(\xi)t^i}{(1+t)^{\dim\xi}}$$

depends only on the stable class of ξ. Letting $u = t/(1+t)^2$, so that $u = t(1-t+t^2-t^3+\dots)^2$ is a power series over Z with leading term t and t is a power series over Z in u with leading term u, one defines classes

$\pi_{\mathbb{R}}^{i}(\xi) \in KO(X)$ by

$$\sum u^i \pi_{\mathbb{R}}^{i}(\xi) = \pi_u^{\mathbb{R}}(\xi) = \lambda_t^{\mathbb{R}}(\xi - \dim\xi)$$

and calls $\pi_{\mathbb{R}}^{i}(\xi)$ the i-th KO theory Pontrjagin class of ξ.

<u>Note</u>: $\Lambda_{\mathbb{C}}^{i}(\xi \otimes \mathbb{C}) = \Lambda_{\mathbb{R}}^{i}(\xi) \otimes \mathbb{C}$, so $\sum u^i \pi_{\mathbb{R}}^{i}(\xi) \otimes \mathbb{C} = \lambda_t(\xi \otimes \mathbb{C} - \dim\xi_{\mathbb{C}})$. Replacing t by $t/(1-t)$ gives

$$\begin{aligned}\lambda_{t/(1-t)}(\xi \otimes \mathbb{C} - \dim\xi_{\mathbb{C}}) &= \sum \left[\frac{t/(1-t)}{(1+t/(1-t))^2}\right]^i (\pi_{\mathbb{R}}^{i}(\xi) \otimes \mathbb{C}),\\ &= \sum [t(1-t)]^i (\pi_{\mathbb{R}}^{i}(\xi) \otimes \mathbb{C}),\\ &= \sum s^i (\pi_{\mathbb{R}}^{i}(\xi) \otimes \mathbb{C})\end{aligned}$$

where $s = t - t^2$. Thus the complexification of $\pi_{\mathbb{R}}^{i}(\xi)$ is the class $\pi^i(\xi)$ previously referred to as the K-theory Pontrjagin class.

This gives the somewhat curious phenomenon that for i odd, $\pi^i(\xi)$ comes from both $KSp(X)$, as $-p(1)^{2i}\psi(\pi_i^s(\xi \otimes \mathbb{C}))$, and $KO(X)$, as $\pi_{\mathbb{R}}^{i}(\xi) \otimes \mathbb{C}$, provided ξ is a complex bundle. This gives:

<u>Theorem</u>: Let M be a stably almost complex manifold, with $s_\omega(e_{\mathscr{P}})$, $\mathscr{J} \in H^*(M;Q)$ the classes given by the s_ω symmetric function of the variables $e^{x_j} + e^{-x_j} - 2$ and the product of the classes $x_j/(e^{x_j} - 1)$, when $c(M)$ is expressed formally as $\Pi(1+x_j)$.

Then $(s_\omega(e_{\mathscr{P}})\mathscr{J})[M]$ is an integer and is an even integer if M is an SU manifold and either:

1) $\dim M \equiv 4 \pmod 8$, or

2) $\dim M \equiv 0 \pmod 8$ and ω is not of the form (ω', ω').

<u>Proof</u>: Since $ch(\pi^j(\tau))$ is the j-th elementary symmetric function in the $e^x + e^{-x} - 2$, $s_\omega(e_{\mathcal{P}}) = ch(s_\omega(\pi))$ is the Chern character of the s_ω symmetric function class for the K theory Pontrjagin class. Thus, up to a periodicity, $(s_\omega(e_{\mathcal{P}})\mathcal{J})[M]$ is the value of the K-theory characteristic number $s_\omega(\pi)[M]$, and hence is an integer.

If M^{4r} is an SU manifold, imbed M in a sphere S^{8k} with SU normal bundle ν, and let $c : S^{8k} \longrightarrow T(\nu)$ be the collapse, $p : \nu \longrightarrow M$ being the projection. Then

$$(s_\omega(e_{\mathcal{P}})\mathcal{J})[M] = ch\cdot c^*(p^*s_\omega(\pi)\cdot\tilde{U}(\nu))[S^{8k}].$$

If $\dim M \equiv 4(8)$, this is given by

$$ch\psi c^*(p^*s_\omega(\pi_R)\cdot u(\nu))[S^{8k}].$$

but $c^*(p^*s_\omega(\pi_R)u(\nu)) \in \widetilde{KO}^{8k-4r}(S^{8k})$, with $8k-4r \equiv 4(8)$, and hence this Chern character has even value.

If $\dim M \equiv 0(8)$ and $\omega \neq (\omega',\omega')$, then s_ω belongs to the ideal generated by 2 and the odd order elementary symmetric functions. [To see this consider $H^*(BO;Z_2)$ as symmetric classes mod 2. The ideal generated by the odd classes is the kernel of the homomorphism sending w_i to $\sum w_j w_{i-j}$ induced by classifying $\gamma \oplus \gamma$. Under this homomorphism, s_ω goes to zero if $\omega \neq (\omega',\omega')$ and if $\omega = (\omega',\omega')$, s_ω maps to $s_{\omega'}{}^2$.] Thus one may write

$$s_\omega(\pi) = -\sum_{j,\omega_j} a_{\omega_j}\pi^{2j+1}\cdot s_{\omega_j}(\pi) + 2\sum b_\lambda s_\lambda(\pi),$$

with a_{ω_j}, $b_\lambda \in Z$. Then $(s_\omega(e_{\mathcal{P}})\mathcal{J})[M]$ is given by

$$ch\psi c^*(p^*\sum_{j,\omega_j}\pi^s_{2j+1}\cdot a_{\omega_j}s_{\omega_j}(\pi_R)u(\nu))[S^{8k}] + 2chc^*(\sum b_\lambda s_\lambda(\pi)\tilde{U}(\nu))[S^{8k}].$$

The last term is even, being twice a K theory characteristic number, while the first is a sum of terms of the form $ch\psi(x)[S^{8k}]$ with $x \in \widetilde{KO}^{8p+4}(S^{8k})$, and so is even. **

From this, one may complete the argument for $\theta^3 = 0$ in Lemma 3, by noting:

Corollary: The integer $\mathcal{S}[M]$ for M a 4 dimensional SU manifold is always even.

Next, one should note that for M an SU manifold of dimension $4r$, one has

$$s_\omega(\pi^S(\tau \otimes \mathbb{C}))[M] = \beta\sigma$$

where $\sigma \in KO^*(pt)$ is the standard generator $z^\varepsilon \bar{p}(1)^{-s}$, $\varepsilon = 0,1$, of dimension $4n(\omega)-4r = 8s-4\varepsilon$, and $\beta \in Z$. Applying $ch\psi$ gives

$$\beta = (-1)^{n(\omega)}(s_\omega(e_{\mathcal{G}})\mathcal{S})[M]/2^\varepsilon.$$

Next, it should be noted in the above that only SU manifolds of dimension congruent to zero mod 4 were considered. The classes $s_\omega(e_{\mathcal{G}})$ have nonzero components only in dimensions congruent to zero mod 4 and one has

Proposition: The $\mathcal{S}$ class is given by $e^{-c_1/2}\cdot\hat{A}$, so the classes $\mathcal{S}_{4j+2}$ are divisible by c_1. In particular, the $\mathcal{S}$ class coincides with the $\hat{A}$ class in SU manifolds.

Proof: $x/(e^x-1) = e^{-x/2}x/(e^{x/2} - e^{-x/2}) = e^{-x/2}\cdot((x/2)/\sinh(x/2))$ so $\mathcal{S} = e^{-c_1/2}\hat{A}$. **

To make use of KO* theoretic characteristic numbers, one must know their values. For this define

$$\rho(\pi) : \Omega_*^U \longrightarrow Z[\alpha_i] : [M] \longrightarrow \sum (s_\omega(e_{\mathcal{P}})\mathcal{S})[M]\cdot\alpha_\omega,$$

with $\rho_2(\pi)$ denoting the mod 2 reduction. One has:

Proposition: $\rho_2(\pi)(z_2') = 1$ and $\rho_2(\pi)(z_{4n}')$, $n \geq 2$, has largest monomial given by:

1) α_n if n is not a power of 2,

2) $(\alpha_{2^{s}-1})^2$ if $n = 2^s$, $s > 0$.

Proof: First, $z_2' = [\mathbb{C}P(1)]$ so that $s_\omega(e_{\mathcal{P}})$ is zero if $n(\omega) > 0$, giving $\rho(\pi)(z_2') = \mathcal{S}[\mathbb{C}P(1)] = -1$.

If n is not a power of 2, z_{4n}' is the class of the submanifold of $\mathbb{C}P(1) \times \mathbb{C}P(2^p) \times \mathbb{C}P(2^{p+1}q)$ where $2n = 2^p(2q+1)$ dual to $\bar{\alpha}_1 + (2^p+1)\bar{\alpha}_2 + (2^{p+1}q+1)\bar{\alpha}_3$. Then $s_n(e_{\mathcal{P}})\mathcal{S}[z_{4n}'] = s_n(\mathcal{P})[z_{4n}'] = s_{2n}(c)[z_{4n}']$ and this s-number was known to be nonzero in the choice of z_{4n}'.

If $n = 2^s$, $s > 0$, z_{4n}' is the class of the submanifold of $\mathbb{C}P(1) \times \mathbb{C}P(2^s) \times \mathbb{C}P(2^s)$ dual to $\bar{\alpha}_1 + (2^s+1)(\bar{\alpha}_2 + \bar{\alpha}_3)$, and it was noted that $z_{4n}' = [\mathbb{C}P(2^s)]^2 + c$, where c belongs to the ideal generated by 2 and generators b_{2^t-1} of $\Omega_*^U \otimes Z_2$. Since $\rho_2(\pi)(\mathbb{C}P(2^s)) = \alpha_{2^s-1}$ + lower terms by s-classes, with $\rho_2(\pi)(2) = 0$, $\rho_2(\pi)(b_{2^t-1}) = 0$ for $t > 1$ since b_{2^t-1} may be taken to be an SU manifold of dimension nonzero mod 8, and $b_1 = z_2'$ gives $\rho_2(\pi)(b_1) = 1$, which decreases the degree of the terms involving b_1, one has the asserted monomial. **

Lemma: (Conner and Landweber [42]) The homomorphism $\rho_2(\pi)$ sends the image of $\rho\partial : \mathcal{W}_*(\mathbb{C},2) \longrightarrow \mathcal{W}_*(\mathbb{C},2)$ into zero.

Proof: Since $\operatorname{im}\rho\partial$ consists of SU manifolds, it suffices to consider the case $[M] = \rho\partial[N]$ with $\dim M = 8k$. Then since $[M] \in \Omega_*^{TSU}$, $\rho\partial[M \times \mathbb{C}P(1)] = 2[M]$ as in Lemma 1, so $a = 2[N] - [M \times \mathbb{C}P(1)]$ is in the kernel of $\rho\partial$ and has dimension $8k+2$. As noted, the component of $s_\omega(e_{\mathcal{P}})\mathcal{S}$ of dimension $8k+2$ is divisible by c_1, so this number vanishes for a. Thus

$$-\rho(\pi)[M] = \rho(\pi)[M \times \mathbb{C}P(1)] = 2\rho(\pi)[N]$$

and $\rho_2(\pi)[M] = 0$. **

Since clearly $\rho_2(\pi)$ sends $2\mathcal{W}_*(\mathbb{C},2)$ to zero, this shows that one has induced a homomorphism

$$\rho_2(\pi) : H_*(\mathcal{W} \otimes Z_2) \longrightarrow Z_2[\alpha_i].$$

Then $\rho_2(\pi)(h_2) = 1$ and $\rho_2(\pi)(h_{8n})$ has largest monomial α_n^2 if n is not a power of 2, or $(\alpha_{2^{s-1}})^4$ if $n = 2^s$, $s > 0$. [Note: $h_{8n} = (z_{4n})^2 + z_2 z_{4n-2} z_{4n}$, but $\rho_2(\pi)(z_{4n-2}) = 0$ since $\partial z_{4n} = z_{4n-2}$; thus $\rho_2(\pi)(h_{8n}) = \rho_2(\pi)(z_{4n})^2$]. Since these classes have distinct largest monomials, one has:

Proposition: $\alpha \in H_n(\mathcal{W} \otimes Z_2)$ is zero if and only if $\rho_2(\pi)(\alpha) = 0$.

[Note the analogy with oriented cobordism in which the mod 2 numbers $s_\omega(\mathcal{P})$ detected $\ker\partial/\operatorname{im}\partial$, while here the numbers $s_\omega(\pi)$ detect.]

Theorem: All relations among the Chern numbers of n dimensional SU manifolds are given by the relations

a) $c_1 c_\omega[M] = 0$ for all ω,

b) $s_\omega(e)\mathcal{S}[M] \in Z$ for all ω, and

c) If $n \equiv 4 \pmod 8$, $(s_\omega(e_{\mathcal{P}})\mathcal{S})[M] \in 2Z$ for all ω.

Proof: These relations have been shown to hold for SU manifolds. If $a \in H_n(BU;Q)$ with $c_1 c_\omega[a] = 0$, $s_\omega(e)\mathcal{J}[a] \in Z$ and for $n \equiv 4(8)$, $s_\omega(e_{\mathcal{P}})\mathcal{J}[a] \in 2Z$, then from the knowledge of Ω_*^U, a is represented by a complex manifold having all numbers divisible by c_1 zero. Thus $a \in \tau\Omega_*^{TSU}$. For $n \not\equiv 4(8)$, this suffices to prove $a \in \tau\Omega_n^{SU}$. If $n \equiv 4(8)$, then the class of a in $H_*(\mathcal{W} \otimes Z_2)$ is zero since $s_\omega(e_{\mathcal{P}})\mathcal{J}[a] \equiv 0(2)$ for all ω, so the class of a in $H_*(\mathcal{W})$ is zero. Hence a belongs to $B(\mathcal{W}_n(\mathbb{C},2),\rho\partial)$ and a is in $\tau\Omega_n^{SU}$. **

Lemma: The KO^* theory characteristic numbers of $\theta \in \Omega_1^{SU}$ are given by:

a) $s_\omega(\pi^S(\tau \otimes \mathbb{C}))[\theta] = 0$ if $n(\omega) > 0$, and

b) $1[\theta] \neq 0$ in $KO^{-1}(pt) \cong Z_2$.

Proof: Let $u : TBSU_{4r+2} \longrightarrow BSp$ be a map defining the orientation class $u(\tilde{\gamma}_{4r+2}) \in \widetilde{KO}^{8r+4}(TBSU_{4r+2})$, r large. Letting $b \in BSU_{4r+2}$, the composite

$$f : S^{8r+4} = Tb \hookrightarrow TBSU_{4r+2} \xrightarrow{u} BSp$$

represents the generator of $\pi_{8r+4}(BSp) \cong \widetilde{KO}^{8r+4}(S^{8r+4})$. Since S^{8r+4} is $8r+3$ connected, this map lifts to the connective cover $BSp(8r+4,\ldots,\infty)$ giving

$$f : S^{8r+4} \xrightarrow{\hat{f}} BSp(8r+4,\ldots,\infty) \xrightarrow{\pi} BSp.$$

Considering S^{8r+4} as the $8r$-fold suspension of S^4 gives

$$h : S^4 \xrightarrow{f'} \Omega^{8r}BSp(8r+4,\ldots,\infty) \xrightarrow{\Omega^{8r}\pi} \Omega^{8r}BSp \xrightarrow{\sigma} BSp$$

where σ is given by the r-fold application of periodicity. Both σ and $\Omega^{8r}\pi$ induce isomorphisms on homotopy in positive dimensions, so h represents the generator of $\pi_4(BSp)$. Thus $h^* : H^4(BSp;Z) \longrightarrow H^4(S^4;Z)$ is an isomorphism and $h^* : H^*(BSp;Z) \longrightarrow H^*(S^4;Z)$ is an isomorphism in dimensions less than 8, so $h_* : \pi_*(S^4) \longrightarrow \pi_*(BSp)$ is an isomorphism in dimensions less than 7. Thus f' induces an isomorphism on homotopy in this range and being in the stable range, $\hat{f}$ induces an isomorphism on homotopy in dimensions less than $8r+7$. Thus $f_* : \pi_*(S^{8r+4}) \longrightarrow \pi_*(BSp)$ is an isomorphism in dimensions $8r+4$ through $8r+6$.

Thus, the composite $\Omega_*^{fr} \xrightarrow{F*} \Omega_*^{SU} \xrightarrow{1[\ \ ]} KO^{-*}(pt)$ is epic in degrees 0 through 2. Since θ is the nonzero elements of Ω_1^{SU}, this shows that $1[\theta] \neq 0$.

Further θ comes from framed cobordism and $s_\omega(\pi^s(\tau \otimes \mathbb{C})) = 0$ for $n(\omega) > 0$ in a framed manifold, so these numbers must vanish on θ. **

Theorem: (Anderson, Brown, and Peterson [6]): Two SU manifolds are cobordant if and only if they have the same integral cohomology and KO^* theory characteristic numbers.

Proof: Suppose M^n is an SU manifold for which all such characteristic numbers are zero. Since all integral characteristic numbers vanish on M, M is a torsion class, and $[M] = \theta^\epsilon[N]$ where $\epsilon = 1,2$ and N is an SU manifold of dimension $8k$ for some k. Then for all ω, $s_\omega(\pi^s(\tau \otimes \mathbb{C}))[M] \in KO^{4n(\omega)-8k-\epsilon}(pt)$ is zero, but

$$s_\omega(\pi^s(\tau \otimes \mathbb{C}))[M] = s_\omega(\pi^s(\tau \otimes \mathbb{C}))[N]\cdot a^\epsilon$$

since $1[\theta] = a$ and all other numbers vanish. Thus for $\omega \in \pi(2t)$, $s_\omega(\pi^s(\tau \otimes \mathbb{C}))[N] = m_\omega \cdot \bar{p}(1)^{k-t}$ with m_ω even, giving $s_\omega(e_{\mathcal{P}})\mathcal{S}[N] \equiv 0(2)$ for $n(\omega) \equiv 0(2)$. Thus $(s_\omega(e_{\mathcal{P}})\mathcal{S})[N] \equiv 0(2)$ for all ω, the other cases being immediate since N is an SU manifold. Thus [N] represents zero in $H_*(\mathcal{W} \otimes Z_2)$ and also in $H_*(\mathcal{W})$, giving $[N] \in \text{image}\,\partial$. Since multiplication by θ annihilates image∂, this gives $[M] = 0$. The remainder of the theorem being obvious, this gives the result. **

The remaining part of the structure of Ω_*^{SU} which is desireable is the multiplicative structure mod torsion. C. T. C. Wall [136] has noted that this may be obtained out of what has already been proved. To begin this analysis, first consider $\mathcal{W}_*(\mathbb{C},2)$.

As a subset of Ω_*^U, $\mathcal{W}_*(\mathbb{C},2)$ is not a ring, as has previously been noted. One may define a product in $\mathcal{W}_*(\mathbb{C},2)$ by the composite

$$* : \mathcal{W}_*(\mathbb{C},2) \otimes \mathcal{W}_*(\mathbb{C},2) \hookrightarrow \Omega_*^U \otimes \Omega_*^U \longrightarrow \Omega_*^U \xrightarrow{\Phi} \mathcal{W}_*(\mathbb{C},2)$$

and in Ω_*^U, $a*b = ab + 2[V^4]\partial a \cdot \partial b$. In particular,

$$(a*b)*c = abc + 2[V^4](a\partial b\partial c + b\partial a\partial c + c\partial a\partial b - [\mathbb{C}P(1)]\partial a\partial b\partial c)$$

since

$$\partial(a*b) = \partial(ab) = a\partial b + b\partial a - [\mathbb{C}P(1)]\partial a\partial b.$$

With this product, $\mathcal{W}_*(\mathbb{C},2)$ becomes a commutative ring with unit. Also, if $\partial a = 0$ then $a*b = a\cdot b$ and in particular, the usual map from Ω_*^{SU} into $\mathcal{W}_*(\mathbb{C},2)$ is a ring homomorphism. Further,
$\partial(a*b) = a*\partial b + b*\partial a - [\mathbb{C}P(1)]*\partial a*\partial b$.

<u>Theorem</u>: Using the product $*$, $\mathcal{W}_*(\mathbb{C},2)$ is the integral polynomial ring on classes x_i, $i \neq 2$, $\dim x_i = 2i$, with $s_i(c)[x_i] = m_i m_{i-1}$ and the operation ∂ is given by

$$\partial x_1 = 2,$$

$$\partial x_{2i} = x_{2i-1}, \quad i > 1$$

with

$$\partial(a*b) = a*\partial b + b*\partial a - x_1*\partial a*\partial b.$$

Note: $m_i = p$ if $i + 1 = p^s$ for some prime p, $s > 0$; $m_i = 1$ otherwise.

Proof: First noting that an element of $\mathcal{W}_*(\mathbb{C},2)$ which is decomposable under $*$ is decomposable under $\cdot$ in Ω_*^U, it is immediate that the s-number detects indecomposables.

One has $s_1(c)[\mathbb{C}P(1)] = 2$ with $m_1 = 2$, $m_0 = 1$, and so one may let $x_1 = [\mathbb{C}P(1)]$.

From the analysis of $\Omega_*^{SU} \otimes Z[1/2]$, one has SU manifolds $M_i \in \Omega_{2i}^{SU}$, $i > 1$, with $s_i(c)[M_i] = 2^{j(i)}m_im_{i-1}$ (multiplying by 2's one may assume $j(i) \geq 0$) which give polynomial generators of $\Omega_*^{SU} \otimes Z[1/2]$. Note: $\Omega_4^{TSU} \cong Z$ with generator $x = 9\mathbb{C}P(1)^2 - 8\mathbb{C}P(2)$, having $c_1^2 = 0$, $c_2 = 12$ and $\mathcal{S}$ number 1. Since $2x = \partial(\mathbb{C}P(1)^3)$ generates the image of Ω_4^{SU}, with $x_1*x_1 = x_1^2\, 2V\cdot 2\cdot 2 = x$, one may assume $M_2 = 2x_1*x_1$.

One also has classes $z'_{2n} \in \mathcal{W}_{2n}(\mathbb{C},2)$, $n > 2$, with $s_n(c)[z'_{2n}]$ odd if $n \neq 2^s$, 2^s-1 and congruent to 2 mod 4 otherwise. Now $s_{2n}(c)[z'_{4n}] \equiv 0(m_{2n})$ being a complex cobordism class, and since $(2n-1)+1 = 2n$, $m_{2n-1} = 1$ or 2 as n is not or is a power of 2. Thus $s_{2n}(c)[z'_{4n}]$ is an odd multiple of $m_{2n}m_{2n-1}$. Since $z'_{4n-2} = \partial z'_{4n}$ is an SU class, $s_{2n-1}(c)[z'_{4n-2}]$ is divisible by m_{2n-2} (which is odd), while $m_{2n-1} = 1$ or 2 as n is not or is a power of 2. Thus $s_n(c)[z'_{2n}]$ is an odd multiple of m_nm_{n-1}.

One may find integers α_n, β_n, with $\alpha_n = 2a_n+1$ (odd) so that $x'_n = \alpha_nz'_{2n} + 2\beta_nM_n$ has s number m_nm_{n-1}. Let

$$x_{2n} = \alpha_{2n}z'_{4n} + 2\beta_{2n}M_{2n} + (a_{2n-1}-a_{2n})x_1z'_{4n-2} + \beta_{2n-1}x_1M_{2n-1},$$

$$x_{2n-1} = \alpha_{2n-1}z'_{4n-2} + 2\beta_{2n-1}M_{2n-1}.$$

Then $s_n(c)[x_n] = s_n(c)[x'_n] = m_n m_{n-1}$ since x_n differs from x'_n by decomposables. Since z'_{4n}, $x_1 \in \mathcal{W}_*(\mathbb{C},2)$ and z'_{4n-2}, M_{2n}, $M_{2n-1} \in \Omega_*^{TSU}$, $x_n \in \mathcal{W}_*(\mathbb{C},2)$. Further,

$$\partial x_{2n} = \alpha_{2n} z'_{4n-2} + 2(a_{2n-1}-a_{2n}) z'_{4n-2} + 2\beta_{2n-1} M_{2n-1} = x_{2n-1}.$$

There is then a homomorphism $\phi : R = Z[x_i | i \neq 2] \longrightarrow \mathcal{W}_*(\mathbb{C},2)$. Under the composite with

$$\psi : \mathcal{W}_*(\mathbb{C},2) \longrightarrow \Omega_*^U \longrightarrow \Omega_*^U \otimes Z_2$$

one has a ring homomorphism ($a*b = a \cdot b$ mod 2), with x_n being sent to z'_{2n} plus decomposables. Thus ϕ is monic (R being torsion free) and $\mathrm{im}\phi$ has odd index.

For any $\alpha \in \mathcal{W}_*(\mathbb{C},2)$, $\partial(x_1 \partial\alpha) = 2\partial\alpha$ so $2\alpha - x_1\partial\alpha \in \ker\partial$ giving $2(2\alpha - x_1\partial\alpha) = \partial(x_1(2\alpha - x_1\partial\alpha))$ or

$$4\alpha = 2x_1\partial\alpha + 2\partial(x_1\alpha)$$

since $\partial(x_1^2) = 0$. Thus

$$\alpha = 1/2(\partial(x_1\alpha)) + 1/2\ x_1\partial\alpha$$

with $\partial(x_1\alpha)$ and $\partial\alpha$ the images of SU classes. <u>Note</u>: Any element of the form $a + x_1 b$, $a, b \in \Omega_*^{SU}$ clearly belongs to $\mathcal{W}_*(\mathbb{C},2)$.

Then $\phi : R \otimes Z[1/2] \longrightarrow \mathcal{W}_*(\mathbb{C},2) \otimes Z[1/2] = \Omega_*^{SU} \otimes Z[1/2]\{1,x_1\}$, the latter being the free $\Omega_*^{SU} \otimes Z[1/2]$ module on 1 and x_1. One may then write $x_{2n} = 1/2\ y_{2n} + 1/2\ x_1 x_{2n-1}$ where $y_{2n} \in \Omega_{4n}^{SU}$ and $s_{2n}(c)[y_{2n}] = 2m_{2n}m_{2n-1}$. The elements $y_{2n} = 2x_{2n} - x_1 x_{2n-1}$, x_{2n-1}, x_1^2 in R map to acceptable generators for $\Omega_*^{SU} \otimes Z[1/2]$, and thus $R \otimes Z[1/2] = Z[1/2][x_1^2, x_{2n-1}, y_{2n}]\{1,x_1\}$ maps isomorphically to

$\Omega_*^{SU} \otimes Z[1/2]\{1,x_1\}$. Thus $\phi : R \longrightarrow \mathcal{W}_*(\mathbb{C},2)$ has 2 primary cokernel, and so must be an isomorphism. **

Using this, one may describe all the rings of interest easily. First, $\partial : \mathcal{W}_*(\mathbb{C},2) \longrightarrow \mathcal{W}_*(\mathbb{C},2)$ extends to $\mathcal{W}_*(\mathbb{C},2) \otimes Z[1/2]$ satisfying the same formulae. If $\partial\alpha = 0$, then $\alpha = 1/2\ \partial(x_1\alpha)$, so $\ker\partial = \mathrm{im}\partial$ in $\mathcal{W}_*(\mathbb{C},2) \otimes Z[1/2]$.

One may write $\mathcal{W}_*(\mathbb{C},2) \otimes Z[1/2] = Z[1/2][x_1, x_{2i-1}, x_{2i}-1/2\ x_1x_{2i-1}]$ with $\partial x_1 = 2$, $\partial x_{2i-1} = 0$, $\partial(x_{2i}-1/2\ x_1x_{2i-1}) = 0$ and $\partial(x_1^2) = 0$. From this it is clear that

$$\mathrm{im}\partial = \ker\partial = Z[1/2][x_1^2, x_{2i-1}, x_{2i}-1/2\ x_1x_{2i-1}] \subset \mathcal{W}_*(\mathbb{C},2) \otimes Z[1/2].$$

Applying this to $\mathcal{W}_*(\mathbb{C},2)$ one has:

<u>Theorem</u>: Letting $\mathcal{W}_*(\mathbb{C},2) = Z[x_i | i \neq 2]$ as above, thought of as a subring of $\mathcal{W}_*(\mathbb{C},2) \otimes Z[1/2] = Z[1/2][x_i | i \neq 2]$, let $A = Z[1/2][x_1^2, x_{2i-1}, x_{2i}-1/2\ x_1x_{2i-1}]$ and then:

a) $\mathrm{kernel}\partial = A \cap \mathcal{W}_*(\mathbb{C},2)$ is the set of $Z[1/2]$ polynomials in x_1^2, x_{2i-1}, $x_{2i}-1/2\ x_1x_{2i-1}$ which have integral coefficients in the x's.

b) $\mathrm{image}\partial = \{u \in A | 1/2\ x_1u \in A + \mathcal{W}_*(\mathbb{C},2)\}$ is the set of $Z[1/2]$ polynomials in x_1^2, x_{2i-1}, $x_{2i}-1/2\ x_1x_{2i-1}$ which when multiplied by $1/2\ x_1$ can be expressed as the sum of such a polynomial and an integral polynomial in the x's.

<u>Note</u>: Since the image of Ω_n^{SU} in $\mathcal{W}_n(\mathbb{C},2)$ is $(\ker\partial)_n$ if $n \not\equiv 4(8)$ and $(\mathrm{im}\partial)_n$ if $n \equiv 4(8)$ this gives a fairly nice description of $\Omega_*^{SU}/\mathrm{Torsion}$ as a subring of A.

For example, one has bases given by

n	$(\ker\partial)_n$	$(\operatorname{im}\partial)_n$	Ω_n^{SU}/Torsion
0	1	2	1
4	x_1^2	$2x_1^2$	$2x_1^2$
6	x_3	x_3	x_3
8	$x_1^4,\ 2(x_4-1/2\ x_1x_3)$	$2x_1^4,\ 2(x_4-1/2\ x_1x_3)$	$x_1^4,\ 2x_4-x_1x_3$
10	$x_1^2x_3,\ x_5$	$x_1^2x_3,\ x_5$	$x_1^2x_3,\ x_5$

An intrinsic description of Ω_*^{SU}/Torsion is extremely complicated, since the square of the 4 dimensional generator is divisible by 4, while the product of the 4 and 6 dimensional generators is divisible by 2.

One should note also that x_1^2 and $x_{2n}^2-x_1x_{2n-1}x_{2n} = (x_{2n}-1/2\ x_1x_{2n-1})^2 -1/4\ x_1^2x_{2n-1}^2$ belong to kernel∂, and are the classes previously chosen to generate $(\ker\partial/\operatorname{im}\partial)$.

<u>Proof</u>: a) If $u \in A \cap \mathcal{W}_*(\mathbb{C},2)$, then $u \in \mathcal{W}_*(\mathbb{C},2)$ and $\partial u = 0$ so $u \in \ker\partial$, while if $u \in \mathcal{W}_*(\mathbb{C},2)$ and $\partial u = 0$ then $u \in A$.

b) If $u = \partial v$, then $u \in A$ since $\partial u = 0$. Further $v \in \mathcal{W}_*(\mathbb{C},2)$ and $\partial(1/2\ x_1u - v) = 0$ so $1/2\ x_1u - v \in A$, giving $1/2\ x_1u \in A + \mathcal{W}_*(\mathbb{C},2)$.

If $u \in A$, $1/2\ x_1u = p + q$, $p \in \mathcal{W}_*(\mathbb{C},2)$, $q \in A$, then $u = \partial(1/2\ x_1u) = \partial p$, so $u \in \operatorname{im}\partial$. **

Relation to Framed Cobordism

The relation of framed cobordism to SU cobordism was first explored by Anderson, Brown, and Peterson [6], who determined the image of $F_* : \Omega_*^{fr} \longrightarrow \Omega_*^{SU}$. The proof given here is due to Conner and Floyd [41]. The hard part of the result is given by:

Proposition: There is a class $M^8 \in \Omega_8^{SU}$ with $\mathcal{S}[M^8] = 1$ and such that $[M^8]^n \times \theta$ is in the image of Ω_{8n+1}^{fr} in Ω_{8n+1}^{SU}.

To prove this a couple of lemmas are convenient:

Lemma 1: There is an element $M^8 \in \Omega_8^{SU}$ with $\mathcal{S}[M^8] = 1$ and such that $[M^8] = 2[B^8]$ in $\Omega_8^{SU,fr}$.

Proof: One has the exact sequence

$$\Omega_8^{SU} = Z \oplus Z \longrightarrow \Omega_8^{SU,fr} \longrightarrow \Omega_7^{fr} \cong Z_{240} \longrightarrow \Omega_7^{SU} = 0$$

(using the fact that $\lim_{n\to\infty} \pi_{n+7}(S^n) \cong Z_{240}$). Consider λ, the canonical quaternionic line bundle over $\mathbb{H}P(n)$. Over $S^4 = \mathbb{H}P(1)$, the tangent bundle of $D(\lambda)$ is isomorphic to $\pi^*(\lambda) \oplus \pi^*(\tau_{S^4})$ and stably $\pi^*(\tau_{S^4})$ is trivial, so $D(\lambda)$ has an Sp structure for which its tangent bundle is stably isomorphic to $\pi^*(\lambda)$. The usual trivialization of $\pi^*(\lambda)$ over $S(\lambda)$ gives $(D(\lambda), S(\lambda))$ an (Sp,fr) structure.

Recalling that $\mathbb{H}P(2)$ is the Thom space of λ over $\mathbb{H}P(1)$, with λ over $\mathbb{H}P(2)$ restricting on $D(\lambda)$ to $\pi^*(\lambda)$ with the standard trivialization over $S(\lambda)$, one has

$$\mathcal{S}[D(\lambda), S(\lambda)] = \mathcal{S}(\lambda)[\mathbb{H}P(2),pt] = \mathcal{S}(\lambda)[\mathbb{H}P(2)].$$

To compute this number, one applies the splitting principle to λ by pulling back to $\mathbb{C}P(5)$ over which λ splits as the sum of the canonical bundle and its conjugate, $\lambda_{\mathbb{C}} \oplus \bar{\lambda}_{\mathbb{C}}$. Then

$$\mathcal{S}(\lambda_{\mathbb{C}} \oplus \bar{\lambda}_{\mathbb{C}}) = \frac{\alpha}{e^{\alpha}-1} \cdot \frac{(-\alpha)}{(e^{-\alpha}-1)} = \frac{\alpha^2}{e^{\alpha}+e^{-\alpha}-2}$$

but

$$\begin{aligned} e^{x}+e^{-x}-2 &= \sum \{\frac{x^i}{i!} + \frac{(-x)^i}{i!}\} - 2, \\ &= 2(1 + \frac{x^2}{2!} + \frac{x^4}{4!} + \frac{x^6}{6!} + \dots) - 2, \\ &= x^2 + x^4/12 + x^6/360 + \dots \end{aligned}$$

so

$$\begin{aligned} \frac{\alpha^2}{e^{\alpha}+e^{-\alpha}-2} &= \frac{1}{1+\alpha^2/12+\alpha^4/360}, \qquad (\alpha^6 = 0) \\ &= 1 - (\alpha^2/12 + \alpha^4/360) + (\alpha^2/12)^2, \\ &= 1 - \alpha^2/12 + 3\alpha^4/720, \end{aligned}$$

Since the cohomology map is monic, $S(\lambda) = 1 + \hat{\alpha}/12 + \hat{\alpha}^2/240$ giving $\mathcal{S}[D(\lambda), S(\lambda)] = 1/240$.

Since $\mathcal{S}$ is integral on SU manifolds, the extension is completely nontrivial, giving $\Omega_8^{SU,fr} = Z \oplus Z$.

One may then take $M^8 = 240[D(\lambda), S(\lambda)]$ and $B^8 = 120[D(\lambda), S(\lambda)]$, since $240[D(\lambda), S(\lambda)]$ belongs to the image of Ω_8^{SU} and has $\mathcal{S}$ number 1. **

<u>Lemma 2</u>: Let $[V^n] \in \Omega_n^{fr}$ be an element of order 2, $[M^k] \in \Omega_k^{SU}$ an element divisible by 2 in $\Omega_k^{SU,fr}$. There is then a class $[V^{n+k}] \in \Omega_{n+k}^{fr}$ of order 2 with image $[M^k][V^n]$ in Ω_{n+k}^{SU}.

<u>Proof</u>: Let B^k be an (SU,fr) manifold with $2[B] = [M]$ in $\Omega_*^{SU,fr}$. By the (SU,fr) exact sequence, $2[\partial B] = 0$ in Ω_*^{fr} and there is a compact framed manifold C^k with $\partial C^k = (\partial B)_1 \cup (\partial B)_2$ being two disjoint copies of ∂B. Let D^{n+1} be a compact framed manifold with $\partial D = V_1^n \cup V_2^n$ being two disjoint copies of V^n.

Consider the product $C^k \times D^{n+1}$, which is a framed manifold with boundary formed by joining $W_1 = [(\partial B_1 \cup \partial B_2) \times D^{n+1}]$ and $W_2 = [(-1)^k C^k \times (V_1 \cup V_2)]$ along their boundaries $(-1)^{k-1}(\partial B_1 \cup \partial B_2) \times (V_1 \cup V_2)$ and $(-1)^k(\partial B_1 \cup \partial B_2) \times (V_1 \cup V_2)$. One may remove tubular neighborhoods of $\partial B_1 \times V_1$ and $\partial B_2 \times V_2$ in $C^k \times D^{n+1}$ and sew the introduced boundary segments together

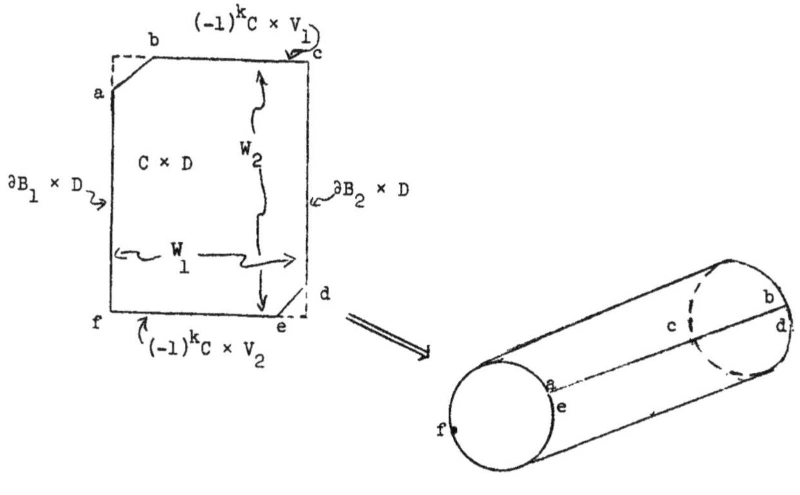

to form a manifold X with two boundary components

$$(-1)^k V_1^{n+k} = (\partial B_1 \times D \cup (-1)^k C \times V_2) \Big/ \begin{array}{ll} (-1)^{k-1}\partial B_1 \times V_2 \equiv (-1)^k \partial B_1 \times V_2 & \text{(at } f\text{)} \\ (-1)^{k-1}\partial B_1 \times V_1 \equiv (-1)^k \partial B_2 \times V_2 & \text{(join along } a\text{–}e\text{)} \end{array}$$

and

$$(-1)^k V_2^{n+k} = (\partial B_2 \times D \cup (-1)^k C \times V_1) \Big/ \begin{array}{ll} (-1)^k \partial B_2 \times V_1 \equiv (-1)^{k-1} \partial B_2 \times V_1 & \text{(at c)} \\ (-1)^{k-1} \partial B_2 \times V_2 \equiv (-1)^k \partial B_1 \times V_1 & \text{(join} \\ \quad \text{along d-b).} & \end{array}$$

Since the framings induced from $C \times D$ along the portions sewn together are compatible, X is given a framing, with $\partial(-1)^k X = V_1^{n+k} \cup V_2^{n+k}$ and clearly $V^{n+k} = V_1^{n+k}$ is isomorphic as a framed manifold with V_2^{n+k}.

This gives a framed manifold V^{n+k} whose class has order 2 in Ω_{n+k}^{fr}. (Note: V^{n+k} depends on a choice of C^k and D^{n+1}, but any choice gives such a class).

Since $[M] = 2[B]$, there is a framed manifold C^k with $\partial C^k = \partial B_1 \cup \partial B_2$ so that

$$M \cup (-B_1) \cup (-B_2) \cup C \,/\, \partial(-B_1) \equiv \partial B_1,\ \partial(-B_2) \equiv \partial B_2$$

is the boundary of an SU manifold W^{k+1}. (Note: This places a demand on C, and hence one may assume C was chosen in this way for the above construction.) Now $\partial(W \times V_1^N)$ contains a copy of $(-B_1) \times V_1 \cup (-B_2) \times V_1$ while $\partial((-1)^k B \times D) = B \times (V_1 \cup V_2) \cup (-1)^k \partial B \times D$ joined along their common boundary so one may form an SU manifold by sewing $(-1)^k B \times D$ to $W \times V_1^n$ by identifying $(-B_i) \times V_1$ to $B \times V_i$. The boundary of the resulting manifold has two components, one of which is $M \times V_1^n$ and the other being formed from $C \times V_1^n \cup (-1)^k \partial B \times D$ by identifying $\partial B_1 \times V_1^n$ to $-\partial B \times V_1$ and $\partial B_2 \times V_1^n$ to $-\partial B \times V_2$, which is just a copy of V^{n+k}. Thus this choice of C will give $[V^{n+k}] = [M^k][V^n]$ in Ω_{n+k}^{SU}. **

Proof of the Proposition: Let M^8 be as in Lemma 1, and let V^1 represent the class of order 2 in Ω_1^{fr} which maps to θ in Ω_1^{SU}. Suppose inductively that $V^{8k+1} \in \Omega_{8k+1}^{fr}$ has order 2 and represents $[M^8]^k \cdot \theta$ in Ω_{8k+1}^{SU}. Applying Lemma 2 to V^{8k+1} and M^8 gives a framed manifold V^{8k+9} with order 2 in Ω_*^{fr} representing $[M^8][V^{8k+1}] = [M^8]^{k+1} \cdot \theta$ in Ω_{8k+9}^{SU}. This completes the induction and gives the Proposition. **

From the Proposition one may completely determine the image of Ω_*^{fr} in Ω_*^{SU}, which is given by:

Theorem: The forgetful homomorphism $F_* : \Omega_*^{fr} \longrightarrow \Omega_*^{SU}$ is a isomorphism in dimension zero, and in positive dimensions the image is zero except in dimensions $8k+1$ and $8k+2$ where it is Z_2 with generator $x_1^{4k} \cdot \theta^\varepsilon$, $\varepsilon = 1$ or 2.

Proof: Clearly $F_0 : \Omega_0^{fr} \xrightarrow{\cong} \Omega_0^{SU} = Z$. For $n > 0$, Ω_n^{fr} is finite, so image(F_n) consists of torsion classes. Thus image$(F_n) = 0$ for n not of the form $8k+1$ or $8k+2$. If $n = 8k+\varepsilon$, $\varepsilon = 1$ or 2 and $\alpha \in \text{im}(F_n)$, then α has all KO* characteristic numbers of the form $s_\omega(\pi)$, $n(\omega) > 0$, zero since these classes vanish in framed manifolds, and so $\alpha \neq 0$ if and only if $1[\alpha] \neq 0$. Thus $\text{im}(F_n)$ is either 0 or Z_2.

From the proposition, $[M^8]^k \times \theta \in \text{im}(F_{8k+1})$ and since $\theta \in \text{im}(F_1)$, $[M^8]^k \cdot \theta^\varepsilon \in \text{im}(F_n)$. Then

$$\begin{aligned} 1[[M^8]^k \cdot \theta^\varepsilon] &= 1[[M^8]^k], \\ &= \mathcal{J}[[M^8]^k] \quad \text{mod } 2, \\ &= (\mathcal{J}[M^8])^k \quad \text{mod } 2, \\ &= 1 \quad \text{mod } 2 \end{aligned}$$

and thus $\text{im}(F_n) \neq 0$, so must be Z_2 with generator $[M^8]^k \cdot \theta^\varepsilon$.

Since $\Omega_8^{SU} = Z \oplus Z$ with base x_1^4 and $\partial(x_1 x_4)$, while θ annihilates image∂, one must have $M^8 \cdot \theta = x_1^4 \theta$ and thus $[M^8]^k \cdot \theta^\varepsilon = x_1^{4k} \theta^\varepsilon$ is the nonzero class in $\mathrm{im}(F_{8k+\varepsilon})$. **

<u>Note</u>: In fact, for any M^8 with odd $\mathcal{S}$ number one has $[M^8]^k \theta^\varepsilon = x_1^{4k} \theta^\varepsilon$. To see this, $\partial(x_1 x_4)\theta = 0$ implies $\mathcal{S}[\partial(x_1 x_4)] \equiv 0(2)$, so $M^8 = (2p+1)x_1^4 + q\partial(x_1 x_4)$ giving $M^8 \theta = x_1^4 \theta$.

Turning to the exact sequence

$$\cdots \longrightarrow \Omega_{n+1}^{fr} \xrightarrow{F_*} \Omega_{n+1}^{SU} \longrightarrow \Omega_{n+1}^{SU,fr} \xrightarrow{\partial} \Omega_n^{fr} \xrightarrow{F_*} \Omega_n^{SU} \longrightarrow \cdots$$

one then knows the nature of F_*, and the remaining question which is within reach is the nature of the extension problem for $\Omega_{n+1}^{SU,fr}$:

$$0 \longrightarrow \Omega_{n+1}^{SU}/\mathrm{im}F_* \longrightarrow \Omega_{n+1}^{SU,fr} \xrightarrow{\partial} \ker F_* \longrightarrow 0.$$

This was settled by Conner and Floyd [41].

It is immediate that the construction of a KO^* orientation carries through for an SU manifold V with framed boundary, giving $U \in \widetilde{KO}^*(T\nu)$. For $n(\omega) > 0$, $s_\omega(\pi)$ may be formed in $KO^*(BSU,*)$, giving KO^* theory characteristic numbers for (SU,fr) manifolds.

Since the torsion subgroup of Ω_*^{SU} is detected by KO^* characteristic numbers, with image(F_*) detected by $1(\pi)$, the image of $\mathrm{Torsion}(\Omega_{n+1}^{SU})/\mathrm{im}(F_*)$ in $\Omega_{n+1}^{SU,fr}$ is detected by Z_2 valued KO^* numbers. This defines a splitting of the torsion part.

To study the extension of the free part, the analysis of KO^* theory characteristic numbers gives immediately that an (SU,fr) manifold V has the same Chern numbers as a closed SU manifold if and only if

$$\mathcal{S}[V,\partial V] \in \begin{cases} Z & \dim V \equiv 0(8) \\ 2Z & \dim V \equiv 4(8), \end{cases}$$

all other relations coming from KO* theory relations which are satisfied by $[V,\partial V]$.

One may define a homomorphism $\mathcal{S}' : \Omega_{n+1}^{SU,fr} \longrightarrow Q$ by $\mathcal{S}'[\alpha] = \mathcal{S}[\alpha]$ if $\dim \alpha \not\equiv 4(8)$ and by $\mathcal{S}'[\alpha] = 1/2\,\mathcal{S}[\alpha]$ if $\dim \alpha \equiv 4(8)$. Note: $\mathcal{S}$ is divisible by c_1 in dimensions not congruent to zero mod 4, so $\mathcal{S}'$ vanishes in these dimensions. $\mathcal{S}'$ sends Ω_{n+1}^{SU} to Z and defines

$$\mathcal{S}' : \text{kernel}(F_*) \longrightarrow Q/Z.$$

The homomorphism $\mathcal{S}' : \Omega_{8k+3}^{fr} = \text{kernel}(F_{8k+3}) \longrightarrow Q/Z$ is just the Adams invariant e_R. The previous results concerning the invariant $\mathcal{S}$ give:

The homomorphism $\mathcal{S}' : \Omega_{4k-1}^{fr} \longrightarrow Q/Z$ maps precisely onto the integral multiples of $1/a_k$ where a_k is the denominator of $B_k/4k$.

Relation to Complex Cobordism

Turning to the relationship with Ω_*^U, one has the exact sequence

$$\cdots \longrightarrow \Omega_n^{SU} \xrightarrow{F_*} \Omega_n^U \longrightarrow \Omega_n^{U,SU} \xrightarrow{\partial} \Omega_{n-1}^{SU} \longrightarrow \cdots$$

in which F_* is completely known. Just as in the (O,SO) sequence, this may be identified with the Ω_*^{SU} bordism sequence of the cofibration

$$\mathbb{C}P(1) \longrightarrow \mathbb{C}P(\infty) \longrightarrow \mathbb{C}P(\infty)/\mathbb{C}P(1)$$

or

$$\cdots \longrightarrow \Omega_n^{SU} \xrightarrow{F_*} \Omega_n^U \xrightarrow{(\partial,d)} \Omega_{n-2}^{SU} \oplus \Omega_{n-4}^U \xrightarrow{(t,0)} \Omega_{n-1}^{SU} \longrightarrow \cdots .$$

From this one knows the complete structure of $\Omega_n^{U,SU}$. The interesting part of the sequence is

$$\Omega_{8n+2}^{SU} \xrightarrow{F_{8n+2}} \Omega_{8n+2}^{U} \longrightarrow \Omega_{8n+2}^{U,SU} \longrightarrow \Omega_{8n+1}^{SU} = Z_2^{|\pi(n)|} \xrightarrow{F_{8n+1}} \Omega_{8n+1}^{U} = 0$$

in which F_{8n+2} maps onto the direct summand of classes for which all numbers divisible by c_1 are zero.

From $\Omega_{8n+2}^{U,SU} \cong \Omega_{8n}^{SU} \oplus \Omega_{8n-2}^{U}$ the extension must be completely nontrivial. Classes in Ω_{8n+1}^{SU} are all of the form $\theta\cdot M^{8n}$, with the characteristic numbers

$$s_{(\omega',\omega')}(e_{\mathcal{P}})\mathcal{S}[M^{8n}] \equiv s_{(\omega',\omega')}(\pi)[\theta\cdot M^{8n}] \quad (\text{in } Z_2)$$

for $\omega' \in \pi(n)$ detecting these classes. Now $\theta \in \Omega_1^{SU}$ comes from $[D^2] \in \Omega_2^{U,SU}$, with D^2 being given the usual framing of its boundary. $[D^2 \subset \mathbb{C}P(1)$ may be realized as the disc bundle of λ over $\mathbb{C}P(0)$, with $\tau(D^2)$ being induced from λ over $\mathbb{C}P(1)$, giving the standard trivialization over S^1, which is the unusual framing of S^1. Over $\mathbb{C}P(1)$, $\mathcal{S}(\lambda) = 1 + k\bar{\alpha}$ with $\mathcal{S}(\tau) = \mathcal{S}(\xi)^2 = \mathcal{S}(\lambda)^{-2} = 1 - 2k\bar{\alpha}$ and $\mathcal{S}[\mathbb{C}P(1)] = -1$ gives $k = 1/2$ or $\mathcal{S}[D^2] = 1/2$.]

Since $s_{(\omega',\omega')}(e_{\mathcal{P}})$ has only nonzero components in dimensions a multiple of 4, while $\mathcal{S}_{4\ell+2}$ is divisible by c_1, the numbers $s_{(\omega',\omega')}(e_{\mathcal{P}})\mathcal{S}[V]$ are meaningful invariants of $8k+2$ dimensional (U,SU) manifolds, and

$$\begin{aligned} s_{(\omega',\omega')}(e_{\mathcal{P}})\mathcal{S}[D^2 \times M^{8n}] &= s_{(\omega',\omega')}(e_{\mathcal{P}})\mathcal{S}[M^{8n}]\cdot\mathcal{S}[D^2], \\ &= 1/2\ s_{(\omega',\omega')}(e_{\mathcal{P}})\mathcal{S}[M^{8n}] \end{aligned}$$

so that the numbers $s_{(\omega',\omega')}(e_{\mathcal{P}})\mathcal{S}$ map $\Omega_{8n+2}^{U,SU}$ onto a subgroup of odd index in $(1/2\ Z)^{|\pi(n)|}$ while sending Ω_{8n+2}^{U} into a subgroup of odd index in $Z^{|\pi(n)|}$. Thus by means of the examples $\{D^2 \times M^{8n}\}$ one also gets complete nontriviality of the extension.

Conner and Floyd [41] have noted that the use of cobordism theory proves the Adams result:

<u>Theorem</u>: The homomorphism

$$e_{\mathbb{C}} : \Omega_n^{fr} \longrightarrow Q/Z$$

maps precisely onto the integral multiples of

$$\begin{array}{lll} 1 & \text{if} & n = 8k+5 \\ 1/2 & \text{if} & n = 8k+1. \end{array}$$

Since this gives a cobordism theoretic proof of a homotopy result on framed cobordism, their argument will be reproduced here.

<u>Proof</u>: Let M be a framed $8k+5$ manifold. From $\Omega_{8k+5}^{SU} = 0$, $M = \partial V$ with V an SU manifold. Then $e_{\mathbb{C}}[M]$ is the reduction mod Z of $\mathcal{J}[V,M] = (\mathcal{J})_{8k+6}[V,M]$ but $\mathcal{J}_{8k+6}$ is divisible by c_1 and $c_1(V,M) = 0$ since V is an SU manifold. Thus $\mathcal{J}[V,M] \in Z$ and $e_{\mathbb{C}}[M] = 1 \in Q/Z$.

If M is a framed $8k+1$ manifold, $2[M] = 0$ in $\Omega_{8k+1}^{SU} = Z_2^{|\pi(k)|}$, so $2M = \partial V$ with V an SU manifold. Thus $2e_{\mathbb{C}}[M] = e_{\mathbb{C}}[2M]$ is the reduction mod Z of $\mathcal{J}[V,\partial V]$, which is zero as above. This makes $e_{\mathbb{C}}[M]$ a multiple of 1/2.

Now let V^{8k+1} be the framed manifold of the last section with V^{8k+1} cobordant to $(M^8)^k \cdot \theta$ in Ω_{8k+1}^{SU}, and with $\partial W = V \cup (-(M^8)^k \times \theta)$, and let $\bar{W} = W \cup (M^8)^k \times D^2/(M^8)^k \times S^1$. Then $\bar{W}$ is a U manifold with boundary V, so $e_{\mathbb{C}}[V^{8k+1}] = \mathcal{J}[W,V]$ mod Z. Since $\mathcal{J}$ is an invariant of (U,SU) cobordism in dimension $8k+2$, $\mathcal{J}[W,V] = S[(M^8)^k \times D^2] = 1/2\{\mathcal{J}(M^8)\}^k = 1/2$. Thus $e_{\mathbb{C}} : \Omega_{8k+1}^{fr} \longrightarrow Q/Z$ maps precisely onto the multiples of 1/2. **

Relation to Unoriented Cobordism

The nature of the forgetful homomorphism $F_* : \Omega_*^{SU} \longrightarrow \mathfrak{N}_*$ has been studied by P. G. Anderson [9], Stong [118, 121], and Conner and Landweber [42]. One way to phrase the result is:

Theorem: One may choose generators x_i of $\mathfrak{N}_*$, $\dim x_i = i$, $i \neq 2^s-1$, so that:

1) $\mathfrak{N}_* = Z_2[x_i]$,

2) $\mathcal{W}_*(R,2) = Z_2[x_k, x_{2^j}^2 \mid k$ not a power of $2]$,

3) There is a derivation $\partial_1 : \mathcal{W}_i(R,2) \longrightarrow \mathcal{W}_{i-1}(R,2)$ with $\partial_1^2 = 0$, $\partial_1 x_{2k} = x_{2k-1}$, k not a power of 2, $\partial_1(x_{2^j}^2) = 0$, and the image of Ω_*^{SO} in $\mathfrak{N}_*$ is additively generated by image ∂_1 and the Z_2 polynomials in the classes x_{2t}^2 (t any integer).

4) The image of Ω_*^U in $\mathfrak{N}_*$ is the squares of the classes in $\mathfrak{N}_*$; i.e. $Z_2[x_i^2]$.

5) The image of $\mathcal{W}_*(\mathbb{C},2)$ in $\mathfrak{N}_*$ is the squares of the classes in $\mathcal{W}_*(R,2)$.

6) The image of Ω_*^{SU} in $\mathfrak{N}_*$ is the subring additively generated by the squares of classes in image ∂_1 and the Z_2 polynomials in the classes x_{2k}^4 (k not a power of 2) and the $x_{2^j}^8$.

Proof: The classes x_i needed were defined in Chapter VIII and satisfy properties 1-5. To verify property 6, consider a $\mathcal{W}_*(\mathbb{C},2)$ manifold M, with $[M] = [M']^2$ in $\mathfrak{N}_*$, M' belonging to $\mathcal{W}_*(R,2)$. The formulae for computing the Chern numbers of ∂M are the same as those for computing Stiefel-Whitney numbers of $\partial_1 M'$ so $[\partial M] = [\partial_1 M']^2$ in $\mathfrak{N}_*$. Thus image ∂ maps precisely onto the squares of classes in image ∂_1.

Now kernel ∂ is additively generated by image ∂ and the polynomials in the classes

$$c_4 = \Phi(\mathbb{C}P(1)^2) = 9\mathbb{C}P(1)^2 - 8\mathbb{C}P(2)$$

and

$$c_{8n} = \Phi(z_{4n}'^2) - z_{4n-2}'\Phi(z_2'z_{4n}') = z_{4n}'^2 + 2V^4 z_{4n-2}'^2 - z_2'z_{4n}'z_{4n-2}' - 4V^4 z_{4n-2}'^2 .$$

Since $z_2' = [\mathbb{C}P(1)]$ bounds in $\mathfrak{N}_*$, the image of kernel ∂ in $\mathfrak{N}_*$ is additively generated by the squares of classes in image ∂_1 and the polynomials in the classes $z_{4n}'^2$. However, z_{4n}' reduces to x_{2n}^2 if n is not a power of 2 and to x_n^4 if n is a power of 2. In addition, the classes c_{8n} contain SU manifolds since kernel ∂ = image Ω_*^{SU} in dimensions congruent to zero mod 8, so the images of Ω_*^{SU} and kernel ∂ in $\mathfrak{N}_*$ coincide and are given by (6). **

The image of Ω_*^{SU} in $\mathfrak{N}_*$ may be described in a slightly different fashion following Conner and Landweber [42], as:

Theorem: The image of Ω_*^{SU} in $\mathfrak{N}_*$ consists of the unoriented cobordism classes $[M]^2$, where M is an oriented manifold for which all Pontrjagin numbers with $\mathcal{P}_1$ as a factor are even.

Proof: Let $A_* \subset \mathfrak{N}_*$ be the cobordism classes of oriented manifolds for which all Pontrjagin numbers with $\mathcal{P}_1$ as a factor are even.

Since Pontrjagin numbers vanish on image $\partial_1 \subset$ image Ω_*^{SO}, image $\partial_1 \subset A_*$. Further z_{4n}' is the class of a complex manifold for which all Chern numbers divisible by c_1^2 are zero. But in mod 2 cohomology c_1 reduces to w_2 and $\mathcal{P}_1$ reduces to w_2^2, so that since Pontrjagin numbers are given by Chern numbers, or Stiefel-Whitney numbers (when reduced mod 2), one has $z'_{4n} \in A_*$. Thus the squares of classes in A_* contains image Ω_*^{SU}.

Letting $B_* \subset A_*$ be the subring additively generated by image ∂_1 and the polynomials in the z'_{4n}, one has the image of Ω_*^{SU} given precisely by the squares of classes in B_*.

Now consider the ring $\ker\partial/\mathrm{im}\partial = \mathrm{im}\Omega_*^{SO}/\mathrm{im}\partial$, whose dual is the space of Pontrjagin numbers reduced mod 2. $A_*/\mathrm{im}\partial$ then has dual space the space of Pontrjagin numbers mod 2 which do not have $\mathcal{P}_1$ as a factor, so $A_*/\mathrm{im}\partial$ has dimension equal to that of $Z_2[\mathcal{P}_i | i > 1]$. Since $B_*/\mathrm{im}\partial = Z_2[z'_{4n} | n \geq 2]$, this gives $B_* = A_*$ completing the proof. **

This characterizes image Ω_*^{SU} in $\mathcal{N}_*$ as those cobordism classes for which all Stiefel-Whitney numbers divisible by an odd w_i, or by w_2 are zero and for which all Stiefel-Whitney numbers of the form $w_4^2 w_{4\omega}^2$ are zero. [Such being the square (odd w_i numbers zero) of a class with w_1 numbers zero (hence oriented) for which the numbers $w_2^2 w_{2\omega}^2 = \mathcal{P}_1 \mathcal{P}_\omega$ are zero mod 2.]

Relation to Oriented Cobordism

Having studied the 2 primary part of the relationship of Ω_*^{SU} with Ω_*^{SO}, one may turn to the composite

$$f : \Omega_*^{SU} \xrightarrow{F_*} \Omega_*^{SO} \xrightarrow{\pi} \Omega_*^{SO}/\mathrm{Torsion}.$$

Writing the universal Pontrjagin class $\mathcal{P} \in H^*(BSO;Q)$ as $\Pi(1 + x_j^2)$, $\dim x_j = 2$, one defines classes $s_\omega(e_{\mathcal{P}})$ as the s_ω symmetric functions of the variables $e^{x_j} + e^{-x_j} - 2$ and lets $\hat{A}$ be the product of the classes $x_j/2 \sinh(x_j/2)$. Let

$$\rho : H_*(BSO;Q) \longrightarrow Q[\alpha_i] : x \longrightarrow \sum s_\omega(e_{\mathcal{P}})\hat{A}[x]\cdot\alpha_\omega$$

and let $B_n^{SO} = \{x \in H_n(BSO;Q) | \rho(x) \in Z[\alpha_i]\}$, with $B_*^{SO} = \bigoplus_n B_n^{SO}$.

Note: If $x \in B_n^{SO}$, and $\omega \in \pi(n/4)$, $s_\omega(e_{\mathcal{P}})\hat{A}[x] = s_\omega(\mathcal{P})[x] \in Z$ so x is the reduction of an integral homology class. Since $\rho(x) \in Z[1/2][\alpha_i]$, x is in fact the image of the fundamental class of an oriented manifold, so $B_*^{SO} \subset \tau\Omega_*^{SO}$.

Since $\rho(x) \in Z[1/2][\alpha_i]$ for all $x \in \tau\Omega_*^{SO}$, one also sees that $B_n^{SO} \subset \tau\Omega_n^{SO}$ has 2 primary index.

Lemma: If M is a manifold with "$P(\mathbb{C}^2)$ structure", then $\tau[M] \in B_*^{SO}$; i.e. the image of $\mathcal{W}_*(\mathbb{C},2)$ in $\tau\Omega_*^{SO}$ is contained in B_*^{SO}.

Proof: If $\dim M \not\equiv 0(4)$, $s_\omega(e_{\mathcal{P}})\hat{A}[M] = 0$, while if $\dim M \equiv 0(4)$,

$$\begin{aligned} s_\omega(e_{\mathcal{P}})\hat{A}[M] &= s_\omega(e_{\mathcal{P}})\hat{A} \cdot \sum \frac{(-c_1)^i}{2^i i!}[M], \\ &= s_\omega(e_{\mathcal{P}})\mathcal{S}[M], \\ &\in Z \end{aligned}$$

for every nonzero component of $s_\omega(e_{\mathcal{P}})\hat{A}$ has dimension congruent to zero mod 4 and is a polynomial in the Chern classes of M so is annihilated by c_1^{2k}, the value being integral since $s_\omega(e_{\mathcal{P}})\mathcal{S}[M]$ is the value of a K theory number. **

Lemma: Let $P_* \subset \mathcal{W}_*(\mathbb{C},2) \otimes Z[1/2]$ be the integral polynomial subring $Z[x_1^2, x_{2i}-1/2\, x_1x_{2i-1}]$. Under the natural homomorphism of groups

$$\tau : \mathcal{W}_*(\mathbb{C},2) \otimes Z[1/2] \longrightarrow B_*^{SO} \otimes Z[1/2]$$

the subring P_* maps into B_*^{SO}, and $\tau|P_*$ is a ring homomorphism.

Proof: Since $P_* \subset$ kernel ∂, the usual and unusual products coincide on P_*, making $\tau | P_*$ a ring homomorphism. Since $x_1^2 \in \mathcal{W}_*(\mathbb{C},2)$, $\tau(x_1^2) \in B_*^{SO}$, while $\tau(x_{2i}-1/2\; x_1 x_{2i-1}) = \tau(x_{2i})$ since $\tau x_1 = 0$, and $\tau(x_{2i}) \in B_*^{SO}$ since $x_{2i} \in \mathcal{W}_*(\mathbb{C},2)$. Since B_*^{SO} is a ring, the result is clear. **

Proposition: B_*^{SO} is an integral polynomial ring on classes y_{4i}, $i \geq 1$.

Proof: Let $y_{4i} = \tau(x_{2i}-1/2\; x_1 x_{2i-1})$, $y_4 = \tau(x_1^2)$. Then $s_i(\mathcal{P})[y_{4i}] = m_{2i} m_{2i-1}$ if $i > 1$, $s_1(\mathcal{P})[y_4] = -8s_1(\mathcal{P})[\mathbb{C}P(2)] = -8.3$ and $\Omega_*^{SO} \otimes Z[1/2]$ is generated by these. Thus τP_* is contained in B_*^{SO} with 2 primary index.

One also notes that $\rho_2(y_{4i})$, the mod 2 reduction of $\rho(y_{4i})$ has largest monomial

a) α_i if $i \neq 2^s$ for any s,

b) $(\alpha_{2^{s-1}})^2$ if $i = 2^s$ for some $s > 0$, and

c) 1 if $i = 1$

as computed in the KO^* analysis of the z's. Thus τP_* is contained in B_*^{SO} with odd index, making $\tau : P_* \longrightarrow B_*^{SO}$ an isomorphism. **

Lemma: For any sequence $(i_1,\ldots,i_r)$, $2y_{4i_1}\cdots y_{4i_r}$ is the image of an SU manifold and y_4 and y_{4i}^2 (for all i) are the images of complex manifolds for which all Chern numbers divisible by c_1 are zero.

Proof: Let y_{4i} be the class of a $\mathcal{W}_*(\mathbb{C},2)$ manifold M_{4i}, and write $M_{4i} = 1/2\,(N_{4i} - \mathbb{C}P(1)N_{4i-2})$ where N_{2j} is an SU manifold (N_2 = empty).

For any sequence $(i_1,\ldots,i_r)$

$$\partial(\mathbb{C}P(1) \times M_{4i_1} \times\ldots\times M_{4i_r}) = \frac{1}{2^r}\,\partial\{N_{4i_1} \times\ldots\times N_{4i_r} \times \mathbb{C}P(1) + \text{terms}$$

$$\mathbb{C}P(1)^t \times \Pi N_{4i} \times \Pi N_{4j-2}\},$$

$$= \frac{1}{2^{r-1}} \cdot N_{4i_1} \times\ldots\times N_{4i_r} + \text{terms divisible by}$$

$$\text{an } N_{4j-2}.$$

As oriented manifolds, the terms with N_{4j-2} factors are zero, so $2y_{4i_1}\cdots y_{4i_r}$ is the class of the SU manifold $\partial(\mathbb{C}P(1) \times M_{4i_1} \times\ldots\times M_{4i_r})$.

Consider the classes in $\Omega_*^U \otimes Q$ given by

$$A = 1/4(N_{4i} \times N_{4i} - M_4 \times N_{4i-2} \times N_{4i-2}) = 1/4(N_{4i}^2 - [9\mathbb{C}P(1)^2-8\mathbb{C}P(2)]N_{4i-2}^2)$$

and

$$B = M_{4i}^2 + \mathbb{C}P(1)N_{4i-2}M_{4i} - 2[\mathbb{C}P(1)^2-\mathbb{C}P(2)]N_{4i-2}^2,$$

$$= 1/4N_{4i}^2-1/2N_{4i}\mathbb{C}P(1)N_{4i-2}+1/4\mathbb{C}P(1)^2N_{4i-2}^2+1/2N_{4i}\mathbb{C}P(1)N_{4i-2}-2/4\mathbb{C}P(1)^2N_{4i-2}^2$$

$$-1/4[8\mathbb{C}P(1)^2 - 8\mathbb{C}P(2)]N_{4i-2}^2.$$

From the expansion one clearly has $A = B$, with B being given as the class of a complex manifold and A having all Chern numbers divisible by c_1 zero. Since N_{4i-2} is zero in oriented cobordism mod torsion, B is clearly the class y_{4i}^2. **

From the Conner and Landweber lemma that KO theory numbers detect $H_*(\mathcal{W} \otimes Z_2)$, one has $\rho_2(x) = 0$ if $x \in \text{image } \partial$, so $\tau : \ker\partial \longrightarrow B_*^{SO}$ takes image ∂ into $2B_*^{SO}$. Thus $H_*(\mathcal{W})$ maps isomorphically into the algebra

$B_*^{SO}/2B_*^{SO}$ with image $Z_2[y_4, y_{4i}^2 | i > 1]$. Since Ω_*^{SU} has image in Ω_*^U coinciding with $\ker\partial$ in dimensions congruent to zero mod 8 and with $\mathrm{image}\,\partial$ in dimensions congruent to 4 mod 8, this gives:

<u>Theorem</u>: Under the forgetful homomorphism to $\Omega_*^{SO}/\mathrm{Torsion}$, $\mathcal{W}_*(\mathbf{C},2)$ is mapped onto an integral polynomial subring $Z[y_{4i}] = B_*^{SO}$. Further, $\mathrm{image}\,\partial$ maps onto $2B_*^{SO}$, $\mathrm{kernel}\,\partial$ maps onto the span of $2B_*^{SO}$ and $Z[y_4, y_{4i}^2 | i > 1]$ and Ω_*^{SU} maps onto the span of $2B_*^{SO}$ and $Z[y_{4i}^2]$.

Chapter XI

Spin, Spin^c and Similar Nonsense

Among the (B,f) cobordism theories, the most interesting examples arise from the classical groups. The most difficult of these which have been successfully analyzed are the theories given by the groups Spin and Spin^c. The group Spin arose classically in the study of Lie groups, being the simply connected covering group of the special orthogonal group.

To justify the study of these theories, another attack will be used here. Briefly, one may consider the cobordism classification problem for manifolds which are orientable for the bundle cohomology theories KO^* and K^*.

Reference material for this is: Atiyah, Bott, and Shapiro [16] for a discussion of the groups, and Anderson, Brown, and Peterson [8] for calculation of the cobordism theories.

To begin, one may return to the construction of K theoretic orientation classes for complex bundles. For this, one considered the vector space $\Lambda(\mathbb{C}^k)$, treating this as a representation space for the unitary group $U(k)$ in order to construct bundles. Clearly one may ask: Is it possible to find a larger group, acting on both $\mathbb{C}^k$ and $\Lambda(\mathbb{C}^k)$ which will possess all of the properties used in the construction?

<u>Lemma 1</u>: The ring of endomorphisms of $\Lambda(\mathbb{C}^k)$ is an algebra over $\mathbb{C}$ and is generated by the endomorphisms F_v and F_v^* for $v \in \mathbb{C}^k$.

<u>Proof</u>: $\Lambda(\mathbb{C}^k)$ has a base consisting of the elements $e_I = e_{i_1} \wedge \dots \wedge e_{i_r}$, $i_1 < \dots < i_r$. Let I,J be any two sequences of this type, $I = (i_1,\dots,i_r)$, $J = (j_1,\dots,j_s)$. There is a sequence K formed from $\{1,\dots,k\} - I$ and $e_K \wedge e_I = \pm \sigma$, $\sigma = e_1 \wedge \dots \wedge e_k$, and say $e_K \wedge e_I = (-1)^t \sigma$. Consider the

operation

$$T = (-1)^t F_{e_{j_1}} \cdots F_{e_{j_s}} F^*_{e_k} \cdots F^*_{e_1} F_{e_{k_1}} \cdots F_{e_{k_p}}$$

where $K = (k_1, \ldots, k_p)$. Then $T(e_L) = 0$ if $L \neq I$ and $T(e_I) = e_J$. Since operations of this type form a base of $\mathrm{End}(\Lambda(\mathbb{C}^k))$, this completes the proof. **

<u>Lemma 2</u>: The operations $\phi_v = F_v + F^*_v$, for $v \in \mathbb{C}^k$ satisfy the identities:

a) $\phi_v^2(x) = \|v\|^2 \cdot x$,

b) $\phi_{iv} = i(F_v - F^*_v)$,

c) If $v, w \in \mathbb{C}^k$ are orthogonal, then $\phi_v\phi_w + \phi_w\phi_v = 0$ and $\phi_v\phi_{iv} + \phi_{iv}\phi_v = 0$.

<u>Proof</u>: a) was verified in Chapter IX. For b) one has $F_{iv} = iF_v$, while

$$< F^*_{iv}y, z > \; = \; < y, iv \wedge z > \; = -i < y, F_v z > \; = \; < -iF^*_v y, z >$$

so $F^*_{iv} = -iF^*_v$. For c) one has $\phi_{v+w} = \phi_v + \phi_w$, so

$$\begin{aligned}(\phi_{v+w})^2(x) &= \|v+w\|^2 \cdot x, \\ &= (\phi_v^2 + \phi_v\phi_w + \phi_w\phi_v + \phi_w^2)(x), \\ &= (\|v\|^2 + \|w\|^2)x + (\phi_v\phi_w + \phi_w\phi_v)x.\end{aligned}$$

If v, w are orthogonal, $\|v+w\|^2 = \|v\|^2 + \|w\|^2$ giving $\phi_v\phi_w + \phi_w\phi_v = 0$. If $w = iv$, $\|v+w\|^2 = |1+i|^2 \cdot \|v\|^2 = 2\|v\|^2$, while $\|w\|^2 = \|v\|^2$, so $\phi_v\phi_w + \phi_w\phi_v = 0$. **

<u>Note</u>: c) may be rephrased: If v and w are orthogonal under the real inner product, $\mathrm{Re} < \, , \, >$, then $\phi_v\phi_w + \phi_w\phi_v = 0$.

Definition: Let V be a real inner product space. The Clifford algebra of V, $\mathrm{Cliff}(V)$, is a pair (A,f) where A is a real algebra with unit and $f : V \longrightarrow A$ is a linear transformation such that $f(v)^2 = \|v\|^2 \cdot 1$ and such that for any pair (B,g) with these properties there is a unique algebra homomorphism $\lambda : A \longrightarrow B$ with $g = \lambda \circ f$.

Note: $\mathrm{Cliff}(V)$ is, of course, unique up to natural isomorphism. If $v_1,\ldots,v_p$ is an orthonormal base of V, the algebra A is the real algebra with unit generated by $v_1,\ldots,v_p$ with relations $v_i^2 = 1$ and $v_iv_j + v_jv_i = 0$ if $i \neq j$. A has dimension 2^p with base given by the monomials $v_{i_1}\ldots v_{i_s}$ with $1 \leq i_1 <\ldots< i_s \leq p$.

Proposition: The linear transformation $\phi : \mathbb{C}^k \longrightarrow \mathrm{End}(\Lambda(\mathbb{C}^k))$ induces an algebra homomorphism $\phi : \mathrm{Cliff}(\mathbb{C}^k) \longrightarrow \mathrm{End}(\Lambda(\mathbb{C}^k))$ which extends to an isomorphism

$$\phi : \mathrm{Cliff}(\mathbb{C}^k) \otimes_{\mathbb{R}} \mathbb{C} \xrightarrow{\cong} \mathrm{End}(\Lambda(\mathbb{C}^k)).$$

Proof: Clearly, $\phi : \mathrm{Cliff}(\mathbb{C}^k) \otimes_{\mathbb{R}} \mathbb{C} \longrightarrow \mathrm{End}(\Lambda(\mathbb{C}^k))$ is defined and is a homomorphism of complex algebras. Both algebras have dimension 2^{2k} over $\mathbb{C}$, so it suffices to show ϕ is epic. Since $\phi(v) = F_v + F_v^*$ and $\phi(iv) = i(F_v - F_v^*)$, image ϕ contains $F_v = (\phi(v) - i\phi(iv))/2$ and $F_v^* = (\phi(v) + i\phi(iv))/2$. By Lemma 1, these generate $\mathrm{End}(\Lambda(\mathbb{C}^k))$ as algebra over $\mathbb{C}$ as v runs over $\mathbb{C}^k$. Thus ϕ is an isomorphism. **

Note: The critical point of this is simply that one has a very easy way to express the endomorphisms. $\mathrm{Cliff}(\mathbb{C}^k)$ really enters through:

Proposition: ϕ identifies $\mathrm{Cliff}(\mathbb{C}^k)$ with the real subalgebra of $\mathrm{End}(\Lambda(\mathbb{C}^k))$ consisting of those operators which commute with μ.

<u>Proof</u>: Since $\mu \circ \phi_v = \phi_v \circ \mu$ (Lemma 8, Chapter IX), $\mathrm{Cliff}(\mathbb{C}^k)$ commutes with μ. Since $\mu i = -i\mu$ (Lemma 3, Chapter IX), $\mathrm{Cliff}(\mathbb{C}^k)$ is precisely the subset of $\mathrm{Cliff}(\mathbb{C}^k) \otimes \mathbb{C}$ consisting of operators which commute with μ. **

It is now clear that one may perform most of the construction of an orientation by considering the group G^c (or G) consisting of elements of $\mathrm{Cliff}(\mathbb{C}^k) \otimes \mathbb{C}$ (or $\mathrm{Cliff}(\mathbb{C}^k)$) which are:

1) Invertible (i.e. are automorphisms of $\Lambda(\mathbb{C}^k)$);

2) Satisfy $x^*x = 1$, where x^* is the conjugate linear antiautomorphism of $\mathrm{Cliff}(C^k) \otimes C$ defined by the adjoint. (<u>Note</u>: $\phi(x)$ preserves the inner product on $\Lambda(C^k)$ if and only if, for all y,z, one has

$$< y,z > = < \phi(x)y,\phi(x)z > = < \phi(x)^*\phi(x)y,z >$$

or if and only if $\phi(x)^*\phi(x) = 1$ or $x^*x = 1$. $*$ is then clearly a conjugate linear antiautomorphism and is the identity on $\mathbb{C}^k \subset \mathrm{Cliff}(\mathbb{C}^k)$ since $(\phi_v)^* = F_v^* + F_v^{**} = \phi_v$.)

3) Preserve the even-odd decomposition of $\Lambda(\mathbb{C}^k)$. (<u>Note</u>: $\mathrm{Cliff}(\mathbb{C}^k)$ is Z_2 graded; this takes the elements of even grading.)

Corresponding to any principal G^c (or G) bundle one may then form an associated complex vector bundle with fiber $\Lambda(\mathbb{C}^k)$, decomposing into even-odd summands (and admitting a bundle map μ). One also requires a vector bundle with fiber $\mathbb{C}^k$ or $\mathbb{R}^{2k}$, with each fiber acting on $\Lambda(\mathbb{C}^k)$ to define $\phi : \pi^*(\Lambda^{ev}(\mathbb{C}^k)) \longrightarrow \pi^*(\Lambda^{od}(\mathbb{C}^k))$. This is the analogue of Lemma 7, Chapter IX, in that one desires an action of G^c (or G) or a subgroup thereof on $\mathbb{C}^k$.

Letting $g \in G^c$, one wishes to find for $v \in \mathbb{R}^{2k}$ an element $gv \in \mathbb{R}^{2k}$ so that $g\circ\phi_v = \phi_{gv}\circ g$. Now letting $g = \phi(x)$,

$$\phi(x)\circ\phi_v(y) = \phi_{gv}\circ\phi(x)(y)$$

gives

$$\phi_{gv} = \phi(x)\circ\phi_v\circ\varphi(x)^{-1} = \phi(xvx^{-1})$$

or

$$gv = x\cdot v\cdot x^{-1}.$$

Note: The subgroup consisting of elements $g = \phi(x)$ for which $xvx^{-1} \in \mathbb{C}^k$ if $v \in \mathbb{C}^k$ clearly acts on $\mathbb{C}^k$, giving an associated bundle with fiber $\mathbb{C}^k$. The orientation construction may then be made for this bundle.

Definition: Let $\mathrm{Spin}^c(k)$ (or $\mathrm{Spin}(k)$) denote the subgroup of $\mathrm{Cliff}(\mathbb{R}^k) \otimes \mathbb{C}$ (or $\mathrm{Cliff}(\mathbb{R}^k)$) consisting of invertible elements g for which:

1) $gvg^{-1} \in \mathbb{R}^k$ for all $v \in \mathbb{R}^k$;

2) $g^*g = 1$ where $*$ is the conjugate linear antiautomorphism extending the identity map on $\mathbb{R}^k$; and

3) the Z_2 grading is zero.

The above analysis shows that Spin^c and Spin bundles are K^* and KO^* orientable respectively. (Note: Restricting to a fiber, the bundle further reduces to U or SU, where it is known that this construction gives a generator for the cohomology of the sphere). The orientation is clearly multiplicative, for $\Lambda(\mathbb{C}^k \oplus \mathbb{C}^\ell) \cong \Lambda(\mathbb{C}^k) \otimes \Lambda(\mathbb{C}^\ell)$ and all constructions are compatible with this decomposition.

To relate this to the classical treatment, one may analyze the groups $\mathrm{Spin}^c(k)$ and $\mathrm{Spin}(k)$ in a group theoretic fashion.

Denote by Γ_k the subgroup of invertible elements x in $\mathrm{Cliff}(\mathbb{R}^k) \otimes C$ (or $\mathrm{Cliff}(\mathbb{R}^k)$) satisfying:

1) $xyx^{-1} \in \mathbb{R}^k$ for all $y \in \mathbb{R}^k$;

2) $x^*x = 1$; and

3) x is a homogeneous element in the Z_2 grading.

Define a representation of Γ_k on $\mathbb{R}^k$ by

$$\rho : \Gamma_k \longrightarrow \mathrm{Aut}(\mathbb{R}^k)$$

with $\rho(x)(y) = (-1)^{\deg x} xyx^{-1}$, where $\deg x$ is the integer mod 2 giving the grading to which x belongs.

Lemma 3: The homomorphism $\rho : \Gamma_k \longrightarrow \mathrm{Aut}(\mathbb{R}^k)$ has kernel precisely the scalar multiples $r \cdot 1$ where $\|r\| = 1$.

Proof: If $x \in \ker \rho$, $xy = (-1)^{\deg x} yx$ for all $y \in \mathbb{R}^k$. Let $e_1, \ldots, e_k$ be the standard base of $\mathbb{R}^k$ and write $x = a + e_1 b$, where a, b do not involve e_1 and $\deg a = \deg x$, $\deg b = \deg x + 1$. Then $xe_1 = ae_1 + e_1 b e_1 = ae_1 + (-1)^{\deg b} b$, and $(-1)^{\deg x} e_1 x = (-1)^{\deg x} e_1 a + (-1)^{\deg x} b$, so $b = 0$. Similar analysis with the other e_j shows that x cannot involve any e_j, so $x = r \cdot 1$. Since $x^* x = \bar{r} r = 1$, $\|r\| = 1$. **

Lemma 4: $\rho(\Gamma_k)$ is contained in the isometries of $\mathbb{R}^k$.

Proof: $\|\rho(x)y\|^2 = (\rho(x)y)^*(\rho(x)y) = (-1)^{\deg x}(x^{-1})^* y^* x^* \cdot (-1)^{\deg x} xyx^{-1}$ $= y^* y = \|y\|^2$ so $\rho(x)$ preserves norms and hence also inner products. **

Lemma 5: $\rho : \Gamma_k \longrightarrow O(k)$ is epic.

Proof: Let $v \in S^{k-1} \subset \mathbb{R}^k$ be any unit vector, and extend to an orthonormal base $e_1 = v$, $e_2, \ldots, e_k$. Then $e_1 \in \Gamma_k$ and

$$\rho(e_1)(e_i) = -e_1 e_i e_1 = \begin{cases} -e_1 & i = 1, \\ e_i & i \neq 1. \end{cases}$$

Thus $\rho(v)$ is the reflection in the hyperplane orthogonal to v. Since $O(k)$ is generated by these reflections, ρ is epic. **

Lemma 6: The homomorphism $\rho : \Gamma_k \longrightarrow O(k)$ maps the subgroup consisting of elements of degree zero precisely onto $SO(k)$.

Proof: Let $x \in \Gamma_k$. Then $\rho(x) = R_1 \circ \ldots \circ R_q$ for some reflections R_j and letting $x_j \in S^{k-1}$ with $\rho(x_j) = R_j$, one has $x = r \cdot x_1 \ldots x_q$. Then $\det\rho(x) = (-1)^q$ and $\deg x = q$, so $\rho(x) \in SO(k)$ if and only if $\deg x = 0$. **

Definition: Γ_k is denoted $Pin^c(k)$ in the case of $Cliff(\mathbb{R}^k) \otimes \mathbb{C}$ and is denoted $Pin(k)$ in the case of $Cliff(\mathbb{R}^k)$.

Theorem: There are exact sequences

$$1 \longrightarrow U(1) \longrightarrow Pin^c(k) \longrightarrow O(k) \longrightarrow 1,$$

$$1 \longrightarrow U(1) \longrightarrow Spin^c(k) \longrightarrow SO(k) \longrightarrow 1,$$

$$1 \longrightarrow Z_2 \longrightarrow Pin(k) \longrightarrow O(k) \longrightarrow 1, \quad \text{and}$$

$$1 \longrightarrow Z_2 \longrightarrow Spin(k) \longrightarrow SO(k) \longrightarrow 1$$

and isomorphisms

$$Pin^c(k) \cong Pin(k) \times_{Z_2} U(1)$$

$$Spin^c(k) \cong Spin(k) \times_{Z_2} U(1)$$

where $U(1), Z_2$ denote the scalars of norm 1.

Proof: All has been proved except the isomorphisms. For these one has inclusions $Pin(k) \longrightarrow Pin^c(k)$ and $U(1) \longrightarrow Pin^c(k)$ and as in Lemma 6, $Pin^c(k)$ consists of all $rx_1 \ldots x_q$, $x_j \in S^{k-1}$, $r \in U(1)$ and $Pin(k)$ consists of all $\pm x_1 \ldots x_q$. This gives the isomorphism for $\mathrm{Pin}^c(k)$, while that for $\mathrm{Spin}^c(k)$ follows by taking the elements of degree zero. **

To complete the analysis of these groups, one has:

Proposition: For $k \geq 2$, $\rho : \mathrm{Spin}(k) \longrightarrow SO(k)$ is the nontrivial double cover.

Proof: It suffices to show that $+1$ and -1, the kernel of ρ, can be connected by a path in $\mathrm{Spin}(k)$. One such path is

$$\lambda : [0,\pi] \longrightarrow \mathrm{Spin}(k) : t \longrightarrow \cos(t) + \sin(t) \cdot e_1 e_2 . \quad **$$

Corollary: For $k \geq 2$, $\mathrm{Spin}(k)$ is connected, and for $k \geq 3$, $\mathrm{Spin}(k)$ is simply connected.

Proof: $\pi_0(SO(k)) = 0$; $\pi_1(SO(k)) = Z_2$ for $k \geq 3$. **

Now one may form the classifying spaces for these groups.

First one notes that Spin is the simply connected cover of SO, so that BSpin is the 2-connective cover of BSO. Thus one may identify BSpin with the total space of the fibration over BSO induced from the path fibration over $K(Z_2,2)$ via the map $f : BSO \longrightarrow K(Z_2,2)$ realizing w_2 (inducing an isomorphism on π_2). Thus one has

$$\begin{array}{ccc} BSpin & \longrightarrow & PK(Z_2,2) \\ {\scriptstyle\pi}\downarrow & & \downarrow \\ BSO & \xrightarrow{\ f\ } & K(Z_2,2) \end{array}$$

and the fiber of π is a $K(Z_2,1)$.

One may compute the cohomology algebra $H^*(BSpin;\mathcal{S})$ for any ring $\mathcal{S}$ containing 1/2 easily, since $H^*(K(Z_2,1);\mathcal{S})$ is trivial, making π^* an isomorphism.

To compute $H^*(BSpin;Z_2)$ is much more difficult, requiring some additional background.

<u>Lemma 7</u>: In $H^*(BO;Z_2)$, $Sq^{n-1}w_n = w_{2n-1}$ + decomposables.

<u>Proof</u>: From the Adem formulae,

$$Sq^{n-1}Sq^n = Sq^{2n-1} + \sum_{i=1}^{s} a_i Sq^{2n-1-i}Sq^i, \quad \text{with} \quad s = [(n-1)/2],$$

so in $\tilde{H}^*(\underset{\sim}{TBO};Z_2)$

$$\begin{aligned} Sq^{n-1}Sq^n U &= w_{2n-1}U + \sum_{i=1}^{s} a_i Sq^{2n-1-i}(w_i U), \\ &= w_{2n-1}U + \sum_{i=1}^{s} a_i \sum_{j=0}^{2n-1-i} (Sq^{2n-1-i-j}w_i)\cdot w_j U \end{aligned}$$

and the terms with $j = 0$ vanish, since for $2n-1-i > i$, one has $Sq^{2n-1-i}w_i = 0$. Thus $Sq^{n-1}Sq^n U = (w_{2n-1}$ + decomposables$)\cdot U$. Further

$$\begin{aligned} Sq^{n-1}Sq^n U &= Sq^{n-1}(w_n U), \\ &= Sq^{n-1}w_n\cdot U + \sum_{j=1}^{n-1} (Sq^{n-1-j}w_n)\cdot w_j U, \\ &= (Sq^{n-1}w_n + \text{decomposables})\cdot U. \end{aligned}$$

Equating these expressions gives the result. **

<u>Corollary</u>: $f^* : H^*(K(Z_2,2);Z_2) \longrightarrow H^*(BSO;Z_2)$ is monic.

<u>Proof</u>: $H^*(K(Z_2,2);Z_2)$ is the Z_2 polynomial ring on the classes $Sq^I \iota_2$, where $I = (2^r,2^{r-1},\ldots,1)$ and by the Lemma, $Sq^I w_2 = w_{2^{r+1}+1}$ + decomposables. Thus f^* is monic. **

This gives:

<u>Proposition</u>: $\pi^* : H^*(BSO;Z_2) \longrightarrow H^*(BSpin;Z_2)$ is epic, with kernel the ideal generated by w_2 over $\mathcal{A}_2$. Thus $Z_2[w_i | i \neq 1, 2^r+1; r \geq 0]$ maps isomorphically onto $H^*(BSpin;Z_2)$.

<u>Proof</u>: Let E^* denote the spectral sequence of the fibration π, E'^* that of the path fibration of $K(Z_2,2)$, with f^* the induced map. Since E^* is an $H^*(BSO;Z_2)$ module, one has induced a map of spectral sequences

$$Z_2[w_i | i \neq 1, 2^r+1; r \geq 0] \otimes E'^* \longrightarrow E^*.$$

This is an isomorphism on the E^2 level, hence also on the E^∞ level. Thus $\pi^* : Z_2[w_i | i \neq 1, 2^r+1; r \geq 0] \longrightarrow H^*(BSpin;Z_2)$ is an isomorphism. **

To form the classifying space $BSpin^c$, one considers the map $g : BSO \times K(Z,2) \longrightarrow K(Z_2,2)$ with $g^*(\iota_2) = w_2 \otimes 1 + 1 \otimes \iota$ and denotes by $BSpin^c$ the total space of the fibration induced via g from the path fibration over $K(Z_2,2)$. This gives a diagram

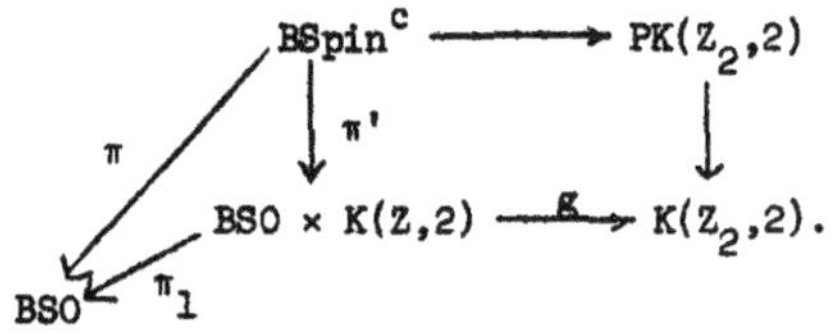

One then has the fibration $\pi : BSpin^c \longrightarrow BSO$ with fiber $K(Z,2) = BU(1)$, corresponding to the exact sequence

$$1 \longrightarrow U(1) \longrightarrow Spin^c \longrightarrow SO \longrightarrow 1.$$

<u>Notes</u>: 1) One has a homomorphism $Spin(k) \times U(1) \longrightarrow SO(k) \times U(1)$: $(x,y) \longrightarrow (\rho(x),y^2)$ inducing $\theta : Spin^c(k) \cong Spin(k) \times_{Z_2} U(1) \longrightarrow SO(k) \times U(1)$

and θ gives an exact sequence

$$1 \longrightarrow Z_2 \longrightarrow \mathrm{Spin}^c(k) \longrightarrow SO(k) \times U(1) \longrightarrow 1.$$

The construction given for $BSpin^c$ is the fibration associated with this sequence.

2) An oriented bundle admits a Spin structure if and only if its second Stiefel-Whitney class is zero; it admits a $Spin^c$ structure if and only if its second Stiefel-Whitney class is the reduction of an integral class.

The cohomology of $BSpin^c$ is now readily computable. Since $H^*(K(Z_2,2);\mathcal{S})$ is trivial if $1/2 \in \mathcal{S}$, $\pi'^* : H^*(BSO \times BU(1);\mathcal{S}) \longrightarrow H^*(BSpin^c;\mathcal{S})$ is an isomorphism. Using Z_2 cohomology, g^* is monic, and using the same spectral sequence argument, π'^* is epic with kernel the ideal generated by $w_2 + \acute{\iota}$ over $\mathcal{A}_2$. In particular, $\pi^* : H^*(BSO;Z_2) \longrightarrow H^*(BSpin^c;Z_2)$ is epic, with kernel the ideal generated by $w_3(Sq^1(w_2 + \dot{\iota}) = w_3$ in $H^*(BSO \times BU(1);Z_2))$.

<u>Note</u>: Let $f : BSpin \times BU(1) \longrightarrow BSO$ classify the sum of the canonical bundles over these spaces. Let $k : BSpin \times BU(1) \longrightarrow K(Z,2)$ be projection on the second factor. Then $f \times k : BSpin \times BU(1) \longrightarrow BSO \times K(Z,2)$ lifts to $BSpin^c$; say $\hat{f} : BSpin \times BU(1) \longrightarrow BSpin^c$ is the lift. It is immediate that $\hat{f}$ induces isomorphisms of homotopy and hence is a homotopy equivalence. (<u>Beware</u>: The multiplication on $BSpin^c$ is not the product multiplication on $BSpin \times BU(1)$.) Thus the cohomology of $BSpin^c$ follows from a knowledge of that of BSpin.

From the knowledge of the rational cohomology of BSpin and $BSpin^c$, one may evaluate the Chern character of the bundle theoretic orientation class constructed above.

<u>Proposition</u>: Let ξ be a Spin (or Spin^c) bundle over a space X and let $U(\xi) \in \widetilde{KO}^*(T\xi)$ (or $\tilde{K}^*(T\xi)$) be the orientation of ξ constructed above. Let $\Phi_H^{-1} : \tilde{H}^*(T\xi;Q) \longrightarrow H^*(X;Q)$ be the inverse of the Thom isomorphism defined by the standard orientation associated with the SO structure of ξ. Then:

a) $\Phi_H^{-1}(\text{ch}(\psi U(\xi))) = \hat{A}(-\xi)$ if ξ is a Spin bundle, (ψ being the usual homomorphism from KO^* to K^*)

b) $\Phi_H^{-1}(\text{ch}(U(\xi))) = e^{c_1(\xi)/2} \cdot \hat{A}(-\xi)$ if ξ is a Spin^c bundle, where $c_1(\xi) \in H^2(X;Z)$ is the integral class reducing to $w_2(\xi)$ which is defined by the Spin^c structure.

<u>Proof</u>: By stability and naturality of the orientation class, $\Phi_H^{-1}(\text{ch}(\psi U(\xi)))$ and $\Phi_H^{-1}(\text{ch}U(\xi))$ come from $H^*(B\text{Spin};Q)$ and $H^*(B\text{Spin}^c;Q)$ respectively. Corresponding to the inclusions $U(k) \longrightarrow \text{Spin}^c(2k)$ and $SU(k) \longrightarrow \text{Spin}(2k)$ one has maps $t : BU \longrightarrow B\text{Spin}^c$, $u : BSU \longrightarrow B\text{Spin}$. (<u>Note</u>: The natural map $BSU \xrightarrow{u'} BSO$ lifts to $B\text{Spin}$ since BSU is 3 connected; the map $BU \xrightarrow{t'} BSO \times BU(1)$, whose projection to BSO classifies the universal bundle and whose projection to $BU(1)$ classifies the determinant bundle clearly lifts to $B\text{Spin}^c$.) Both t^* and u^* are monic on rational cohomology (since t'^*, u'^* are monic and π'^* and π^* are isomorphisms), and since the orientation class is sent to the previously constructed orientation for U and SU bundles one may apply the previous computations.

For a U bundle, one has $\Phi_H^{-1}(\text{ch}U(\xi)) = \Phi_H^{-1}(\text{ch}\tilde{U}(\xi))$ (<u>Note</u>: To form $U(\xi)$ one had to introduce a sign depending on $\dim \xi$ in order to get an orientation: This was given by $\hat{U}(\xi) = (-1)^n\tilde{U}(\xi)$ if ξ is a complex n-plane bundle. See Chapter X), and this is $\mathscr{J}(-\xi) = e^{-c_1(-\xi)/2}\hat{A}(-\xi) = e^{c_1(\xi)/2}\hat{A}(-\xi)$.

For an SU bundle, $\Phi_H^{-1}(\text{ch}\psi U(\xi)) = \mathscr{J}(-\xi)$, exactly as for a complex bundle. Since $c_1(\xi) = 0$, this reduces to $\hat{A}(-\xi)$. **

In the computations which will follow, it will be necessary to have the following:

Proposition: Let $BO(k,\ldots,\infty)$, $BU(k,\ldots,\infty)$ denote the $(k-1)$ connective covers of BO and BU. Let

$$f : BO(k,\ldots,\infty) \longrightarrow K(\pi_k(BO),k)$$

and

$$g : BU(k,\ldots,\infty) \longrightarrow K(\pi_k(BU),k)$$

be the maps realizing the least possibly nonzero homotopy group. Then:

a) If $k \equiv 0,1,2,4 \pmod 8$, $f^* : H^i(K(\pi_k(BO),k);Z_2) \longrightarrow H^i(BO(k,\ldots,\infty);Z_2)$ is epic for $i < 2k$, and in these dimensions $H^*(BO(k,\ldots,\infty);Z_2)$ is isomorphic to

1) $(\mathcal{A}_2/\mathcal{A}_2Sq^1 + \mathcal{A}_2Sq^2)f^*(\iota_k)$ $\quad k \equiv 0\ (8)$,

2) $(\mathcal{A}_2/\mathcal{A}_2Sq^2)f^*(\iota_k)$ $\quad k \equiv 1\ (8)$,

3) $(\mathcal{A}_2/\mathcal{A}_2Sq^3)f^*(\iota_k)$ $\quad k \equiv 2\ (8)$,

4) $(\mathcal{A}_2/\mathcal{A}_2Sq^1 + \mathcal{A}_2Sq^5)f^*(\iota_k)$ $\quad k \equiv 4\ (8)$.

b) If k is even, $g^* : H^i(K(\pi_k(BU),k);Z_2) \longrightarrow H^i(BU(k,\ldots,\infty);Z_2)$ is epic for $i < 2k$, and in these dimensions $H^*(BU(k,\ldots,\infty);Z_2)$ is isomorphic to $(\mathcal{A}_2/\mathcal{A}_2Sq^1 + \mathcal{A}_2Sq^3)g^*(\iota_k)$.

Proof: The main step in the proof will be an induction on the statement:

$P(j)$: With the given hypotheses, f^* and g^* are epic for $i < 2k$ and $k + j$, and in these dimensions, the groups $H^*(BO(k,\ldots,\infty);Z_2)$ and $H^*(BU(k,\ldots,\infty);Z_2)$ are as asserted.

This is clearly true for $j = 1$, since f^* and g^* are isomorphisms in dimension k. One may then suppose $P(j)$ is true, and try to prove $P(j+1)$.

Consider the fibrations

$$\begin{array}{ccc} BO(k+1,\ldots,\infty) \longrightarrow BO(k,\ldots,\infty) & & BU(k+1,\ldots,\infty) \longrightarrow BU(k,\ldots,\infty) \\ \downarrow & & \downarrow \\ K(\pi_k(BO),k) & & K(\pi_k(BU),k). \end{array}$$

The spectral sequences of these fibrations give exact sequences

$$\ldots \longrightarrow H^i(X(k+1,\ldots,\infty)) \xrightarrow{\tau} H^{i+1}(K(\pi_k(X),k)) \longrightarrow H^{i+1}(X(k,\ldots,\infty)) \longrightarrow \ldots$$

$$\ldots \longrightarrow H^{2k}(X(k+1,\ldots,\infty)) \xrightarrow{\tau} H^{2k+1}(K(\pi_k(X),k)) \longrightarrow H^{2k+1}(X(k,\ldots,\infty))$$

in which $\tau : H^i(X(k+1,\ldots,\infty)) \longrightarrow H^{i+1}(K(\pi_k(X),k))$ is the transgression.

By the inductive assumption $H^*(X(k+1,\ldots,\infty)) = (\mathcal{A}_2/\mathcal{J}_\ell)\cdot x_\ell$ in dimensions less than $2(k+1)$ and $k + 1 + j$, where $x_\ell \in H^\ell(X(k+1,\ldots,\infty))$ is the nonzero class of least positive dimension and $\mathcal{J}_\ell$ is the appropriate ideal. τ is a homomorphism over $\mathcal{A}_2$ and hence it suffices to prove that the sequence of $\mathcal{A}_2$ module homomorphisms

$$0 \longrightarrow \mathcal{A}_2\mathcal{J}_\ell \xrightarrow{t} H^*(K(\pi_k(X),k)) \longrightarrow \mathcal{A}_2\mathcal{J}_k \longrightarrow 0$$

is exact, where $\tau x_\ell = t(1)\cdot \iota_k$, since then the exact sequence of the fibration will prove the isomorphism, giving $P(j+1)$.

Thus one must determine $t(1)$ and prove appropriate exact sequences.

<u>Lemma</u>: $\tau : H^{2\ell+2}(BU(2\ell+1,\ldots,\infty)) \longrightarrow H^{2\ell+3}(K(Z,2\ell))$ sends $x_{2\ell+2}$ into $Sq^3\iota_{2\ell}$.

<u>Proof</u>: Noting that $H^{2\ell+3}(K(Z,2\ell))$ has a generator $Sq^3\iota_{2\ell}$, it suffices to show that $\tau x_{2\ell+2} \neq 0$. By periodicity, $\Omega^{2\ell-4}BU(2\ell+1,\ldots,\infty) = BU(5,\ldots,\infty)$,

and by looping, it suffices to show $\tau x_6 = Sq^3 \iota_4$ in the fibration $BU(5,\ldots,\infty) \longrightarrow BSU = BU(4,\ldots,\infty) \longrightarrow K(Z,4)$. If $\tau x_6 = 0$, $Sq^3 x_4 \neq 0$ in $H^7(BSU)$, but one knows that $H^7(BSU) = 0$. Thus $\tau x_6 = Sq^3 \iota_4$. **

Lemma:

$$\tau : H^{8k+1}(BO(8k+1,\ldots,\infty)) \longrightarrow H^{8k+2}(K(Z,8k)) : x_{8k+1} \longrightarrow Sq^2 \iota_{8k},$$

$$\tau : H^{8k+2}(BO(8k+2,\ldots,\infty)) \longrightarrow H^{8k+3}(K(Z_2,8k+1)) : x_{8k+2} \longrightarrow Sq^2 \iota_{8k+1},$$

$$\tau : H^{8k+4}(BO(8k+3,\ldots,\infty)) \longrightarrow H^{8k+5}(K(Z_2,8k+2)) : x_{8k+4} \longrightarrow Sq^3 \iota_{8k+2},$$

$$\tau : H^{8k+8}(BO(8k+5,\ldots,\infty)) \longrightarrow H^{8k+9}(K(Z,8k+4)) : x_{8k+8} \longrightarrow Sq^5 \iota_{8k+4}.$$

Proof: Clearly one may write

$$\tau x_{8k+1} = aSq^2 \iota_{8k}, \qquad \tau x_{8k+2} = bSq^2 \iota_{8k+1},$$

$$\tau x_{8k+4} = cSq^3 \iota_{8k+2} + dSq^2 Sq^1 \iota_{8k+2}, \qquad \tau x_{8k+8} = eSq^5 \iota_{8k+4},$$

with $a,b,c,d,e \in Z_2$. Since x_{8k+4} is an integral class, $0 = \tau Sq^1 x_{8k+4} = Sq^1 \tau x_{8k+4} = dSq^3 Sq^1 \iota_{8k+2}$ and thus $d = 0$.

Looping $8k-4$ times and applying periodicity, $BO(8k,\ldots,\infty)$ becomes BSp [The case $k = 0$ is trivial and may be ignored.] and one has:

$$\tau : H^5(BSp(5,\ldots,\infty)) \longrightarrow H^6(K(Z,4)) : x_5 \longrightarrow aSq^2 \iota_4,$$

$$\tau : H^6(BSp(6,\ldots,\infty)) \longrightarrow H^7(K(Z_2,5)) : x_6 \longrightarrow bSq^2 \iota_5,$$

$$\tau : H^8(BSp(8,\ldots,\infty)) \longrightarrow H^9(K(Z_2,6)) : x_8 \longrightarrow cSq^3 \iota_6,$$

$$\tau : H^{12}(BSp(12,\ldots,\infty)) \longrightarrow H^{13}(K(Z,8)) : x_{12} \longrightarrow eSq^5 \iota_8.$$

Now consider the fibration $BSp(5,\ldots,\infty) \xrightarrow{i} BSp \xrightarrow{\pi} K(Z,4)$. One has $H^5(BSp) = 0$ so $\tau x_5 \neq 0$ giving $\tau x_5 = Sq^2 \iota_4$; hence $a = 1$.

One may then compute $H^*(BSp(5,\ldots,\infty))$ from the spectral sequence. Specifically, $H^*(BSp(5,\ldots,\infty))$ is the polynomial algebra over Z_2 on the classes $i^*(\mathcal{P}_i^s)$, $i \geq 2$, and the classes $Sq^I x_5$ with I given by:

$I = (0)$ - giving $x_5^{2^s}$ which transgresses to $Sq^{2^{s-1}\cdot 5} \ldots Sq^5 Sq^2 \iota_4$,

$I = (2^k,\ldots,4)$ - giving $(Sq^{2^k} \ldots Sq^4 x_5)^{2^s}$ which transgresses to

$$Sq^{2^{s-1}(2^{k+1}+1)} \ldots Sq^{2^{k+1}+1} Sq^{2^k} \ldots Sq^4 Sq^2 \iota_4,$$

$I = (1)$ - giving $(Sq^1 x_5)^{2^s}$ which transgresses to $Sq^{2^k\cdot 3} \ldots Sq^6 Sq^3 \iota_4$.

[Note: The given transgressions, together with ι_4 form polynomial generators for $H^*(K(Z,4))$].

Now consider the fibration $BSp(6,\ldots,\infty) \xrightarrow{i'} BSp(5,\ldots,\infty) \xrightarrow{\pi'} K(Z_2,5)$. Then $H^6(BSp(5,\ldots,\infty))$ has base $Sq^1 x_5$ and $i'^*(Sq^1 x_5) = 0$ so $\tau x_6 \neq 0$, giving $\tau x_6 = Sq^2 \iota_5$; hence $b = 1$.

Assertion: $\pi'^*(Sq^2 Sq^1 \iota_5) = i^*(\mathcal{P}_2^s)$.

[If not $\tau : H^7(BSp(6,\ldots,\infty)) \longrightarrow H^8(K(Z_2,5)) = Z_2 \oplus Z_2$ is an isomorphism since $\pi'^*(Sq^2 Sq^1 \iota_5) = \pi'^*(Sq^3 \iota_5) = 0$. Thus the map $BSp(6,\ldots,\infty) \longrightarrow K(Z_2,6)$ realizing x_6 is not epic on 7 dimensional cohomology, implying $\pi_7(BSp) \neq 0$.]

Assertion: $Sq^{2^r} \ldots Sq^4 \mathcal{P}_2^s = \mathcal{P}^s_{2^{r-1}+1}$ + decomposables.

[To see this one maps $a : BO \longrightarrow BSp$ by quaternionification so $a^*(\mathcal{P}_i^s) = w_i^4$. Then $a^*(Sq^{2^r} \ldots Sq^4 \mathcal{P}_2^s) = (Sq^{2^{r-2}} \ldots Sq^1 w_2)^4 = (w_{2^{r-1}+1} +$ decomposables$)^4$ giving indecomposability of $Sq^{2^r} \ldots Sq^4 \mathcal{P}_2^s$.]

One may now compute $H^*(BSp(6,\ldots,\infty))$. For later uses, it suffices to know the answer in dimensions ≤ 13. One has π'^* epic in dimensions less than 16 and in dimensions ≤ 14

$$H^*(BSp(5,\dots,\infty)) = Z_2[Sq^I x_5 | I = (0),(1),(2^k,\dots,4),(2^k,\dots,2,1)]$$

and

$$H^*(K(Z_2,5)) = Z_2[Sq^I \iota_5 | I \text{ admissible}, \ e(I) < 5]$$

with the kernel of π'^* being generated by the classes

dim 14: $Sq^6Sq^3\iota_5 = \tau(Sq^6Sq^1x_6), (Sq^2\iota_5)^2 + Sq^6Sq^2Sq^1\iota_5 = \tau(Sq^4Sq^2Sq^1x_6)$

dim 13: $Sq^6Sq^2\iota_5 = \tau(x_6^2), \ Sq^5Sq^2Sq^1\iota_5 = \tau(Sq^4Sq^2x_6)$

dim 12: $Sq^5Sq^2\iota_5 = \tau(Sq^5x_6)$

dim 11: $Sq^5Sq^1\iota_5 = \tau(Sq^3Sq^1x_6), \ Sq^4Sq^2\iota_5 = \tau(Sq^4x_6)$

dim 10: $\iota_5^2 + Sq^4Sq^1\iota_5 = \tau(Sq^2Sq^1x_6)$

dim 9 : $Sq^3Sq^1\iota_5 = \tau(Sq^2x_6)$

dim 8 : $Sq^3\iota_5 = \tau(Sq^1x_6)$

dim 7 : $Sq^2\iota_5 = \tau(x_6)$.

Thus in dimensions ≤ 13, $H^*(BSp(6,\dots,\infty)) \cong Z_2[x_6, Sq^1x_6, Sq^2x_6, Sq^2Sq^1x_6, Sq^4x_6, Sq^3Sq^1x_6, Sq^5x_6, Sq^4Sq^2x_6, Sq^6Sq^1x_6, Sq^4Sq^2Sq^1x_6]$.

Now consider the fibration $BSp(8,\dots,\infty) \xrightarrow{i''} BSp(6,\dots,\infty) \xrightarrow{\pi''} K(Z_2,6)$. Then $H^8(BSp(6,\dots,\infty))$ has base Sq^2x_6 and $i''^*(Sq^2x_6) = 0$ so $\tau x_8 \neq 0$. Thus $\tau x_8 = Sq^3\iota_6$ and hence $c = 1$.

One may now compute $H^*(BSp(8,\dots,\infty))$ in low dimensions, for π''^* is epic in dimensions ≤ 13 and below dimension 14, the spectral sequence reduces to an exact sequence. Thus $\tau : H^i(BSp(8,\dots,\infty)) \cong \text{kernel}(\pi''^*)^{i+1}$ for $i \leq 12$. Now kernel π''^* is given by

dim 9 : $Sq^3\iota_6 = \tau x_8$

dim 10: 0

dim 11: $Sq^5\iota_6 + Sq^4Sq^1\iota_6 = \tau Sq^2 x_8$

dim 12: $Sq^5Sq^1\iota_6 = \tau Sq^3 x_8$

dim 13: $Sq^5Sq^2\iota_6 = \tau Sq^4 x_8$.

Now consider the fibration $BSp(12,\ldots,\infty) \xrightarrow{i'''} BSp(8,\ldots,\infty) \xrightarrow{\pi'''} K(Z,8)$. Then $H^{12}(BSp(8,\ldots,\infty))$ has base Sq^4x_8 and $i'''{}^*(Sq^4x_8) = 0$ so $\tau x_{12} \neq 0$. Thus $\tau x_{12} = Sq^5\iota_8$ and hence $e = 1$. **

To complete the proof of the proposition, one has:

<u>Lemma</u>: The sequences

$$\mathcal{A}_2/\mathcal{A}_2 Sq^1 \xrightarrow{\widetilde{Sq}^3} \mathcal{A}_2/\mathcal{A}_2 Sq^1 \xrightarrow{\widetilde{Sq}^3} \mathcal{A}_2/\mathcal{A}_2 Sq^1$$

and

$$\begin{array}{ccc} \mathcal{A}_2 & \xrightarrow{\widetilde{Sq}^2} & \mathcal{A}_2/\mathcal{A}_2 Sq^1 \\ {\scriptstyle \widetilde{Sq}^2}\uparrow & & \downarrow{\scriptstyle \widetilde{Sq}^5} \\ \mathcal{A}_2 & \xleftarrow{\widetilde{Sq}^3} & \mathcal{A}_2/\mathcal{A}_2 Sq^1 \end{array}$$

are exact, where $\widetilde{Sq}^i(\alpha) = \alpha \circ Sq^i$.

<u>Note</u>: These sequences are due to H. Toda: On exact sequences in Steenrod algebra mod 2, Mem. Coll. Sci. Univ. Kyoto. Ser. A, 31(1958), 33-64. The proof given here is essentially that of C.T.C.Wall: Generators and relations for the Steenrod algebra, Annals of Math., 72 (1960), 429-444.

<u>Proof</u>: Let $\mathcal{A}_2' \subset \mathcal{A}_2$ be the sub-Hopf algebra generated by Sq^1 and Sq^2. $\mathcal{A}_2'$ has eight elements:

$$1,\ Sq^1,\ Sq^2,\ Sq^2Sq^1,\ Sq^3 = Sq^1Sq^2,\ Sq^3Sq^1,\ Sq^5 + Sq^4Sq^1, \text{ and } Sq^5Sq^1.$$

Consider the sequences

$$\mathcal{A}_2'/\mathcal{A}_2'Sq^1 \xrightarrow{\widetilde{Sq}^3} \mathcal{A}_2'/\mathcal{A}_2'Sq^1 \xrightarrow{\widetilde{Sq}^3} \mathcal{A}_2'/\mathcal{A}_2'Sq^1$$

and

$$\begin{array}{ccc} \mathcal{A}_2' & \xrightarrow{\widetilde{Sq}^2} & \mathcal{A}_2'/\mathcal{A}_2'Sq^1 \\ {\scriptstyle \widetilde{Sq}^2}\uparrow & & \downarrow{\scriptstyle \widetilde{Sq^5+Sq^4Sq^1}} \\ \mathcal{A}_2' & \xleftarrow{\widetilde{Sq}^3} & \mathcal{A}_2'/\mathcal{A}_2'Sq^1 . \end{array}$$

It is trivial to verify that these sequences are exact. Since $\mathcal{A}_2'$ is a sub-Hopf algebra of $\mathcal{A}_2$, $\mathcal{A}_2$ is a right $\mathcal{A}_2'$ module and a coalgebra, with comultiplication being a homomorphism of right $\mathcal{A}_2'$ modules. Since $\nu : \mathcal{A}_2' \longrightarrow \mathcal{A}_2 : x \longrightarrow 1\cdot x$ is monic, the Milnor-Moore theorem applies and $\mathcal{A}_2 \cong \mathcal{L} \otimes_{Z_2} \mathcal{A}_2'$ as right $\mathcal{A}_2'$ modules. Tensoring the exact sequences above with $\mathcal{L}$ gives the exact sequences of the lemma. **

Corollary: Let ξ be an n-dimensional vector bundle over X and suppose $U(\xi) \in \tilde{K}^n(T\xi)$ (or $\widetilde{KO}^n(T\xi)$) is an orientation. Then stably ξ admits a Spin^c (or Spin) structure.

Proof: Let n' be such that $n + n' = 8k$ and let $\xi' = \xi \oplus n'\cdot 1$. Then ξ' has an orientation class $U(\xi) \otimes \acute{\imath} \in \widetilde{KG}^{8k}(T(\xi) \wedge S^{n'}) = \widetilde{KG}^{8k}(T\xi')$ $(G = U$ or $O)$. Let $f : T\xi' \longrightarrow BG$ represent this orientation. Since $T\xi'$ is $8k-1$ connected, one has a lifting $f : T\xi' \longrightarrow BG(8k,\ldots,\infty)$ and the inclusion $S^{8k} = T(\text{fiber}) \hookrightarrow T\xi' \longrightarrow BG(8k,\ldots,\infty)$ represents the generator of $\pi_{8k}(BG)$. The class $f^*(x_{8k})$ is an orientation class $U' \in \tilde{H}^{8k}(T\xi';Z)$, so ξ' is an orientable bundle.

If $G = O$, $Sq^2U' = f^*(Sq^2x_{8k}) = 0$, so $w_2(\xi') = 0$ and ξ' admits a Spin structure.

If $G = U$, one notes that there is an integral class $v \in H^{8k+2}(BU(8k,\ldots,\infty))$ with v reducing to $Sq^2 x_{8k}$ mod 2. To see this, the integral spectral sequence of $BU(8k+2,\ldots,\infty) \longrightarrow BU(8k,\ldots,\infty) \longrightarrow K(Z,8k)$ has x_{8k+2} transgressing to the nonzero class of order 2 in $H^{8k+3}(K(Z,8k);Z)$ giving $H^{8k+2}(BU(8k,\ldots,\infty);Z) \cong Z$, with generator v mapping to $2x_{8k+2}$ and with mod 2 reduction $Sq^2 x_{8k}$. Thus $f^*(v) = \pi^*(x)\cdot U'$ with $x \in H^2(X;Z)$ having mod 2 reduction $w_2(\xi')$. Thus ξ' admits a $Spin^c$ structure. **

<u>Note</u>: The orientation defined by the choice of a $Spin^c$ or Spin structure may differ from $U(\xi)$, but clearly differs only by multiplication with an invertible element in $KU(X)$ or $KO(X)$. Thus one knows precisely the fashion in which all K or KO theory orientations arise.

<u>Definition</u>: A class $x \in KO^*(X)$ (or $K^*(X)$) has filtration n if for every finite complex Y of dimension less than n and map $f : Y \longrightarrow X$ one has $f^*(x) = 0$.

If X is a complex of finite type, x has filtration n if and only if $i^*(x) = 0$, where $i : X^{n-1} \longrightarrow X$ is the inclusion of the $(n-1)$-skeleton of X. Additionally, if $f : X \longrightarrow BG$ realizes x, then x has filtration n if and only if f lifts to $BG(n,\ldots,\infty)$. <u>Note</u>: This assumes that $x \in KG(X)$ has positive filtration, or more precisely, the restriction of x to each component of X has virtual dimension zero.

<u>Proposition</u>: $KG(X)$ is a filtered ring.

<u>Proof</u>: If $x,y \in KG(X)$, with filtration $(x) = n$ and filtration $(y) = m$, one needs that $x\cdot y$ has filtration $n+m$. Let Y be a finite complex of dimension less than $n+m$ and $f : Y \longrightarrow X$ any map. Let $g : Y \longrightarrow BG_r(n,\ldots,\infty)$, $h : Y \longrightarrow BG_s(m,\ldots,\infty)$ be maps with $g^*(\gamma_r - \dim\gamma_r) = f^*(x)$, $h^*(\gamma_s - \dim\gamma_s) = f^*(y)$.

Now $f^*(xy)$ is given by $k^*((\gamma_r - \dim\gamma_r) \otimes (\gamma_s - \dim\gamma_s))$ where

$$k : Y \xrightarrow{\Delta} Y \times Y \xrightarrow{g\times h} BG_r(n,\dots,\infty) \times BG_s(m,\dots,\infty).$$

Letting $p \in BG_r(n,\dots,\infty)$, $q \in BG_s(m,\dots,\infty)$ be base points, $u = (\gamma_r - \dim\gamma_r) \otimes (\gamma_s - \dim\gamma_s)$ is trivial over $BG_r(n,\dots,\infty) \vee BG_s(m,\dots,\infty)$ $= BG_r(n,\dots,\infty) \times q \cup p \times BG_s(m,\dots,\infty)$ for $\gamma_r - \dim\gamma_r = 0$ over p and $\gamma_s - \dim\gamma_s = 0$ over q. Thus $u = j^*(v)$ where

$$j : BG_r(n,\dots,\infty) \times BG_s(m,\dots,\infty) \longrightarrow V = BG_r(n,\dots,\infty) \wedge BG_s(m,\dots,\infty)$$

and $v \in \widetilde{KG}(V)$. Now V is $(n+m-1)$ connected $[\tilde{H}^*(V;F) \cong \tilde{H}^*(BG_r(n,\dots,\infty);F) \otimes_F \tilde{H}^*(BG_s(m,\dots,\infty);F)$ for any field F so the least nonzero class has dimension at least $n+m]$ so $j \circ k : Y \longrightarrow V$ is homotopic to zero, and thus $(j\circ k)^*(v) = f^*(xy) = 0$. Thus xy has filtration $n+m$. **

Following Anderson, Brown and Peterson, the analysis of Spin^c or Spin cobordism depends centrally on the knowledge of the filtration of the K and KO theory characteristic classes. The basic result is:

<u>Proposition</u>: Let ξ be an oriented real vector bundle over a space X, and let $\pi_R^i(\xi)$ be the i-th KO-theory Pontrjagin class of ξ (defined by $\pi_u^R(\xi) = \lambda_t^R(\xi - \dim\xi)$ with $u = t/(1+t)^2$). For $I = (i_1,\dots,i_r)$, let $\pi_R^I(\xi) = \pi_R^{i_1}(\xi)\dots\pi_R^{i_r}(\xi)$. Then $\pi_R^I(\xi)$ has filtration

$$F(\pi_R^I(\xi)) = \begin{cases} 4n(I) & \text{if } n(I) \equiv 0\ (2), \\ 4n(I)-2 & \text{if } n(I) \equiv 1\ (2) \end{cases}$$

where $n(I) = i_1 + \dots + i_r$, and $\pi_R^I(\xi) \otimes \mathbb{C}$ has filtration $4n(I)$ in $K(X)$.

If $\mathcal{P}_I(\xi) = \mathcal{P}_{i_1}(\xi)\dots\mathcal{P}_{i_r}(\xi)$ is nonzero in $H^{4n(I)}(X;Q)$, and if $n(I)$ is odd the class $\mathcal{P}_I(\xi)$ is not divisible by 2 in $\rho_Q H^{4n(I)}(X;Z)$, then these are the precise filtrations of $\pi_R^I(\xi)$ and its complexification.

Further, the lifting $\hat{f} : X \longrightarrow BO(F(\pi_R^I(\xi)),\ldots,\infty)$ may be chosen with $f^*(x_{4n(I)}) = \mathcal{P}_I(\xi) + \delta Sq^2Sq^1\alpha_I$ (δ being Bockstein, α_I a polynomial in Stiefel-Whitney classes) if $n(I)$ is even, or $Sq^2f^*(x_{4n(I)-2}) = \rho_2(\mathcal{P}_I(\xi))$ if $n(I)$ is odd, and $f : X \longrightarrow BU(4n(I),\ldots,\infty)$ may be chosen so that $f^*(x_{4n(I)}) = \mathcal{P}_I(\xi)$. (Note: x_i generates $H^i(BG(i,\ldots,\infty);\pi_i(BG))$.)

The proof of this given by Anderson, Brown, and Peterson makes use of the KO^*-theory spectral sequence. The proof given here is fairly involved but will be just standard obstruction theory.

Let Y be a space with $y \in KG(Y)$ a class having filtration n. Then if $f : Y \longrightarrow BG$ realizes y, there is a lifting $\hat{f} : Y \longrightarrow BG(n,\ldots,\infty)$, and $\hat{f}$ lifts to $BG(n+1,\ldots,\infty)$ if and only if $f^*(x_n) = 0$ in $H^n(Y,\pi_n(BG))$. Denoting by $[y]$ the subset of $H^n(Y,\pi_n(BG))$ consisting of the classes $\hat{f}^*(x_n)$ for all lifts $\hat{f}$, y has filtration $n+1$ if and only if $0 \in [y]$.

If $G = 0$, there are four cases to be considered, depending on the class of n mod 8.

Case I: Suppose $y \in KO(Y)$ has filtration $8k$, with $\hat{f} : Y \longrightarrow BO(8k,\ldots,\infty)$ one lifting so that $\hat{f}^*(\gamma) = y$ where γ is the universal class. Letting $g : S^{8k} \longrightarrow BO(8k,\ldots,\infty)$ generate $\pi_{8k}(BO)$ one has $g^*(x_{8k}) = \dot{\iota}$ and $g^*(\gamma) = \dot{\iota}$, with $g^*(ch(\gamma \otimes \mathbb{C})) = ch(g^*\gamma \otimes \mathbb{C}) = \dot{\iota}$ so $ch(\gamma \otimes \mathbb{C}) = x_{8k}$ + higher terms. Thus $ch(y \otimes \mathbb{C}) = \hat{f}^*(ch(\gamma \otimes \mathbb{C})) = \hat{f}^*(x_{8k})$ + higher terms. Further $Sq^2\rho_2\hat{f}^*(x_{8k}) = \hat{f}^*(Sq^2\rho_2x_{8k}) = 0$ from the analysis of the Z_2 cohomology of $BO(8k,\ldots,\infty)$. This gives: If $\alpha \in [y]$, then $Sq^2\rho_2\alpha = 0$ and $\rho_Q(\alpha)$ + higher terms $= ch(y \otimes \mathbb{C})$.

If $\hat{f} : Y \longrightarrow BO(8k,\ldots,\infty)$ is one lift, then one has

$$\begin{array}{ccc} & & K(Z,8k-5) \\ & & \downarrow i \\ Y & \xrightarrow{\hat{f}} & BO(8k,\dots,\infty) \\ & & \downarrow \pi \\ & & BO(8k-4,\dots,\infty) \end{array}$$

and π being a principal fibration, the lifts $f' : Y \longrightarrow BO(8k,\dots,\infty)$ covering $\pi\circ\hat{f}$ are classified by maps into $K(Z,8k-5)$ or classes $x \in H^{8k-5}(Y;Z)$. Since $\rho_2 i^*(x_{8k}) = i^*(\rho_2 x_{8k}) = Sq^5 \dot{\iota}_{8k-5}$, one has $i^*(x_{8k}) = \delta Sq^4 \rho_2 \dot{\iota}_{8k-5}$ (ρ_2 being monic), where δ is the integral Bockstein. Thus $[y]$ is a union of cosets of $\delta Sq^4 \rho_2 H^{8k-5}(Y;Z)$.

<u>Case II</u>: Suppose $y \in KO(Y)$ has filtration $8k+1$, with $\hat{f} : Y \longrightarrow BO(8k+1,\dots,\infty)$ one lifting. From the Z_2 cohomology analysis, one has $Sq^2 \hat{f}^*(x_{8k+1}) = \hat{f}^*(Sq^2 x_{8k+1}) = 0$. This gives: If $\alpha \in [y]$, then $Sq^2\alpha = 0$.

If one considers

$$\begin{array}{ccc} & & K(Z,8k-1) \\ & & \downarrow i \\ Y & \xrightarrow{\hat{f}} & BO(8k+1,\dots,\infty) \\ & & \downarrow \pi \\ & & BO(8k,\dots,\infty) \end{array}$$

in which $i^*(x_{8k+1}) = Sq^2 \dot{\iota}_{8k-1}$, one has: $[y]$ is a union of cosets of $Sq^2 \rho_2 H^{8k-1}(Y;Z)$.

<u>Case III</u>: Suppose y has filtration $8k+2$ with $\hat{f} : Y \longrightarrow BO(8k+2,\dots,\infty)$ one lifting. As above one has: If $\alpha \in [y]$, then $Sq^3\alpha = 0$. Further, $[y]$ is a union of cosets of $Sq^2 H^{8k}(Y;Z_2)$.

Note: If y has filtration $8k + 2$, one may consider $ch(y \otimes \mathbb{C}) = \alpha_{8k+4} +$ higher terms. For $BO(8k+2,\ldots,\infty)$, the integral spectral sequence of the fibration

$$BO(8k+4,\ldots,\infty) \xrightarrow{i} BO(8k+2,\ldots,\infty) \xrightarrow{\pi} K(Z_2,8k+2)$$

gives $\tau x_{8k+4} = \delta Sq^2 \hat{\iota}_{8k+2}$ and thus an integral class v_{8k+4} with $i^*(v_{8k+4}) = 2x_{8k+4}$. Since $H^{8k+4}(BO(8k+2,\ldots,\infty);Z_2) \cong Z_2$ with base $Sq^2 x_{8k+2}$, one has $\rho_2 v_{8k+4} = Sq^2 x_{8k+2}$. Letting $g : S^{8k+4} \longrightarrow BO(8k+2,\ldots,\infty)$ be a generator for $\pi_{8k+4}(BO)$ one has $g^*(\gamma) = \dot{\iota}$ so $g^*(ch(\gamma \otimes \mathbb{C})) = ch(\dot{\iota} \otimes \mathbb{C}) = 2\iota$. Since x_{8k+4} pulls back to ι in S^{8k+4}, $ch(\gamma \otimes \mathbb{C})$ must pull back to $2x_{8k+4}$ + higher terms in $BO(8k+4,\ldots,\infty)$. Thus $ch(\gamma \otimes \mathbb{C}) = v_{8k+4}$ + higher terms.

In particular, if $\alpha \in [y]$ is $\hat{f}^*(x_{8k+2})$, then $Sq^2\alpha$ is the reduction of the integral class $\hat{f}^*(v_{8k+4})$ and $ch(y \otimes \mathbb{C}) = \rho_Q(\hat{f}^*(v_{8k+4}))$ + higher terms.

Case IV: Suppose $y \in KO(Y)$ has filtration $8k + 4$ with $\hat{f} : Y \longrightarrow BO(8k+4,\ldots,\infty)$ one lifting. Then, if $\alpha \in [y]$, $Sq^5\rho_2\alpha = 0$ and $ch(y \otimes \mathbb{C}) = 2\rho_Q(\alpha)$ + higher terms. Further, $[y]$ is a union of cosets of $\delta Sq^2 H^{8k+1}(Y;Z_2)$.

Using these facts in looking at the Proposition, consider first the upper bound condition on the filtration. If ξ is an oriented vector bundle over X, then $ch(\pi_R^i(\xi) \otimes \mathbb{C}) = \mathcal{P}_i(\xi)$ + higher terms. This gives

$$ch(\pi_R^I(\xi) \otimes \mathbb{C}) = \mathcal{P}_I(\xi) + \text{higher terms.}$$

Thus if $\mathcal{P}_I(\xi) \neq 0$ in $H^{4n(I)}(X;Q)$ one must have filtration $(\pi_R^I(\xi)) \leq 4n(I)$ and filtration $(\pi_R^I(\xi) \otimes \mathbb{C}) \leq 4n(I)$.

If $n(I) \equiv 1 \bmod 2$ and filtration $(\pi_R^I(\xi)) = 4n(I) = 8k+4$, there is a lifting $\hat{f} : X \longrightarrow BO(8k+4,\ldots,\infty)$ with $\hat{f}^*(\gamma) = \pi_R^I(\xi)$. This gives an integral class $x' = \hat{f}^*(x_{8k+4})$ such that $ch(\pi_R^I(\xi) \otimes \mathbb{C}) = 2\rho_Q(x')$ + higher terms. Thus $\mathcal{P}_I(\xi) = 2\rho_Q(x')$. Thus, if $\mathcal{P}_I(\xi)$ is not divisible by 2 in $\rho_Q H^{4n(I)}(X;Z)$, $\pi_R^I(\xi)$ has filtration less than $4n(I)$, so less than or equal to $4n(I)-2$.

To prove that $\pi_R^I(\xi)$ has the asserted filtration, with the given classes defined by the lift, one considers $g : X \longrightarrow BSO$ with $g^*(\gamma) = \xi - \dim\xi$, γ being the universal stable bundle. Then $\pi_R^I(\xi) = g^*\pi_R^I(\gamma)$, and it suffices to prove the remainder of the result in the special case in which $X = BSO$ and ξ is the universal bundle. This may be accomplished by a careful study of BSO. From the obstruction analysis, it is clear that one is interested in the action of Sq^1 and Sq^2 in $H^*(BSO;Z_2)$.

<u>Lemma</u>: $Sq^2 w_i = \binom{i+3}{2} w_{i+2} + \binom{i+2}{1} w_{i+1} w_1 + \binom{i+1}{0} w_i w_2$.

<u>Proof</u>:

$$Sq^2 w_i = Sq^2(\Sigma x_1 \ldots x_i) = \Sigma x_1 \ldots x_j^2 \ldots x_k^2 \ldots x_i,$$

$$w_2 w_i = (\Sigma x_1 \ldots x_i)(\Sigma x_1 x_2),$$

$$= \binom{i+2}{2} \Sigma x_1 \ldots x_{i+2} + \binom{i}{1} \Sigma x_1 \ldots x_j^2 \ldots x_{i+1} + \Sigma\, x_1 \ldots x_j^2 \ldots x_k^2 \ldots x_i,$$

$$w_1 w_{i+1} = (\Sigma x_1 \ldots x_{i+1})(\Sigma x_1)$$

$$= \binom{i+2}{1}\, \Sigma\, x_1 \ldots x_{i+2} + \Sigma\, x_1 \ldots x_j^2 \ldots x_{i+1},$$

so

$$-Sq^2 w_i + w_2 w_i = (i)[w_1 w_{i+1} - (i+2) w_{i+2}] + \binom{i+2}{2} w_{i+2},$$

$$= \left[\frac{(i+2)(i+1)}{2} - \frac{(i+2)(2i)}{2}\right] w_{i+2} + i w_1 w_{i+1},$$

$$= \frac{(i+2)(i-1)}{2}\, w_{i+2} + i w_1 w_{i+1},$$

giving the desired formula. **

Recall that one has operations $Q_0 = Sq^1$ and $Q_1 = Sq^1Sq^2 + Sq^2Sq^1$ in $\mathcal{A}_2$ satisfying:

$$Q_0Q_1 = Q_1Q_0, \qquad Q_i^2 = 0$$

and

$$Q_i(a\cdot b) = Q_i(a)\cdot b + a\cdot Q_i(b).$$

Thus in any space X one may form the homology groups $H(H^*(X;Z_2);Q_i) = \text{kernel}(Q_i)/\text{image}(Q_i)$. One has a natural map $\phi_i : \ker Q_0 \cap \ker Q_1 \longrightarrow \ker Q_i/\text{im}Q_i$, $(i = 0,1)$ and clearly $\text{im}Q_0 \cap \text{im}Q_1 \subset \ker\phi_i$ $(i = 0,1)$.

<u>Definition</u>: X has <u>isomorphic homologies</u> if the homomorphisms

$$\lambda_i : \frac{\ker(Q_0) \cap \ker(Q_1)}{\text{im}(Q_0) \cap \text{im}(Q_1)} \longrightarrow \frac{\ker(Q_i)}{\text{im}(Q_i)}$$

are isomorphisms. The group $\ker Q_0 \cap \ker Q_1/\text{im}Q_0 \cap \text{im}Q_1$ will be denoted $H(H^*(X;Z_2))$.

<u>Lemma</u>: BSO has isomorphic homologies.

<u>Proof</u>: One has $Q_0 w_i = (i+1)w_{i+1}$ so $H(H^*(BSO);Q_0) \cong Z_2[(w_{2i})^2]$. Also

$$Sq^2Sq^1w_i = (i+1)Sq^2w_{i+1} = (i+1)[\tbinom{i+4}{2}w_{i+3} + w_{i+1}w_2],$$

and

$$Sq^1Sq^2w_i = Sq^1[\tbinom{i+3}{2}w_{i+2} + w_iw_2] = (i+3)\tbinom{i+3}{2}w_{i+3} + (i+1)w_{i+1}w_2 + w_iw_3,$$

giving

$$\begin{aligned} Q_1w_i &= w_iw_3 + [\frac{(i+3)(i+3)(i+2)}{2} + \frac{(i+1)(i+4)(i+3)}{2}]\, w_{i+3}, \\ &= w_iw_3 + \frac{(i+3)}{2}\,[i^2 + 5i + 6 + i^2 + 5i + 4]\, w_{i+3}, \end{aligned}$$

$= w_i w_3 + (i+3)(i^2 + 5i + 5)w_{i+3},$

$= w_i w_3 + (i+3)w_{i+3}.$

In particular, $Q_1 w_{2i} = w_{2i+3} + w_{2i}w_3$, $Q_1 w_3 = w_3^2$ giving $H^*(BSO) \cong Z_2[w_{2i}, Q_1 w_{2i}] \otimes Z_2[w_3]$, so $H(H^*(BSO);Q_1) \cong Z_2[(w_{2i})^2]$.

Then $Z_2[(w_{2i})^2] \subset \ker Q_0 \cap \ker Q_1$ so λ_i is epic for $i = 0,1$.

Suppose $x \in (\ker Q_0 \cap \ker Q_1)^n$ and $\phi_0(x) = 0$. If $n \not\equiv 0$ (4), then $\phi_1(x) = 0$ also, so $x \in \mathrm{im} Q_0 \cap \mathrm{im} Q_1$. Suppose $n \equiv 0$ (4). Then $x = Q_1 y + f((w_{2j})^2)$, where f is a polynomial over Z_2. Since $x = Q_0 z$, this gives $f((w_{2j})^2) = Q_0 z + Q_1 y$, but $\mathrm{im} Q_0 + \mathrm{im} Q_1$ is contained in the ideal generated by the w_{2k+1}. Thus $f((w_{2j})^2) = 0$, giving $x = Q_1 y = Q_0 z \in \mathrm{im} Q_0 \cap \mathrm{im} Q_1$. Thus $\ker \phi_0 = \mathrm{im} Q_0 \cap \mathrm{im} Q_1$.

Thus $\lambda_0 : \ker Q_0 \cap \ker Q_1 / \mathrm{im} Q_0 \cap \mathrm{im} Q_1 \longrightarrow H(H^*(BSO);Q_0)$ is an isomorphism. Since λ_1 is epic, with the two groups having the same rank over Z_2 in each dimension, λ_1 is also an isomorphism. **

Proposition: Let M be a positively graded module over the exterior algebra E generated by Q_0 and Q_1, and suppose M has isomorphic homologies. Then M is the direct sum of a free E module and a trivial E module.

Proof: Let $\pi : M \longrightarrow M/\bar{E}M = N$. Then $\pi(\ker Q_0) = \pi(\ker Q_0 \cap \ker Q_1) = \pi(\ker Q_1)$. [If $Q_i a = 0$, there is an $\ell \in \ker Q_0 \cap \ker Q_1$ with $a + \ell = Q_j b$, and thus $\pi(a) = \pi(\ell) \in \pi(\ker Q_0 \cap \ker Q_1)$]. Let $L \subset \mathrm{kernel} Q_0 \cap \mathrm{kernel} Q_1$ with $\pi : L \longrightarrow \pi(\ker Q_0 \cap \ker Q_1)$ an isomorphism. Let $T \subset M$ with π mapping T isomorphically onto a complementary summand for $\pi(L)$ in N. Then $f : E \otimes (L \oplus T) \longrightarrow M : e \otimes (\ell,t) \longrightarrow e\ell + et$ is epic, inducing an epimorphism

$$\phi : L \oplus (E \otimes T) \longrightarrow M.$$

<u>Assert</u>: $L \subset \ker Q_0 \cap \ker Q_1$ maps isomorphically onto $H(M)$.

[If $a \in \ker Q_0 \cap \ker Q_1$, there is an $\ell \in L$ with $\pi(a) = \pi(\ell)$ so $a = \ell + Q_0x + Q_1y$. Then $Q_1y = a + \ell + Q_0x \in \ker Q_0 \cap \ker Q_1$, and $Q_1y \in \operatorname{im}Q_1$. Thus $a + \ell + Q_0x \in \operatorname{im}Q_0$, and $\phi_0(a) = \phi_0(\ell)$, and hence L maps onto $\ker Q_0/\operatorname{im}Q_0$. Hence $\psi : L \longrightarrow H(M)$ is epic. Since $\ker\psi \subset \ker\pi$, $\psi : L \cong H(M)$.]

Now suppose $Q_0Q_1t = 0$. Then $Q_1(Q_0t) = 0$ and $Q_0(Q_0t) = 0$, so $Q_0t \in \ker Q_0 \cap \ker Q_1$ and $Q_0t \in \operatorname{im}Q_0$, giving $Q_0t = Q_1s$ for some s $(\dim s + 3 = \dim t + 1)$. Then $Q_0Q_1s = Q_0Q_0t = 0$. If $\dim t < 2$, then $\dim s < 0$, so $s = 0$ and $Q_0t = 0$, giving an $\ell \in L$ with $t + \ell = Q_0u$. Suppose inductively that $s = \ell + Q_0a + Q_1b$, with $\ell \in L$. Then $Q_0t = Q_1s = Q_1Q_0a$, so $Q_0(t + Q_1a) = 0$, giving $t + Q_1a \in \ker Q_0$ so $t + Q_1a = \ell' + Q_0b'$, $\ell' \in L$. This proves:

<u>Assertion</u>: If $Q_0Q_1t = 0$, then $t = \ell + Q_0a + Q_1b$, for some $\ell \in L$, $a,b \in M$.

Now suppose $a = (\ell,\ 1 \otimes t + Q_0 \otimes t_0 + Q_1 \otimes t_1 + Q_0Q_1 \otimes t_2) \in L \oplus (E \otimes T)$ with $\phi(a) = 0$. Then $\pi\phi(a) = \pi(\ell) + \pi(t) = 0$ in $\pi(L) \oplus \pi(T)$, so $\pi(\ell) = \pi(t) = 0$. Since π is monic on both L and T, $\ell = t = 0$. Thus $0 = \phi(a) = Q_0t_0 + Q_1t_1 + Q_0Q_1t_2$. Thus $Q_1Q_0t_0 = Q_1\phi(a) = 0$, $Q_0Q_1t_1 = Q_0\phi(a) = 0$, and by the assertion, $\pi(t_0) \in \pi(L)$, $\pi(t_1) \in \pi(L)$, but $\pi(T) \cap \pi(L) = 0$. Thus $\pi(t_0) = \pi(t_1) = 0$, giving $t_0 = t_1 = 0$. Then $0 = \phi(a) = Q_0Q_1t_2$ and similarly $t_2 = 0$.

Thus $\phi : L \oplus (E \otimes T) \longrightarrow M$ is an isomorphism. **

<u>Note</u>: $\ker(Q_0Q_1) = \phi(L \oplus (\bar{E} \otimes T))$ and $\ker Q_0 \cap \ker Q_1 = \phi(L \oplus (Q_0Q_1 \otimes T))$.

<u>Lemma</u>: Let $\mu \in KO(BSO)$ have filtration n, and suppose:

a) If $n = 8k$, $ch(\mu \otimes \mathbb{C})_n = \rho_Q(\mathcal{P})$ where $\mathcal{P} \in H^n(BSO;Z)$ with $Sq^2\rho_2\mathcal{P} = 0$;

b) If $n = 8k+1$, $\mathcal{P} \in H^n(BSO;Z_2)$ with $Sq^2\mathcal{P} = 0$;

c) If $n = 8k+2$, $ch(\mu \otimes \mathbb{C})_{n+2} = \rho_Q(\mathcal{Q})$, $\mathcal{Q} \in H^{n+2}(BSO);Z)$ with $Sq^3\rho_2\mathcal{Q} = 0$;

d) If $n = 8k+4$, $ch(\mu \otimes \mathbb{C})_n = \rho_Q(2\mathcal{P})$, $\mathcal{P} \in H^n(BSO;Z)$ with $Sq^5\rho_2\mathcal{P} = 0$.

Then, for $n = 8k+2$, $\rho_2(\mathcal{Q}) \in im(Sq^2)$ and let $\mathcal{P}$ be any class in $H^{8k}(BSO;Z_2)$ with $Sq^2\mathcal{P} = \rho_2\mathcal{Q}$.

Then for any n, there is a lift $\hat{f} : BSO \longrightarrow BO(n,\ldots,\infty)$ for μ such that $\hat{f}^*(x_n) = \mathcal{P}$ and such that for $n = 8k+2$, $f^*(v_{8k+4}) = \mathcal{Q}$.

<u>Proof</u>: Let $f : BSO \longrightarrow BO(n,\ldots,\infty)$ be any lift for μ and let $f^*(x_n) = \mathcal{P}'$.

a) $n = 8k$: Letting $\alpha = \mathcal{P} - \mathcal{P}'$, one has $\rho_Q(\alpha) = 0$ and $Sq^2\rho_2(\alpha) = 0$. Since $\rho_Q(\alpha) = 0$, α is a torsion class and there is a $\beta \in H^{8k-1}(BSO;Z_2)$ with $\alpha = \delta\beta$ or $\rho_2\alpha = Sq^1\beta$. Further $Sq^2\rho_2\alpha = Sq^2Sq^1\beta = 0$.

Thus $Sq^2Q_0\beta = 0$ and $Q_0Q_0\beta = 0$ so $Q_0\beta \in \ker Q_0 \cap \ker Q_1$ giving $Q_0\beta = \ell + Q_0Q_1t$ so $Q_0\beta = Q_0Q_1t$ [$\ell \in L$ and applying π gives $\pi(\ell) = 0$].

Thus $Q_0Q_1Sq^2t = Sq^2Q_0Q_1t = Sq^2Q_0\beta = 0$ and $Sq^2t = \ell + Q_0u + Q_1v$ [$\dim Sq^2t = 8k-2$, and L is zero in this dimension, so $\ell = 0$] or $Sq^2t = Q_0u + Q_1v$.

Then $Q_0Q_1t = Sq^2Sq^2t = Sq^2(Q_0u + Q_1v) = Sq^2Sq^1u + Sq^2Sq^1Sq^2v = Sq^2Sq^1\gamma$ where $\gamma = u + Sq^2v$. Then $0 = Q_0Q_0Q_1t = Q_0(Sq^2Sq^1\gamma) = Q_0Q_1\gamma$ so $\gamma = \ell + Q_0p + Q_1q$ [$\dim \gamma = 8k-3$, so $\ell = 0$] or $\gamma = Q_0p + Q_1q$.

Thus $\rho_2\alpha = Q_0\beta = Q_0Q_1t = Sq^2Sq^1\gamma = Sq^2Sq^1(Q_0p + Q_1q) = Sq^2Sq^1Q_1q = Sq^5Sq^1q = \rho_2[\delta Sq^4\rho_2\delta_q]$. Since ρ_2 is monic on the torsion subgroup, $\alpha = \delta Sq^4\rho_2\delta q$. The lift from $BO(8k-4,\ldots,\infty)$ may be modified using the class δq, giving a lift $\hat{f}$ such that $\hat{f}^*(x_n) = \mathcal{P}' + \alpha = \mathcal{P}$.

b) $n = 8k+1$: Letting $\alpha = \mathcal{P} - \mathcal{P}'$ one has $Sq^2\alpha = 0$.

Thus $Q_0Q_1\alpha = Sq^2Sq^2\alpha = 0$, so $\alpha = \ell + Q_0x + Q_1y$ [dim α = 8k+1 so $\ell = 0$] or $\alpha = Q_0x + Q_1y = Q_0x + Q_0Sq^2y + Sq^2Sq^1y$.

Letting $\alpha' = \alpha + Sq^2Sq^1y = Q_0(x + Sq^2y)$ one has $Q_0\alpha' = 0$ and $Sq^2\alpha' = 0$, so $\alpha' = \ell + Q_0Q_1\beta$ [dim α' = 8k+1 so $\ell = 0$] or $\alpha' = Q_0Q_1\beta$.

Then $Q_0Q_1Sq^2\beta = Sq^2Q_0Q_1\beta = Sq^2\alpha' = 0$ so $Sq^2\beta = \ell + Q_0a + Q_1b$ [dim β = 8k-1 so $\ell = 0$] or $Sq^2\beta = Q_0a + Q_1b$. Thus $Q_0Q_1\beta = Sq^2Sq^2\beta = Sq^2Sq^1a + Sq^2Sq^1Sq^2b$, giving $\alpha = Sq^2Sq^1(y + a + Sq^2b)$. Hence $\alpha = Sq^2\rho_2\delta(y + a + Sq^2b)$ and modifying the lift from $BO(8k,\ldots,\infty)$ by $\delta(y + a + Sq^2b)$ gives a lift $\hat{f}$ with $\hat{f}^*(x_n) = \mathcal{P}' + \alpha = \mathcal{P}$.

c) $n = 8k+2$: Let $\mathcal{G}' = f^*(v_{8k+4})$, so that $Sq^2\mathcal{P}' = \rho_2\mathcal{G}'$ and $\rho_Q(\mathcal{G}') = ch(\mu \otimes \mathbb{C})_{n+2} = \rho_Q(\mathcal{G})$. Then $\mathcal{G} - \mathcal{G}' = \delta\beta$ with β a torsion class. Then $Sq^2\mathcal{P}' = \rho_2(\mathcal{G}') = \rho_2(\mathcal{G} - \delta\beta) = \rho_2\mathcal{G} + Sq^1\beta$.

Then $Q_1Sq^1\beta = Q_1\rho_2\mathcal{G} + Q_1Sq^2\mathcal{P}' = Q_1Sq^2\mathcal{P}' = Sq^2Sq^1Sq^2\mathcal{P}' = Sq^2Sq^1(\rho_2\mathcal{G} + Sq^1\beta) = 0$ and $Q_0Sq^1\beta = 0$, so $Sq^1\beta \in \ker Q_0 \cap \ker Q_1$ and $Sq^1\beta = \ell + Q_0Q_1\gamma$ [$\ell \in L$ and applying π, $\pi(\ell) = 0$]. Then $Sq^2\mathcal{P}' = \rho_2\mathcal{G} + Sq^1\beta = \rho_2\mathcal{G} + Q_0Q_1\gamma = \rho_2\mathcal{G} + Sq^2Sq^2\gamma$. Thus $\rho_2\mathcal{G} = Sq^2(\mathcal{P}' + Sq^2\gamma)$. Thus $\rho_2\mathcal{G} \in im(Sq^2)$.

Let $\mathcal{P} \in H^{8k}(BSO;Z_2)$ be any class with $Sq^2\mathcal{P} = \rho_2\mathcal{G}$, and let $\bar{\alpha} = \mathcal{P}' + Sq^2\gamma + \mathcal{P}$. Then $Sq^2\bar{\alpha} = 0$.

Then $Q_0Q_1\bar{\alpha} = Sq^2Sq^2\bar{\alpha} = 0$, so $\bar{\alpha} = \ell + Q_0x + Q_1y$ [dim $\bar{\alpha}$ = 8k+2 so $\ell = 0$] or $\bar{\alpha} = Q_0x + Q_1y = Q_0x + Q_0Sq^2y + Sq^2Sq^1y$.

Let $\alpha' = \bar{\alpha} + Sq^2Sq^1y = Q_0x + Q_0Sq^2y$ giving $Q_0\alpha' = 0$ and $Sq^2\alpha' = 0$, so $\alpha' = \ell + Q_0Q_1z$ [dim α' = 8k+2, so $\ell = 0$] or $\alpha' = Q_0Q_1z$.

Thus $\mathcal{P} = \mathcal{P}' + Sq^2\gamma + Sq^2Sq^1y + Sq^2Sq^2z = \mathcal{P}' + Sq^2(\gamma + Sq^1y + Sq^2z)$. By modifying the lift from $BO(8k+1,\ldots,\infty)$ by $\gamma + Sq^1y + Sq^2z$, one has a lift $\hat{f}$ with $\hat{f}^*(x_n) = \mathcal{P}$. Then $f^*(v_{8k+4}) - \mathcal{Q} = \sigma$ is a torsion class and $\rho_2(\sigma) = Sq^2\mathcal{P} + \rho_2\mathcal{Q} = 0$, but ρ_2 is monic on the torsion subgroup so $\sigma = 0$, giving $\hat{f}^*(v_{8k+4}) = \mathcal{Q}$.

d) $n = 8k+4$: Letting $\alpha = \mathcal{P} - \mathcal{P}'$ one has $\rho_Q(\alpha) = 0$ so α is a torsion class, giving $\alpha = \delta\beta$. Since $Sq^5\rho_2(\alpha) = 0$, $Sq^5Sq^1\beta = Sq^2Q_0Q_1\beta = 0$.

Then $Q_0Q_1Sq^2\beta = 0$ gives $Sq^2\beta = \ell + Q_0u + Q_1v$ [dim $Sq^2\beta = 8k+5$ so $\ell = 0$] or $Sq^2\beta = Q_0u + Q_1v$. Thus $Q_0Q_1\beta = Sq^2Sq^2\beta = Sq^2Sq^1u + Sq^2Sq^1Sq^2v = Sq^2Sq^1\gamma$.

Then $Q_0Q_1\gamma = Sq^3Sq^1\gamma = Sq^1Q_0Q_1\beta = 0$ so $\gamma = Q_0p + Q_1q + \ell$. Thus $Q_0Q_1\beta = Sq^2Sq^1\gamma = Sq^2Sq^1Q_1q$. Then $Q_1(Q_0\beta + Q_0Sq^2q) = 0$ and $Q_0(Q_0\beta + Q_0Sq^2q) = 0$ so $Q_0\beta + Q_0Sq^2q = \ell + Q_0Q_1t$ [applying π, $\pi(\ell) = 0$] or $Q_0\beta = Sq^3q + Sq^3Sq^1t = \rho_2\delta Sq^2(q + Sq^1t)$.

Since ρ_2 is monic on the torsion subgroup, $\alpha = \delta Sq^2(q + Sq^1t)$ and by modifying the lift from $BO(8k+2,\ldots,\infty)$ by $(q + Sq^1t)$ one has a lift $\hat{f}$ with $\hat{f}^*(x_n) = \mathcal{P}' + \alpha = \mathcal{P}$. **

To complete the proof of the Proposition, one applies this lemma to the class $\pi_R^I(\gamma) \in KO(BSO)$, noting that $ch(\pi_R^I(\gamma) \otimes \mathbb{C}) = \mathcal{P}_I(\gamma)$ + higher terms. Applying the lemma with $\mathcal{P}$ or $\mathcal{Q}$ both zero shows that $\pi_R^I(\gamma)$ has filtration at least $F(\pi_R^I(\gamma))$. If $n(I)$ is odd, $Sq^3\rho_2(\mathcal{P}_I) = Sq^3(w_{2I}{}^2)$ is zero and the lift may be chosen so that $Sq^2f^*(x_{4n(I)-2}) = \rho_2(\mathcal{P}_I)$. For $n(I)$ even, $Sq^2\rho_2\mathcal{P}_I \neq 0$, but $f^*(x_{4n(I)}) = \mathcal{P}_I + \delta\beta$ for any lift f. Then $Q_0\rho_2\delta\beta = Sq^1Sq^1\beta = 0$ and $Q_1\rho_2\delta\beta = Sq^3\rho_2\delta\beta = Sq^3\rho_2f^*(x_{4n(I)}) + Sq^3\rho_2\mathcal{P}_I = 0$. Thus $\rho_2\delta\beta = Q_0Q_1\alpha$ for some α, or $\delta\beta = \delta Sq^2Sq^1\alpha$. Thus $f^*(x_{4n(I)})$ is given by $\mathcal{P}_I + \delta Sq^2Sq^1\alpha$ where α is a polynomial in the Stiefel Whitney classes.

For the complex case, clearly filtration $(\lambda \otimes \mathbb{C}) \geq$ filtration (λ). If $\lambda \in KO(X)$ has filtration $8k+2$, let $f : X \longrightarrow BO(8k+2,\ldots,\infty)$ classify λ.

Letting $u : BO(8k+2,\ldots,\infty) \longrightarrow BU(8k+2,\ldots,\infty)$ classify $\gamma \otimes \mathbb{C}$, one has $H^{8k+2}(BO(8k+2,\ldots,\infty);Z) = 0$ $(Sq^1 x_{8k+2} \neq 0$ so x_{8k+2} is not the reduction of an integral class) so u lifts to $BU(8k+4,\ldots,\infty)$. Thus $u \circ f$ lifts to $BU(8k+4,\ldots,\infty)$ and filtration $(\lambda \otimes \mathbb{C}) \geq 8k+4$. This gives filtration $(\pi_R^I(\gamma) \otimes \mathbb{C}) \geq 4n(I)$.

Being given a lift $f : BSO \longrightarrow BU(4n(I),\ldots,\infty)$ for $\pi_R^I(\gamma) \otimes \mathbb{C}$, one has $ch(\pi_R^I(\gamma) \otimes \mathbb{C}) = \rho_Q f^*(x_{4n(I)})$ + higher terms $= \rho_Q(\mathcal{P}_I)$ + higher terms. Thus $f^*(x_{4n(I)}) = \mathcal{P}_I + \delta\beta$ for some β. Since Sq^1 and Sq^3 both annihilate $\rho_2 f^*(x_{4n(I)})$ and $\rho_2\mathcal{P}_I$, $\rho_2\delta\beta = Sq^1\beta \in \ker Q_0 \cap \ker Q_1$ and so belongs to image Q_0Q_1. Thus $\delta\beta = \delta Sq^2\rho_2\delta\alpha$ for some α. Since the lift from $BU(4n(I)-2,\ldots,\infty)$ may be modified by any class in $\delta Sq^2 H^{4n(I)-3}(BSO;Z)$, there is a lift with $f^*(x_{4n(I)}) = \mathcal{P}_I$.

This completes the proof of the proposition. **

Turning to the case of a Spin bundle or Spin^c bundle, one has:

<u>Proposition</u>: Let γ denote the universal bundle over BSpin or BSpin^c. Then

a) $\pi_R^I(\gamma)$ for $I = (i_1,\ldots,i_r)$ with $i_j > 1$ for all j, has filtration precisely

$$\begin{cases} 4n(I) & \text{if } n(I) \text{ is even,} \\ 4n(I)-2 & \text{if } n(I) \text{ is odd} \end{cases}$$

in KO(BSpin); and

b) $\pi_R^I(\gamma) \otimes \mathbb{C}$ for all I has filtration $4n(I)$ in KU(BSpin^c).

<u>Proof</u>: Since the induced homomorphism from $H^*(BSO;Q)$ to $H^*(BSpin;Q)$ or $H^*(BSpin^c;Q)$ is monic, this is clear from the proposition, except for

filtration $(\pi_R^I(\gamma))$ with $n(I)$ odd, which is less than or equal to $4n(I)$. One must check that $\mathcal{P}_I(\gamma)$ is not divisible by 2 in $\rho_Q H^*(BSpin;Z)$.

For this one has:

Lemma: Let $v_{2^i} \in H^*(BO;Z_2)$ be the Wu class. Then v_{2^i} is indecomposable, and $Sq^1 v_{2^i}$ belongs to the ideal generated by w_1 and w_3 (over $\mathcal{A}_2$).

Proof: Let λ be the canonical bundle over $RP(\infty)$, so $w(2^i\lambda) = (1+\alpha)^{2^i} = 1 + \alpha^{2^i}$. Then $v(2^i\lambda) = (1 + \alpha + \alpha^2 + \ldots + \alpha^{2^j} + \ldots)^{2^i} = 1 + \alpha^{2^i}$ + higher terms. Thus $v_{2^i}(2^i\lambda) = \alpha^{2^i}$ is nonzero, while all decomposable classes of dimension 2^i have value zero on $2^i\lambda$, making v_{2^i} indecomposable.

To evaluate $Sq^1 v_{2^i}$, one considers a manifold M^n. For any $x \in H^*(M^n;Z_2)$

$$\begin{aligned}(Sq^1 v_{2^i} \cup x)[M] &= [Sq^1(v_{2^i} \cup x) + v_{2^i}\cdot Sq^1 x][M],\\ &= [(v_1 v_{2^i} \cup x) + Sq^{2^i} Sq^1 x][M],\\ &= [(v_1 v_{2^i} \cup x) + (Sq^2 Sq^{2^i-1} + Sq^1 Sq^{2^i})x][M], \quad (i \geq 2)\\ &= [v_1 v_{2^i} \cup x + v_2 Sq^{2^i-1}x + v_1 Sq^{2^i} x][M],\\ &= [v_1 v_{2^i} \cup x + v_1 v_2 Sq^{2^i-2}x + Sq^1 v_2\cdot Sq^{2^i-2}x + v_1 Sq^{2^i} x][M].\end{aligned}$$

Now applying the relation

$$\begin{aligned}a.Sq^j b[M] &= Sq^j(a\cdot b)[M] + \sum_{t=1}^{j} (Sq^t a\cdot Sq^{j-t} b)[M],\\ &= (v_j\cdot a)\cdot b[M] + \sum_{t=1}^{j} (Sq^t a\cdot Sq^{j-t} b)[M]\end{aligned}$$

one may "push Sq^j off of b" since in the right hand side of this expression only operators $Sq^k b$ with $k < j$ occur. Applying this to the above relation gives $Sq^1 v_{2^i}\cdot x[M] = \lambda\cdot x[M]$ where λ belongs to the ideal in $H^*(BO;Z_2)$ generated by $v_1 = w_1$ and $Sq^1 v_2 = Sq^1(w_2 + w_1^2) = w_3 + w_2 w_1$. [Note: If a belongs to this ideal, so do $Sq^t a$ and $v_j a$].

Thus $Sq^1 v_{2^i} - \lambda$ maps to zero in all manifolds. Letting $M = RP(2^{i+1})^{2^i+1}$, $w(M) = \Pi(1+\alpha_j)^{2^{i+1}+1} = \Pi(1+\alpha_j+\alpha_j^{2^{i+1}})$, so $w_k(M)$ is the k-th elementary symmetric function in the α_j for $k \leq 2^i + 1$, making $\tau^* : H^*(BO;Z_2) \longrightarrow H^*(M;Z_2)$ monic in dimensions less than or equal to 2^i+1. Thus $Sq^1 v_{2^i} = \lambda$ and belongs to the ideal generated by w_1 and $Sq^1 v_2$ over $\mathcal{A}_2$. [Since $v_1 = w_1$, $v_2 = w_2 + w_1^2$, and $Sq^1 v_2 = w_3 + w_2 w_1$, this trivially holds for $i < 2$.] **

Lemma: $H(H^*(BSpin;Z_2);Q_0) \cong Z_2[w_{2j}^2, v_{2^i} \mid j$ not a power of 2, $i > 1]$.

Proof: Since v_{2^i} is indecomposable, $H^*(BSpin;Z_2)$ is

$$Z_2[w_{2j}, Q_0 w_{2j}, v_{2^i} \mid j \text{ not a power of } 2, i > 1].$$

Then $Q_0 v_{2^i} = 0$ and applying the Künneth theorem gives the homology. **

Corollary: All torsion in $H_*(BSpin;Z)$ has order 2.

Now returning to the proof of the proposition, let $I = (i_1,\ldots,i_r)$, $i_j > 1$, $n(I) \equiv 1\ (2)$ and suppose $\rho_Q(\mathcal{P}_I) = 2\rho_Q(x)$. Then $\mathcal{P}_I = 2x + \alpha$, where α is a torsion class, so $\alpha = \delta\beta$. This gives $w_{2I}^2 = \rho_2(\mathcal{P}_I) = \rho_2(\alpha) = Sq^1\beta$. Since $v_{2^i} = w_{2^i}$ mod decomposables, one may write

$$v_{2^i} = w_{2^i} + f(w_{2k}) + \sum_{\substack{j \text{ odd} \\ j < 2^i}} w_j \cdot c_j$$

where f is a polynomial in the even w_{2k}, $2k < 2^i$. Then

$$v_{2^i}^2 = w_{2^i}^2 + f(w_{2k}^2) + \sum_{\substack{j \text{ odd} \\ j < 2^i}} Sq^1(w_{j-1} w_j c_j^2).$$

Thus $Z_2[w_{2k}^2 | k \neq 1]$ maps monomorphically into $H(H^*(BSpin;Z_2);Q_0)$, contradicting $w_{2I}^2 = Sq^1\beta$. Thus $\pi_R^I(\gamma)$ has filtration $4n(I)-2$. **

Note: The sequences I involving 1's were eliminated since $\mathcal{P}_1$ reduces to w_2^2 which is zero in BSpin.

Lemma: $H(H^*(BSpin^c;Z_2);Q_0) = Z_2[w_{2j}{}^2, v_{2^i} | j$ not a power of $2, i \geq 1]$.

Proof: $H^*(BSpin^c;Z_2) = Z_2[w_{2j}, Q_0w_{2j}, v_{2^i}]$ with $Q_0v_{2^i} = 0$. **

Corollary: All torsion in $H_*(BSpin^c;Z)$ has order 2.

Note: $v_2^2 = w_2^2 = \rho_2\mathcal{P}_1$ and $H(H^*(BSpin^c;Z_2);Q_0) \supset Z_2[\rho_2\mathcal{P}_i]$.

One now has:

Lemma: Let ξ be an n dimensional vector bundle over a complex X and $U \in \widetilde{KG}^n(T\xi)$ an orientation. Then $\alpha \in KG^*(X)$ has filtration k if and only if $\pi^*(\alpha)\cdot U = \Phi^U(\alpha)$ has filtration $n + k$.

Proof: Since ξ is trivial over cells in X, $T\xi$ has a cellular decomposition in which the $n + r$ skeleton of $T\xi$ is $T(\xi|_{X^r})$, X^r being the r skeleton of X. Further, U restricts to an orientation of $T(\xi|_{X^r})$ so $\Phi^U : KG^*(X^r) \xrightarrow{\cong} \widetilde{KG}^{n+*}(T(\xi|_{X^r}))$. Thus α restricts to zero in $KG^*(X^r)$ if and only if $\Phi^U(\alpha)$ restricts to zero in $\widetilde{KG}^{n+*}((T\xi)^{n+r})$. **

Note: Suppose $\hat{f} : X \longrightarrow BG(k,\dots,\infty)$ is a lifting for α and $\hat{U} : T\xi \longrightarrow BG(n,\dots,\infty)$ is a lift for U (n even if $G = U$, $n \equiv 0 \bmod 8$ if $G = 0$). Then

$$\widehat{Tf} : T\xi \xrightarrow{\emptyset} X \wedge T\xi \xrightarrow{\hat{f}\wedge\hat{U}} BG(k,\dots,\infty) \wedge BG(n,\dots,\infty) \xrightarrow{\beta} BG(n+k,\dots,\infty)$$

is a lift of $\Phi^U(\alpha)$ and $\widehat{Tf}^*(x_{k+n}) = \pi^*\hat{f}^*(x_k)\cdot U'$ where U' is the Thom class of ξ. This is immediate from the fact that $\beta^*(x_{k+n}) = x_k \otimes x_n$. (If $G = 0$, $n = 8\ell$, $k = 8t+2$, then $\beta^*(v_{8t+8\ell+4}) = v_{8t+4} \otimes x_{8\ell}$ for $v_{8t+8\ell+4}$ and $v_{8t+4} \otimes x_{8\ell}$ are bases, and evaluating the Chern character of the canonical bundle over each side gives the result.).

Letting γ denote the canonical $8n$ plane bundle over $BSpin_{8n}$, one then has classes

$$\pi^*(\pi_R^I(\gamma))\cdot U(\gamma) \in \widetilde{KO}^{8n}(TBSpin_{8n}) \cong \widetilde{KO}(TBSpin_{8n})$$

of filtration $8n + 4n(I)$ or $8n + 4n(I)-2$ as $n(I)$ is even or odd for each sequence I having $i_j > 1$, and the choice of liftings defines a map

$$Tf : TBSpin_{8n} \longrightarrow \prod_{n(I)\text{odd}} BO(8n+4n(I)-2,\ldots,\infty) \times \prod_{n(I)\text{even}} BO(8n+4n(I),\ldots,\infty)$$

with $\rho_2 Tf^*(x_{8n+4n(I)}) = (w_{2I}^2 + Sq^3Sq^1\alpha_I)\cdot U$, $Sq^2 Tf^*(x_{8n+4n(I)-2}) = w_{2I}^2$, and thus a homomorphism

$$(Tf)^* : \bigoplus_{n(I)\text{odd}} (\mathcal{A}_2/\mathcal{A}_2 Sq^3)x_{8n+4n(I)-2} \oplus \bigoplus_{n(I)\text{even}} (\mathcal{A}_2/\mathcal{A}_2 Sq^1 + \mathcal{A}_2 Sq^2)x_{8n+4n(I)}$$

$$\longrightarrow \tilde{H}^*(TBSpin_{8n};Z_2).$$

<u>Note</u>: Letting $\underset{\sim}{BO}(k,\ldots,\infty)$ be the spectrum with $(\underset{\sim}{BO}(k,\ldots,\infty))_{8n} = BO(8n+k,\ldots,\infty)$ and with the intermediate spaces given by loop spaces, one has a map

$$Tf : \underset{\sim}{TBSpin} \longrightarrow \Pi\underset{\sim}{BO}(4n(I)-2,\ldots,\infty) \times \Pi\underset{\sim}{BO}(4n(I),\ldots,\infty)$$

defined by a careful choice of lifts for the Thom classes. Up to any given dimension, this may be obtained just by taking n sufficiently large in the above.

In exactly the same fashion, one has a map

$$Tf : TBSpin^c_{2n} \longrightarrow \Pi\ BU(4n(I) + 2n, \ldots, \infty)$$

inducing

$$(Tf)^* : \oplus\ (\mathcal{A}_2/\mathcal{A}_2Sq^1 + \mathcal{A}_2Sq^3)x_{4n(I)+2n} \longrightarrow \tilde{H}^*(TBSpin^c_{2n}; Z_2).$$

The main result of Anderson, Brown, and Peterson is:

<u>Theorem</u>: The homomorphisms

$$(Tf)^*: \bigoplus_{\substack{n(I)\\ \text{odd}}} (\mathcal{A}_2/\mathcal{A}_2Sq^3)x_{4n(I)-2} \oplus \bigoplus_{\substack{n(I)\\ \text{even}}} (\mathcal{A}_2/\mathcal{A}_2Sq^1 + \mathcal{A}_2Sq^2)x_{4n(I)} \to \tilde{H}^*(\underset{\sim}{TBSpin}; Z_2)$$

for I with $i_j > 1$, and

$$(Tf)^* : \oplus\ (\mathcal{A}_2/\mathcal{A}_2Sq^1 + \mathcal{A}_2Sq^3)x_{4n(I)} \longrightarrow \tilde{H}^*(\underset{\sim}{TBSpin}^c; Z_2)$$

are monic and have cokernels which are free $\mathcal{A}_2$ modules. In particular, there exist classes $z_i \in \tilde{H}^*(\underset{\sim}{TBSpin}; Z_2)$ and $z_i' \in \tilde{H}^*(\underset{\sim}{TBSpin}^c; Z_2)$, defining maps g into products of $\underset{\sim}{K}(Z_2)$ spectra, so that

$$Tf\times g: \underset{\sim}{TBSpin} \longrightarrow \prod_{\substack{n(I)\\ \text{odd}}} \underset{\sim}{BO}(4n(I)-2, \ldots, \infty) \times \prod_{\substack{n(I)\\ \text{even}}} \underset{\sim}{BO}(4n(I), \ldots, \infty) \times \prod_i \underset{\sim}{K}(Z_2, \deg z_i)$$

and

$$Tf \times g : \underset{\sim}{TBSpin}^c \longrightarrow \prod_I \underset{\sim}{BU}(4n(I), \ldots, \infty) \times \prod_i \underset{\sim}{K}(Z_2, \deg z_i')$$

are 2 primary homotopy equivalences.

The proof of this result requires a detailed analysis of the Steenrod algebra and the cohomology of these spaces.

Lemma 1: The homomorphisms

$$\nu : \mathcal{A}_2 \longrightarrow \tilde{H}^*(T\underset{\sim}{\mathrm{BSpin}}^c; Z_2) \ : \ a \longrightarrow a(U)$$

and

$$\nu : \mathcal{A}_2 \longrightarrow \tilde{H}^*(T\underset{\sim}{\mathrm{BSpin}}; Z_2) \ : \ a \longrightarrow a(U)$$

have kernels exactly $\mathcal{A}_2 Sq^1 \oplus \mathcal{A}_2 Sq^3$ and $\mathcal{A}_2 Sq^1 \oplus \mathcal{A}_2 Sq^2$ respectively.

Proof: Since $Sq^i U = w_i U$, these are clearly contained in the kernels. One then has

$$\bar{\nu} : \mathcal{A}_2 / \mathcal{A}_2 Sq^1 + \mathcal{A}_2 Sq^3 \longrightarrow \tilde{H}^*(T\underset{\sim}{\mathrm{BSpin}}^c; Z_2)$$

which is clearly a homomorphism of coalgebras. To prove this monic, it suffices to show it is monic on primitive elements. (Note: x is primitive if its diagonal is $x \otimes 1 + 1 \otimes x$.)

Recall that the Steenrod algebra $\mathcal{A}_2$ is a Hopf algebra whose dual is also a Hopf algebra $\mathcal{A}_2^*$. Letting $\xi_k \in (\mathcal{A}_2^*)_{2^k-1}$ be the dual with respect to the base Sq^I, I admissible, of the class $Sq^{2^{k-1}} Sq^{2^{k-2}} \cdots Sq^2 Sq^1$, one has $\mathcal{A}_2^* \cong Z_2[\xi_k]$. Dually, $\mathcal{A}_2$ has a unique nonzero primitive element in each dimension $2^{i+1}-1$, which is the element Q_i.

The Steenrod algebra $\mathcal{A}_2$ also admits a "canonical antiautomorphism" χ, given by $\chi(1) = 1$, and if $\Delta x = x \otimes 1 + \sum x_i' \otimes x_i'' + 1 \otimes x$, then $\chi(x) = x + \sum \chi(x_i') \cdot x_i''$. In particular, $\sum_{j=0}^{i} \chi(Sq^{i-j}) Sq^j = 0$. (Note: $\chi(Q_i) = Q_i$ since χ is an isomorphism of Hopf algebras and hence takes the nonzero primitive into itself.)

Now consider the exact sequence

$$\mathcal{A}_2 \oplus \mathcal{A}_2 \xrightarrow{\tilde{Q}_0 + \tilde{Q}_1} \mathcal{A}_2 \longrightarrow \mathcal{A}_2 / \mathcal{A}_2 Q_0 + \mathcal{A}_2 Q_1 \longrightarrow 0.$$

Applying the canonical antiautomorphism gives

$$\mathcal{A}_2 \oplus \mathcal{A}_2 \xrightarrow{LQ_0+LQ_1} \mathcal{A}_2 \longrightarrow \chi(\mathcal{A}_2/\mathcal{A}_2Q_0+\mathcal{A}_2Q_1) \longrightarrow 0$$

or dually

$$0 \longrightarrow (\chi(\mathcal{A}_2/\mathcal{A}_2Q_0+\mathcal{A}_2Q_1))^* \longrightarrow \mathcal{A}_2^* \xrightarrow{(LQ_0)^*+(LQ_1)^*} \mathcal{A}_2^* \oplus \mathcal{A}_2^*$$

where LQ_i is left multiplication by Q_i.

<u>Assert</u>: $(LSq)^*(\xi_k) = \xi_k + \xi_{k-1}$.

<u>Proof</u>: Let $x \in H^1(RP(\infty);Z_2)$. For I admissible, $Sq^I x = 0$ if $I \neq (2^t,\ldots,1)$ for some t, and $Sq^{2^t}\ldots Sq^1 x = x^{2^{t+1}}$. For I admissible of degree 2^k-1-i, $Sq^i Sq^I = \sum \alpha_J Sq^J$ with J admissible of degree 2^k-1. Then $(LSq^i)^*(\xi_k)(Sq^I) = \xi_k(Sq^i Sq^I) = \alpha_{(2^{k-1},\ldots,1)}$, but $Sq^i Sq^I x = \alpha_{(2^{k-1},\ldots,1)} x^{2^k}$ and $Sq^i Sq^I x = 0$ except for $I = (2^{k-2},\ldots,1)$, $i = 2^{k-1}$. **

Then $(LSq^1)^*(\xi_k) = 0$ if $k \neq 1$ $(2^{k-1}-1+1 = 2^k-1)$ and $(LSq^2Sq^1)^*(\xi_k) = (LSq^1)^*(LSq^2)^*(\xi_k) = 0$ if $k \neq 2$, while $(LSq^1)^*(\xi_1) = 1$, $(LSq^2)^*(\xi_2) = \xi_1$ so $(LSq^1)^*(LSq^2)^*(\xi_2) = 1$. Now $(LSq^3)^*(\xi_k) = 0$ always. Thus $(LQ_0)^*(\xi_k) = 0$ if $k \neq 1$, 1 if $k = 1$, and $(LQ_1)^*(\xi_k) = 0$ if $k \neq 2$, 1 if $k = 2$.

Since $(LQ_0)^*$ and $(LQ_1)^*$ are derivations, $\ker(LQ_0)^* \cap \ker(LQ_1)^*$ is clearly $Z_2[\xi_1^2,\xi_2^2,\xi_3,\ldots]$. Thus $\mathcal{A}_2/\mathcal{A}_2Sq^1 + \mathcal{A}_2Sq^3 = \mathcal{A}_2/\mathcal{A}_2Q_0 + \mathcal{A}_2Q_1$ has dual a polynomial algebra on classes $\xi_1^2,\xi_2^2,\xi_3,\ldots$.

By duality, $\mathcal{A}_2/\mathcal{A}_2Sq^1 + \mathcal{A}_2Sq^3$ has nonzero primitive elements given by:

$$Sq^2,\ Sq^4Sq^2 + Sq^2Sq^4, \text{ and } Q_i \text{ for } i \geq 2.$$

(<u>Note</u>: The image of Q_i is a nonzero primitive for $i \geq 2$, while the others are primitive by a direct computation of the diagonal). Then

$$\bar{\nu}(Sq^2) = Sq^2U = w_2\cdot U,$$

$$\bar{\nu}(Sq^4Sq^2+Sq^2Sq^4) = Sq^4(w_2U) + Sq^2(w_4U) = (w_6+w_4w_2+w_2^3)U$$

$$\bar{\nu}(Q_i) = s_{2^{i+1}-1}(w)\cdot U, \quad i \geq 2.$$

[To see the latter, $\nu(Q_i) = \alpha\cdot U \in \tilde{H}^*(TBO;Z_2)$ and α is a nonzero primitive of dimension $2^{i+1}-1$. The only such class is $s_{2^{i+1}-1}(w)$.] Now $s_{2k+1}(w)$ is indecomposable, and the map $BSpin^c \longrightarrow BSO$ sends indecomposables to zero only in dimensions of the form 2^j+1, so these are nonzero classes. Thus $\bar{\nu}$ is monic.

[<u>Note</u>: To see that $s_{2k+1}(w)$ is indecomposable one has the formula for $s_j = \sum_{i=1}^{n} x_i^j$ given by

$$s_j - s_{j-1}\sigma_1 + s_{j-2}\sigma_2 - \ldots + (-1)^{j-1}s_1\sigma_{j-1} + (-1)^j j\sigma_j = 0$$

if $j \leq n$, where σ_k is the k-th elementary symmetric function in the x_i. Thus $s_j \equiv (-1)^{j+1}j\sigma_j$ mod decomposables (See: Van der Waerden, vol. I, §26).]

One may prove the result for Spin in the same way, or consider the map $f : BSpin^c_{n-2} \longrightarrow BSpin_n$ classifying $\gamma \oplus \xi$, where ξ is the complex line bundle given by the $Spin^c$ structure. $f^*(w_n) = w_{n-2}w_2 = Sq^2w_{n-2}$ and thus, on the Thom space level $f^*U = Sq^2U'$. Thus it suffices to show that $\mathcal{A}_2 \xrightarrow{\widetilde{Sq}^2} \mathcal{A}_2/\mathcal{A}_2Sq^1 + \mathcal{A}_2Sq^3$ has kernel precisely $\mathcal{A}_2Sq^1 + \mathcal{A}_2Sq^2$. If $a\cdot Sq^2 = b\cdot Sq^1 + c\cdot Sq^3$, then $(a+cSq^1)Sq^2 = bSq^1$ but $\mathcal{A}_2 \xrightarrow{\widetilde{Sq}^2} \mathcal{A}_2 \xrightarrow{\widetilde{Sq}^2} \mathcal{A}_2/\mathcal{A}_2Sq^1$ is exact, so $a + cSq^1 = dSq^2$ giving $a = cSq^1 + dSq^2 \in \mathcal{A}_2Sq^1 + \mathcal{A}_2Sq^2$. **

For later purposes, it is desirable to have the forms for $\mathcal{A}_2/\mathcal{A}_2Sq^1 + \mathcal{A}_2Sq^2$ and $\mathcal{A}_2/\mathcal{A}_2Sq^3$. First one has:

$$\mathcal{A}_2 \oplus \mathcal{A}_2 \xrightarrow{\widetilde{Sq}^1+\widetilde{Sq}^2} \mathcal{A}_2 \longrightarrow \mathcal{A}_2/\mathcal{A}_2Sq^1 + \mathcal{A}_2Sq^2 \longrightarrow 0$$

and

$$\mathcal{A}_2 \xrightarrow{\widetilde{Sq}^3} \mathcal{A}_2 \longrightarrow \mathcal{A}_2/\mathcal{A}_2 Sq^3 \longrightarrow 0$$

giving:

$$0 \longrightarrow (\chi(\mathcal{A}_2/\mathcal{A}_2 Sq^1 + \mathcal{A}_2 Sq^2))^* \longrightarrow \mathcal{A}_2^* \xrightarrow{(LSq^1)^* + (LSq^2)^*} \mathcal{A}_2^* \oplus \mathcal{A}_2^*$$

and

$$0 \longrightarrow (\chi(\mathcal{A}_2/\mathcal{A}_2 Sq^3))^* \longrightarrow \mathcal{A}_2^* \xrightarrow{(LSq^2 Sq^1)^*} \mathcal{A}_2^*.$$

Then $(LSq^1)^*(\xi_k) = 0$ *or* 1 *as* $k \neq 1$ *or* $k = 1$; $(LSq^2)^*(\xi_k) = 0$ *or* ξ_1 as $k \neq 2$ or $k = 2$; and $(LSq^2)^*(\xi_1^2) = ((LSq^1)^*(\xi_1))^2 = 1$ since $(LSq^2)^*(a \cdot b) = (LSq^2)^* a \cdot b + (LSq^1)^* a \cdot (LSq^1)^* b + a \cdot (LSq^2)^* b$. It is then immediate that $(\chi(\mathcal{A}_2/\mathcal{A}_2 Sq^1 + \mathcal{A}_2 Sq^2))^* \supset Z_2[\xi_1^4, \xi_2^2, \xi_3, \ldots] = A$ and also $(\chi(\mathcal{A}_2/\mathcal{A}_2 Sq^3))^* \supset A$. $\mathcal{A}_2^*$ is a free A module with base $\xi_1^i \xi_2^j$ for $0 \leq i \leq 3$, $0 \leq j \leq 1$, with $(LSq^1)^*(\xi_1^i \xi_2^j) = i\xi_1^{i-1}\xi_2^j$ and $(LSq^2)^*(\xi_1^i \xi_2^j) = j\xi_1^{i+1} + \binom{i}{2}\xi_1^{i-2}\xi_2^j$ giving

a	$(LSq^1)^* a$	$(LSq^2)^* a$	$(LSq^2 Sq^1)^* a$
1	0	0	0
ξ_1	1	0	0
ξ_1^2	0	1	0
ξ_1^3	ξ_1^2	ξ_1	1
ξ_2	0	ξ_1	1
$\xi_1\xi_2$	ξ_2	ξ_1^2	0
$\xi_1^2\xi_2$	0	$\xi_1^3+\xi_2$	ξ_1^2
$\xi_1^3\xi_2$	$\xi_1^2\xi_2$	$\xi_1^4+\xi_1\xi_2$	ξ_2

Thus $(LSq^1)^* + (LSq^2)^*$ is monic, giving $(\chi(\mathcal{A}_2/\mathcal{A}_2Sq^1+\mathcal{A}_2Sq^2))^* = A$, and $(\chi(\mathcal{A}_2/\mathcal{A}_2Sq^3))^*$ is the free A module on 1, ξ_1, ξ_1^2, $\xi_1^3 + \xi_2$, and $\xi_1\xi_2$.

<u>Lemma 2</u>: $(Tf)^*$ induces isomorphisms on $H(\ ;Q_i)$, $i = 0,1$.

<u>Proof</u>: This requires considerable work. First, one needs the homology of each of the cyclic $\mathcal{A}_2$ modules. Applying χ and dualization, one needs the action of Q_0 and Q_1 in $\mathcal{A}_2^*$, by right action $[(aSq^i)(\lambda) = a(\lambda Sq^i)$ for $a \in \mathcal{A}_2^*$, $\lambda \in \mathcal{A}_2]$, given by $(RQ_i)^*$.

<u>Sublemma 1</u>: $(RSq)^*(\xi_k) = \xi_k + \xi_{k-1}^2$.

<u>Proof</u>: $(RSq^i)^*(\xi_k)(Sq^I) = \xi_k(Sq^I Sq^i) = \alpha_{(2^{k-1},\ldots,1)}$ where $Sq^I Sq^i = \sum \alpha_J Sq^J$ but $\alpha_{(2^{k-1},\ldots,1)} x^{2^k} = Sq^I Sq^i x$ which is zero except for $i = 1$, $I = (2^{k-1},\ldots,2)$, where it is x^{2^k}. Thus $(RSq^i)^*(\xi_k) = 0$ if $i > 1$ and $(RSq^1)^*(\xi_k)(Sq^I) = 0$ except for $I = (2^{k-1},\ldots,2)$. Now $\xi_{k-1}^2(Sq^I) = (LSq^I)^*(\xi_{k-1}^2) = 0$ if I has any odd entry, and $(LSq^{2I'})^*(\xi_{k-1}^2)$ is $((LSq^{I'})^*\xi_{k-1})^2$, which vanishes if $I' \neq (2^{k-2},\ldots,1)$. Thus $(RSq^1)^*(\xi_k) = \xi_{k-1}^2$. **

<u>Sublemma 2</u>: $(RQ_0)^*(\xi_k) = \xi_{k-1}^2$, $(RQ_1)^*(\xi_k) = \xi_{k-2}^4$.

<u>Proof</u>: $(RQ_0)^* = (RSq^1)^*$ gives the first. Since $(RSq^2)^*(\xi_k) = 0$, $(RQ_1)^*(\xi_k) = (RSq^2)^*(RSq^1)^*(\xi_k) = (RSq^2)^*(\xi_{k-1}^2) = ((RSq^1)^*\xi_{k-1})^2 = \xi_{k-2}^4$. **

<u>Sublemma 3</u>: $H((\chi(\mathcal{A}_2/\mathcal{A}_2Sq^1+\mathcal{A}_2Sq^3))^*;RQ_0^*) = Z_2[\xi_1^2]$,

$H((\chi(\mathcal{A}_2/\mathcal{A}_2Sq^1+\mathcal{A}_2Sq^2))^*;RQ_0^*) = Z_2[\xi_1^4]$,

$H((\chi(\mathcal{A}_2/\mathcal{A}_2Sq^3))^*;RQ_0^*) = \xi_1^2 \cdot Z_2[\xi_1^4]$.

Proof: $(\chi(\mathcal{A}_2/\mathcal{A}_2Sq^1+\mathcal{A}_2Sq^3))^* = Z_2[\xi_1^2,\xi_2^2,\xi_3,\ldots]$ and $(RQ_0)^*(\xi_1^2) = (RQ_0)^*(\xi_2^2) = 0$ while $(RQ_0)^*(\xi_k) = \xi_{k-1}^2$ for $k \geq 3$. Thus $\ker(RQ_0)^* = Z_2[\xi_1^2,\xi_2^2,\xi_3^2,\ldots]$ and $\operatorname{im}(RQ_0)^*$ is the ideal generated by $Z_2[\xi_2^2,\xi_3^2,\ldots]$.

Turning to $(\chi(\mathcal{A}_2/\mathcal{A}_2Sq^1+\mathcal{A}_2Sq^2))^* = Z_2[\xi_1^4,\xi_2^2,\xi_3,\ldots]$, $\ker(RQ_0)^* = Z_2[\xi_1^4,\xi_2^2,\xi_3^2,\ldots]$ and $\operatorname{im}(RQ_0)^*$ is the ideal generated by $Z_2[\xi_2^2,\xi_3^2,\ldots]$.

Now $(\chi(\mathcal{A}_2/\mathcal{A}_2Sq^3))^* = Z_2[\xi_1^4,\xi_2^2,\xi_3,\ldots]\ \{1,\ \xi_1,\ \xi_1^2,\ \xi_1^3+\xi_2,\ \xi_1\xi_2\}$ with $(RQ_0)^*(1) = 0$, $(RQ_0)^*(\xi_1) = 1$, $(RQ_0)^*(\xi_1^2) = 0$, $(RQ_0)^*(\xi_1^3+\xi_2) = 0$, $(RQ_0)^*(\xi_1\xi_2) = \xi_1^3 + \xi_2$. Then let $a = \alpha + \beta\xi_1 + \gamma\xi_1^2 + \delta(\xi_1^3+\xi_2) + \varepsilon\ \xi_1\xi_2$.

$$(RQ_0)^*(a) = (RQ_0)^*\alpha + [(RQ_0)^*\beta]\xi_1 + [(RQ_0)^*\gamma]\xi_1^2 + [(RQ_0)^*\delta](\xi_1^3+\xi_2)$$

$$+ [(RQ_0)^*\varepsilon](\xi_1\xi_2) + \beta + \varepsilon\ (\xi_1^3+\xi_2),$$

giving $a \in \ker(RQ_0)^*$ if and only if $\beta = (RQ_0)^*\alpha$, $(RQ_0)^*\beta = 0$, $(RQ_0)^*\gamma = 0$, $(RQ_0)^*\delta = \varepsilon$, $(RQ_0)^*\varepsilon = 0$, but

$$(RQ_0)^*[\alpha\xi_1+\delta\xi_1\xi_2] = [(RQ_0)^*\alpha]\xi_1 + [(RQ_0)^*\delta](\xi_1\xi_2) + \alpha + \delta(\xi_1^3+\xi_2),$$

$$= \alpha + \beta\xi_1 + \delta(\xi_1^3+\xi_2) + \varepsilon\ \xi_1\xi_2$$

if $a \in \ker(RQ_0)^*$. Thus $\ker(RQ_0)^*/\operatorname{im}(RQ_0)^* = (\ker(RQ_0)^*|_A/\operatorname{im}(RQ_0)^*|_A)\cdot\xi_1^2$. **

Thus $H(\mathcal{A}_2/\mathcal{A}_2Sq^1+\mathcal{A}_2Sq^3;Q_0)$ is isomorphic to Z_2 in each dimension of the form $2k$ and is zero in odd dimensions. This gives a class $\alpha_{2k} \in (\mathcal{A}_2/\mathcal{A}_2Sq^1+\mathcal{A}_2Sq^3)^{2k}$ with $Q_0\alpha_{2k} = 0$ and $\chi(\alpha_{2k})$ evaluates to 1 on ξ_1^{2k}.

It is clear that $\bar{\nu}(\alpha_{2k}) \in \tilde{H}^*(\underset{\sim}{TBSpin}^c;Z_2)$ belongs to $\ker Q_0$, so $\bar{\nu}(\alpha_{2k}) = \pi^*(u_{2k})\cdot U$, wwith $u_{2k} \in H^{2k}(BSpin^c;Z_2)$ and $Q_0u_{2k} = 0$ since $Q_0U = 0$. [Since $Q_0U = Q_1U = 0$, the Thom isomorphism induces isomorphisms on Q_i homology.]

Assert: u_{2k} is indecomposable if $k = 2^s$.

Proof: One may write $\chi(\alpha_{2k}) = \sum a_J Sq^J$ with J admissible and $(\sum a_J Sq^J)$ evaluates to 1 on ξ_1^{2k}, but $\xi_1^{2k}(\sum a_J Sq^J)\cdot x^{4k} = (\sum a_J Sq^J)(x^{2k})$, and $Sq^J x^{2k} = 0$ if $J \neq (2k)$, since k is a power of two. Thus $\chi(\alpha_{2k}) = Sq^{2k} + \sum a_{J'} Sq^{J'} = Sq^{2k}$ + decomposable operations, giving $\alpha_{2k} = Sq^{2k}$ + decomposable operations.

On the other hand, one may consider the bundle $2^{s+1}\lambda$ over $RP(\infty)$. This is a Spin^c bundle, of course. Then $Sq^i U = w_i(2^{s+1}\lambda)\cdot U$ is zero for $i \neq 2^{s+1}$, but $Sq^{2^{s+1}} U \neq 0$, so $\alpha_{2k}U = w_{2k}U$ in this bundle. Since all decomposable classes map to zero, this makes u_{2k} indecomposable. **

Thus $H^*(B\mathrm{Spin}^c;Z_2) = Z_2[w_{2j}, Q_0 w_{2j}, u_{2^s} \mid j$ not a power of 2, $s \geq 1]$ with $Q_0 u_{2^s} = 0$, giving $H(H^*(B\mathrm{Spin}^c;Z_2);Q_0) = Z_2[w_{2j}^2, u_{2^s}]$. Now $\mathcal{P}_i$ reduces to w_{2i}^2 and as noted $Z_2[\mathcal{P}_i]$ maps monomorphically into this homology (sending $\mathcal{P}_{2^s}$ into $(u_{2^s}$ + decomposables$)^2$) so $H(H^*(B\mathrm{Spin}^c;Z_2);Q_0)$ is a free $Z_2[\mathcal{P}_i]$ module with base formed by monomials $u_{2^{s_1}}\cdots u_{2^{s_n}}$, $1 \leq s_1 < \cdots < s_n$. (Such a monomial will be denoted u_S).

Partially order the base $\pi^*(\mathcal{P}_I u_S)\cdot U$ of $H(\tilde{H}^*(TB\mathrm{Spin}^c;Z_2);Q_0)$ by $\pi^*(\mathcal{P}_I u_S)\cdot U < \pi^*(\mathcal{P}_{I'} u_{S'})\cdot U$ if $\dim I < \dim I'$. [Note: $4 \dim I$ is the dimension of the "squared" factor in $\mathcal{P}_I u_S$.]

Recalling that $(Tf)^*(x_{4n(I)}) = \pi^*(\mathcal{P}_I)\cdot U$, one has

Assertion: $(Tf)^*(\alpha_{2k} \otimes x_{4n(I)}) = \pi^*(\mathcal{P}_I u_S)\cdot U + \sum \pi^*(\mathcal{P}_{I'} u_{S'})\cdot U$ with $\dim I' > \dim I$, where $S = (2^{s_1}, \ldots, 2^{s_n})$ is the dyadic expansion of $2k$.

Proof: Clearly, $(Tf)^*(\alpha_{2k} \otimes x_{4n(I)}) = \alpha_{2k}[\pi^*(\mathcal{P}_I)\cdot U]$ and $\pi^*(\mathcal{P}_I)\cdot \alpha_{2k}U$ both belong to kernelQ_0, with their difference being of the form $\sum \alpha' \pi^*(\mathcal{P}_I)\cdot \alpha'' U$ with $\deg \alpha' > 0$, and each term $\alpha'\pi^*(\mathcal{P}_I)\cdot\alpha'' U$ has a larger squared factor, when expressed as an element of $\tilde{H}^*(\underset{\sim}{\text{TBSpin}}^c;Z_2)$. Thus one may write

$$\sum \alpha'\pi^*(\mathcal{P}_I)\cdot\alpha''U = \sum \pi^*(\mathcal{P}_{I''}u_{S''})\cdot U + Q_0V\cdot U.$$

Now Q_0V belongs to the ideal generated by odd w_i, so every term $\mathcal{P}_{I''}$ occuring must have larger "squared" term.

Thus it suffices to consider $\alpha_{2k}U$, and since for degree $2k$, every term $\pi^*(\mathcal{P}_{I'}u_{S'})U$ has squared term larger than $\pi^*(u_S)\cdot U$, it suffices to prove the coefficient of $\pi^*(u_S)\cdot U$ is nonzero.

For this one notes that the map $\bar{\nu} : \mathcal{A}_2/\mathcal{A}_2Sq^1 + \mathcal{A}_2Sq^3 \longrightarrow \tilde{H}^*(\underset{\sim}{\text{TBSpin}}^c;Z_2)$ is a map of coalgebras, inducing a map of coalgebras on $H(\ \ ;Q_0)$. The product rule $\xi_1^{2k}\cdot\xi_1^{2\ell} = \xi_1^{2k+2\ell}$ translates to the dual statement $\Delta(\alpha_{2k}) = \sum_{i+j=k} \alpha_{2i} \otimes \alpha_{2j}$. Thus $\Delta^n(\alpha_{2k})$ has a term $\alpha_{2^{s_1}} \otimes\ldots\otimes \alpha_{2^{s_n}}$ or $\bar{\nu}\Delta^n(\alpha_{2k}) = \Delta^n\bar{\nu}(\alpha_{2k})$ has a term $u_{2^{s_1}}U \otimes\ldots\otimes u_{2^{s_n}}U$. On the other hand $\Delta^n(\mathcal{P}_i)$ always contains a squared term in at least one factor, since $\Delta^n(\mathcal{P}_i) = \Delta^n(w_{2i}^2) = [\Delta^n(w_{2i})]^2$. Thus $\Delta^n(\pi^*(\mathcal{P}_{I'}u_{S'})\cdot U)$ never contains a term $u_{2^{s_1}}U \otimes\ldots\otimes u_{2^{s_n}}U$, showing that the coefficient of $\pi^*(u_S)U$ in $\bar{\nu}(\alpha_{2k})$ is nonzero. **

This proves that $(Tf)^*$ induces an isomorphism on $H(\ \ ;Q_0)$ for the Spin^c case. For the Spin case one may cheat slightly.

<u>Sublemma 4</u>: $H(H^*(BSpin;Z_2);Q_0) = Z_2[w_{2j}^2, u_{2^s} | s > 1,\ j \text{ not a power of } 2]$ and the homomorphism given by inclusion

$$\tilde{H}^*(\underset{\sim}{\text{TBSpin}}^c;Z_2) \longrightarrow \tilde{H}^*(\underset{\sim}{\text{TBSpin}};Z_2)$$

induces an epimorphism on $H(\ ;Q_0)$. Further, $H(\tilde{H}^*(\underset{\sim}{\mathrm{TBSpin}};Z_2);Q_0)$ and $H(\tilde{H}^*(\underset{\sim}{\mathrm{BO}};Z_2);Q_0)$ have the same rank in each dimension, where $\underset{\sim}{\mathrm{BO}}$ denotes the product of truncated BO spectra corresponding to Tf.

Proof: Noting that $H^*(BSpin^c;Z_2)$ maps onto $H^*(BSpin;Z_2)$ with u_2 sent to zero, $H^*(BSpin;Z_2) = Z_2[w_{2j}, Q_0 w_{2j}, u_{2^s} \mid s > 1]$ with $Q_0 u_{2^s} = 0$. This gives the homology and epimorphism easily. A dimension count gives the asserted equality easily. **

Now let $BO(8n+4n(I),\ldots,\infty) \xrightarrow{h} BU(8n+4n(I),\ldots,\infty)$ and $BO(8n+4n(I)-2,\ldots,\infty) \xrightarrow{h} BU(8n+4n(I),\ldots,\infty)$ be maps classifying $\gamma \otimes C$, with $h^*(x_{4n(I)+8n}) = x_{4n(I)+8n}$ or $v_{4n(I)+8n}$ in the cases $n(I)$ even or $n(I)$ odd.

Letting $\underset{\sim}{\mathrm{BU}} = \Pi \underset{\sim}{\mathrm{BU}}(4n(I),\ldots,\infty)$ be the spectrum used in realizing the map Tf for $\underset{\sim}{\mathrm{TBSpin}}^c$ and $\underset{\sim}{\mathrm{BO}}$ the spectrum used in realizing Tf for $\underset{\sim}{\mathrm{TBSpin}}$ one has a diagram

$$\begin{array}{ccc} \underset{\sim}{\mathrm{TBSpin}} & \xrightarrow{Ti} & \underset{\sim}{\mathrm{TBSpin}}^c \\ {\scriptstyle Tf}\downarrow & & \downarrow{\scriptstyle Tf} \\ \underset{\sim}{\mathrm{BO}} & \xrightarrow{h} & \underset{\sim}{\mathrm{BU}} \end{array}$$

where h is obtained by the product of the above maps h and point maps to the factors $\underset{\sim}{\mathrm{BU}}(4n(I),\ldots,\infty)$ if I contains a 1. This diagram does not, of course, commute, but

Sublemma 5: After applying the functor $H(\tilde{H}^*(\ ;Z_2);Q_0)$ the above diagram commutes.

Proof: It suffices to consider the summands $(\mathcal{A}_2/\mathcal{A}_2 Sq^1 + \mathcal{A}_2 Sq^3)x_{4n(I)}$ in $\tilde{H}^*(\underset{\sim}{\mathrm{BU}};Z_2)$ individually. There are three cases.

1) I contains a 1. Then $(Ti)^*(Tf)^*(x_{4n(I)}) = \mathcal{P}_I U$, but $\mathcal{P}_1$ divides $\mathcal{P}_I$ and $\mathcal{P}_1$ is zero in $H^*(BSpin;Z_2)$. Since h is a point map to this

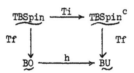

factor, one also has $h^*(x_{4n(I)}) = 0$.

2) *I contains no* 1's *and* $n(I)$ *is odd. Then* $h^*(x_{4n(I)}) = v_{4n(I)}$ which is $Sq^2 x_{4n(I)-2}$ mod 2. Thus $(Tf)^*h^*(x_{4n(I)}) = \mathcal{P}_I U$. This coincides with $(Ti)^*(Tf)^*(x_{4n(I)})$.

3) I contains no 1's and $n(I)$ is even. Then $(Tf)^*h^*(x_{4n(I)}) = (Tf)^*(x_{4n(I)}) = (\mathcal{P}_I + Sq^3Sq^1\alpha_I)\cdot U$. Now in $\tilde{H}^*(\underset{\sim}{TBSpin^c};Z_2)$, one has for any $a \in \mathcal{A}_2/\mathcal{A}_2Sq^1 + \mathcal{A}_2Sq^3$ such that $Q_0a = 0$, $a(Sq^3Sq^1\alpha_I\cdot U) \in \ker Q_0$. This class belongs to the ideal generated by the odd w_i (Since $Sq^1\alpha_I$ belongs to this ideal) and thus $a(Sq^3Sq^1\alpha_I U)$ is zero in $H(\tilde{H}^*(\underset{\sim}{TBSpin^c};Z_2);Q_0)$. Thus $(Tf)^*h^*$ and $(Ti)^*(Tf)^*$ induce the same homomorphism on homology with respect to Q_0. **

This establishes that $(Tf)^*$ induces an isomorphism on $H(\ ;Q_0)$ for the Spin case.

<u>Note</u>: If desired one may be more specific in the choice of a representative α_{2k}. Since $Q_1Sq^{2k-2} = Q_0Sq^{2k} + Sq^{2k}Q_0$ one may let $\alpha_{2k} = \chi(Sq^{2k})$. Now $\chi(Sq^i)U = \pi^*(v_i)U$, where v_i is the Wu class. Thus one may take $u_{2k} = v_{2k}$ if desired, coinciding with the choice in the first calculation of $H(H^*(BSpin^c;Z_2);Q_0)$.

It is interesting and crucial to note that one did not need to find a specific α_{2k} in the above.

<u>Sublemma 6</u>: $H((\chi(\mathcal{A}_2/\mathcal{A}_2Sq^1+\mathcal{A}_2Sq^3))^*;RQ_1^*) = E[\xi_i^2 | i \geq 1]$,

$$H((\chi(\mathcal{A}_2/\mathcal{A}_2Sq^1+\mathcal{A}_2Sq^2))^*;RQ_1^*) = E[\xi_i^2 | i > 1], \text{ and}$$

$$H((\chi(\mathcal{A}_2/\mathcal{A}_2Sq^3))^*;RQ_1^*) = \xi_1^2\cdot E[\xi_i^2 | i > 1],$$

where E denotes the exterior algebra over Z_2.

Proof: By Sublemma 2, $(RQ_1)^*(\xi_k) = \xi_{k-2}^4$. Since $\chi(\mathcal{A}_2/\mathcal{A}_2Sq^1+\mathcal{A}_2Sq^3)^* = Z_2[\xi_1^2,\xi_2^2,\xi_3,\ldots]$, $\ker(RQ_1)^* = Z_2[\xi_1^2,\xi_2^2,\xi_3^2,\ldots]$ and $\mathrm{im}(RQ_1)^*$ is the ideal generated by $Z_2[\xi_1^4,\xi_2^4,\xi_3^4,\ldots]$. Thus $\ker(RQ_1)^*/\mathrm{im}(RQ_1)^* = E[\xi_i^2]$. Since $\chi(\mathcal{A}_2/\mathcal{A}_2Sq^1+\mathcal{A}_2Sq^2)^* = Z_2[\xi_1^4,\xi_2^2,\xi_3,\ldots]$, $\ker(RQ_1)^* = Z_2[\xi_1^4,\xi_2^2,\xi_3^2,\ldots]$ and $\mathrm{im}(RQ_1)^*$ is generated by $Z_2[\xi_1^4,\xi_2^4,\ldots]$.

Letting $a = \alpha + \beta\xi_1 + \gamma\xi_1^2 + \delta(\xi_2+\xi_1^3) + \varepsilon\,\xi_1\xi_2$ with $\alpha,\ldots,\varepsilon$ in A,

$$(RQ_1)^*a = RQ_1^*\alpha + RQ_1^*\beta\cdot\xi_1 + RQ_1^*\gamma\cdot\xi_1^2 + RQ_1^*\delta\cdot(\xi_2+\xi_1^3) + RQ_1^*\varepsilon\cdot\xi_1\xi_2$$

$$+ \delta + \varepsilon\,\xi_1,$$

and $\ker(RQ_1)^*/\mathrm{im}(RQ_1)^*$ is the $E[\xi_i^2|i > 1]$ module on ξ_1^2. **

If $p = (2^{t_1}-2) + (2^{t_2}-2) +\ldots+ (2^{t_n}-2)$ with $2 \le t_1 < t_2 <\ldots< t_n$, then $p < 2^{t_n+1}-2$. Thus if $2^k-2 \le p < 2^{k+1}-2$, one must have $t_n = k$, and p has a unique expression in this form. Let P be the set of integers which are so expressible.

Thus $H(\mathcal{A}_2/\mathcal{A}_2Sq^1+\mathcal{A}_2Sq^3;Q_1)$ is isomorphic to Z_2 in each dimension p belonging to P and is zero in other dimensions. There is then a class $\beta_p \in (\mathcal{A}_2/\mathcal{A}_2Sq^1+\mathcal{A}_2Sq^3)^p$ with $Q_1\beta_p = 0$ and such that writing $p = (2^{t_1}-2) +\ldots+ (2^{t_n}-2)$ one has $\chi(\beta_p)$ evaluating to 1 on $\xi_{t_1-1}^2\cdots\xi_{t_n-1}^2$.

One then defines $\tilde{u}_p \in H^p(BSpin^c;Z_2)$ by $\pi^*(\tilde{u}_p)\cdot U = \bar{\nu}(\beta_p)$ and then $Q_1u_p = 0$.

Assert: $\tilde{u}_{2^t-2}$ for $t \ge 2$ is indecomposable.

Proof: Writing $\chi(\beta_p) = \sum a_J Sq^J$ with J admissible, $p = 2^t-2$, one has $\sum a_J Sq^J$ evaluating to 1 on ξ_{t-1}^2. Now $(\xi_{t-1}^2)(Sq^J) \ne 0$ if and only if $\Delta(Sq^J) = \sum Sq^U \otimes Sq^V$ contains the term $Sq^{J'} \otimes Sq^{J'}$ where $J' = (2^{t-2},\ldots,1)$ which holds if and only if $Sq^J(x^2) \ne 0$ ($\dim x = 1$) which holds if and only if $J = (2^{t-1},\ldots,2)$. Thus $\chi(\beta_p) = Sq^{2^{t-1}}\cdots Sq^2 + \sum a_{J'}Sq^{J'}$.

Letting J be admissible of degree 2^t-2, $J=(j_1,\dots,j_s)$ cannot have excess 1, so $e(J)=(j_1-2j_2)+\dots+(j_{s-1}-2j_s)+j_s=2j_1-\deg J\geq 2$, giving $2j_1\geq 2^t$ or $j_1\geq 2^{t-1}$, and $j_1=2^{t-1}$ if and only if $J=I$ where $I=(2^{t-1},\dots,2)$. Now suppose $Sq^J U=\alpha U$ where α is indecomposable. Letting $J=(j,J')$, $J'\neq\emptyset$, $Sq^J U=Sq^j(Sq^{J'}U)=Sq^j(\alpha' U)$ so $\alpha=Sq^j\alpha'$ mod decomposables, but $j\geq 2^{t-1}$ and $\deg\alpha'\leq 2^{t-1}-2<j$ so $Sq^j\alpha'=0$. Thus $Sq^J U=\alpha U$ with α indecomposable if and only if $J=(2^t-2)$. Thus it suffices to show that the coefficient of Sq^{2^t-2} in β_p is nonzero.

Let $y\in H^2(\mathbb{C}P(2^t-1);Z_2)$ be the generator, with J admissible of degree 2^t-2 and consider $Sq^J y^{2^{t-1}-1}$. Writing $J=(j,J')$, suppose $Sq^J y^{2^{t-1}-1}\neq 0$. Then $Sq^{J'}y^{2^{t-1}-1}=y^{2^t-2-j/2}$ [being nonzero] or $Sq^J y^{2^{t-1}-1}=\binom{2^t-2-j/2}{j/2}y^{2^t-2}$, but $\binom{2^r-2-k}{k}\neq 0\ (2)$ implies $k=2^s-1$, $j=2^{s+1}-2\geq 2^{t-1}$ and $j=2^t-2$. Thus it suffices to show $\beta_p(y^{2^{t-1}-1})\neq 0$.

<u>Sublemma 7</u>: If $a,b\in H^*(\mathbb{C}P(2^t-1);Z_2)$ and $\theta\in\mathcal{A}_2$ with $\deg\theta+\dim a+\dim b=2^{t+1}-2$, then $[\chi(\theta)a]\cdot b=a\cdot\theta(b)$.

<u>Proof</u>: It suffices to prove this by induction on $i=$ degree θ, the case $i=0$ being trivial. Suppose the result true for degrees less than i. Since $w(\mathbb{C}P(2^t-1))=(1+y)^{2^t}=1+y^{2^t}=1$, $v(CP(2^t-1))=1$ and all operations into the top degree are zero. Letting $\Delta(\theta)=\theta\otimes 1+\sum\theta'\otimes\theta''+1\otimes\theta$,

$$\begin{aligned}0=\theta(a\cdot b)&=\theta(a)\cdot b+\sum\theta'(a)\cdot\theta''(b)+a\cdot\theta(b),\\&=\theta(a)+\sum[\chi(\theta'')\theta'(a)]\cdot b+a\cdot\theta(b),\\&=[\chi(\theta)a]\cdot b+a\cdot\theta(b).\quad **\end{aligned}$$

Thus $\beta_p(y^{2^{t-1}-1})\neq 0$ if and only if $\chi(\beta_p)y\neq 0$. Since $\chi(\beta_p)=Sq^{2^{t-1}}\dots Sq^2+\sum a_{J'}Sq^{J'}$ and $Sq^{J'}y=0$, $\chi(\beta_p)y=y^{2^{t-1}}\neq 0$, which completes the proof. **

Thus $H^*(BSpin^c;Z_2) = Z_2[w_{2j}, Q_1w_{2j}, \tilde{u}_{2^t-2} \mid j \neq 2^s-1,\ t \geq 2]$ with $Q_1\tilde{u}_{2^t-2} = 0$. [Note: $Q_1w_{2j} = w_{2j+3}$, so these are generators.]

From this point onward, the case Q_1 is formally identical with the Q_0 case, showing that $(Tf)^*$ induces isomorphisms on homology with respect to Q_1 for both the $Spin^c$ and Spin cases.

This completes the proof of Lemma 2. **

Lemma 3: Let M be a connected coalgebra over Z_2 with counit $1 \in M_0$ and a left module over $\mathcal{A}_2$ such that the diagonal map is a map of $\mathcal{A}_2$ modules. Let $f : N \longrightarrow M$ be a map $\mathcal{A}_2$ modules with either:

1) $N = (\mathcal{A}_2/\mathcal{A}_2Sq^1 + \mathcal{A}_2Sq^3) \otimes X$ and $\ker \nu = \mathcal{A}_2Sq^1 + \mathcal{A}_2Sq^3$, or

2) $N = (\mathcal{A}_2/\mathcal{A}_2Sq^1 + \mathcal{A}_2Sq^2) \otimes X + (\mathcal{A}_2/\mathcal{A}_2Sq^3) \otimes Y$ and $\ker \nu = \mathcal{A}_2Sq^1 + \mathcal{A}_2Sq^2$, and such that f induces isomorphisms on homology with respect to Q_0 and Q_1. Then f is monic and cokernel (f) is a free $\mathcal{A}_2$ module.

Proof: Let $\pi : M \longrightarrow M/\bar{\mathcal{A}}_2M$ be the projection and let $T \subset M$ be a Z_2 subspace mapped by π isomorphically onto a complementary summand for $\pi f(N)$ in $M/\bar{\mathcal{A}}_2M$. Let $e : N \oplus (\mathcal{A}_2 \otimes T) \longrightarrow M : (n, a \otimes t) \longrightarrow f(n) + a(t)$. Then e is epic and induces isomorphisms on $H(\ ;Q_i)$, $i = 0,1$.

For any $\mathcal{A}_2$ module B, let B_n denote the n-th degree part of B and $B^{(n)}$ the $\mathcal{A}_2$ submodule generated by the elements of degree less than or equal to n. [Note: $f : B \longrightarrow C$ epic implies $f^{(n)} : B^{(n)} \longrightarrow C^{(n)}$ epic.]

Clearly $e^{(-1)} : (N \oplus (\mathcal{A}_2 \otimes T))^{(-1)} \longrightarrow M^{(-1)}$ is an isomorphism, both being zero. Suppose then that $e^{(n-1)} : (N \oplus (\mathcal{A}_2 \otimes T))^{(n-1)} \longrightarrow M^{(n-1)}$ is an isomorphism.

Let $y \in (N \oplus (\mathcal{A}_2 \otimes T))^{(n)}$ and suppose $e(y) = 0$. One may then write $y = \sum_i b_i \otimes x_i^n + \sum_j c_j \otimes y_j^n + \sum_k d_k \otimes t_k^n + z$ where $b_i \in \mathcal{A}_2/\ker \nu$, $c_j \in \mathcal{A}_2/\mathcal{A}_2Sq^3$ (zero in case 1)), and $d_k \in \mathcal{A}_2$, x_i^n, y_j^n, t_k^n being linearly independent elements of X_n, Y_n, and T_n respectively, and $z \in (N \oplus (\mathcal{Q}_2 \otimes T))^{(n-1)}$.

Now $\mathcal{Q}_2$ is a free right module over $\mathcal{C}$ ($\mathcal{C} = E$ in case 1), $\mathcal{Q}_2'$ in case 2)) and let a_α be a base of $\mathcal{Q}_2$ over $\mathcal{C}$. One may then write

$$y = \sum u_{i\alpha} a_\alpha \otimes x_i^n + \sum a_\alpha v_{j\alpha} \otimes y_j^n + \sum a_\alpha w_{k\alpha} \otimes t_k^n + z$$

with $u_{i\alpha} \in Z_2$, $v_{j\alpha} \in \mathcal{Q}_2'/\mathcal{Q}_2' Sq^3$, $w_k \in \mathcal{C}$. Let $m = \sup\{\deg a_\alpha \mid$ the coefficient of a_α in this expression is nonzero$\}$.

Let $\psi : M \xrightarrow{\Delta} M \otimes M \longrightarrow M \otimes (M/M^{(n-1)})$ where Δ is the diagonal and the last map is the obvious quotient. If $m \in M^{(n-1)}$, then $\Delta(m) = \sum m' \otimes m''$ and each $m'' \in M^{(n-1)}$, so $\psi(m) = 0$. Similarly, if $\deg q = n$, then $\psi(q) = 1 \otimes q$. If $b \in \mathcal{C}$, then $\psi(bq) = b\psi(q) = b(1 \otimes q) = 1 \otimes bq$, for $\mathcal{C}$ is a subHopf algebra annihilating 1, and thus $\psi(a_\alpha bq) = a_\alpha(1 \otimes bq) = a_\alpha(1) \otimes bq$ + terms in which the first factor has lesser degree.

Composing with $\pi' : M \otimes (M/M^{(n-1)}) \longrightarrow M_m \otimes (M/M^{(n-1)})$ one has

$$0 = \pi'\psi e(y) = \sum_\alpha a_\alpha(1) \otimes \{\sum_i u_{i\alpha} x_i^n + \sum_j v_{j\alpha} y_j^n + \sum_k w_{k\alpha} t_k^n\}$$

the sum being over those α such that $\deg(a_\alpha) = m$. Since kernel ν is the ideal $\mathcal{Q}_2 \bar{\mathcal{C}}$, these classes $a_\alpha(1)$ are linearly independent, and the right hand factors must all be zero.

Thus $e : X_n \oplus (\mathcal{Q}_2'/\mathcal{Q}_2' Sq^3) \otimes Y_n \oplus (\mathcal{C} \otimes T_n) \longrightarrow M/M^{(n-1)}$ must have nontrivial kernel. Now

$$\bar{e} : \frac{N \oplus (\mathcal{Q}_2 \otimes T)}{(N \oplus (\mathcal{Q}_2 \otimes T))^{(n-1)}} \longrightarrow \frac{M}{M^{(n-1)}}$$

is epic and induces isomorphisms on $H(\ ;Q_i)$ (since both e and $e^{(n-1)}$ do so).

Case 1): Thus

$$\bar{e} : \bigoplus_{j \geq n} [(\mathcal{A}_2/\mathcal{A}_2 Sq^1 + \mathcal{A}_2 Sq^3) \otimes X_j + (\mathcal{A}_2 \otimes T_j)] \longrightarrow M/M^{(n-1)}$$

induces isomorphisms on $H(\ ;Q_i)$, $i = 0,1$.

a) $\bar{e}$ is monic on $X_n \oplus T_n$. If $\bar{e}(x_n+t_n) = 0$, then $\pi(x_n+t_n) = 0$ for $\pi : M \longrightarrow M/\overline{\mathcal{A}}_2 M$ factors through $M/M^{(n-1)}$, and so $\pi(t_n) \in \pi f(N)$, making $t_n = 0$. Thus $\bar{e}(x_n) = 0$, but x_n is a Q_0 cycle and represents zero in $H(M/M^{(n-1)};Q_0)$, so $x_n \in \operatorname{im} Q_0$. Thus $x_n = 0$.

b) $\bar{e}$ is monic on $Q_i \otimes T_n$. If $\bar{e}(Q_i t_n) = 0$, then $\bar{e}(t_n)$ represents a class in $H(M/M^{(n-1)};Q_i)$ and there must be an x_n with $\bar{e}(x_n) = \bar{e}(t_n) + Q_i u = \bar{e}(t_n)$. Applying a), $t_n = 0$.

c) $\bar{e}$ is monic on $Q_0Q_1 \otimes T_n$. If $\bar{e}(Q_0Q_1t_n) = 0$, then $Q_1\bar{e}(Q_0t_n) = 0$, so $\bar{e}(Q_0t_n)$ represents a class in $H(M/M^{(n-1)};Q_1)$. Thus for some $x_{n+1} \in X_{n+1}$, $\bar{e}(Q_0t_n) + \bar{e}(x_{n+1}) \in (\operatorname{im} Q_1)_{n+1} = 0$. Then x_{n+1} is a Q_0 cycle and $\bar{e}(x_{n+1}) = Q_0\bar{e}(t_n)$, which by monicity on $H(\ ;Q_0)$ makes $x_{n+1} \in \operatorname{im} Q_0$. Thus $x_{n+1} = 0$, so $\bar{e}(Q_0t_n) = 0$ and by b) $t_n = 0$.

Case 2): Thus

$$\bar{e} : \bigoplus_{j \geq n} [(\mathcal{A}_2/\mathcal{A}_2 Sq^1 + \mathcal{A}_2 Sq^2) \otimes X_j \oplus (\mathcal{A}_2/\mathcal{A}_2 Sq^3) \otimes Y_j \oplus \mathcal{A}_2 \otimes T_j] \longrightarrow M/M^{(n-1)}$$

induces isomorphisms on $H(\ ;Q_i)$, $i = 0,1$.

a) $\bar{e}$ is monic on $X_n \oplus Y_n \oplus T_n$. If $\bar{e}(x_n+y_n+t_n) = 0$, then $\pi(x_n+y_n+t_n) = 0$ so $\pi(t_n) \in \pi f(N)$ and $t_n = 0$. Thus $\bar{e}(x_n+y_n) = 0$, so $0 = Sq^2(\bar{e}(x_n+y_n)) = \bar{e}(Sq^2x_n) + \bar{e}(Sq^2y_n) = \bar{e}(Sq^2 \otimes y_n)$. Since $Sq^2 \otimes y_n$ represents a nonzero class in $H(\ ;Q_1)$, this gives $y_n = 0$. Then $\bar{e}(x_n) = 0$, but x_n represents a nonzero class in $H(\ ;Q_0)$, so $x_n = 0$.

b) $\bar{e}$ is monic on $Q_0 \otimes Y_n \oplus Q_0 \otimes T_n$. If $\bar{e}(Q_0y_n+Q_0t_n) = 0$, then $\bar{e}(y_n+t_n)$ represents a class in $H(M/M^{(n-1)};Q_0)$, so $\bar{e}(y_n+t_n) = \bar{e}(x_n)$ for some x_n, and by a), $x_n = y_n = t_n = 0$.

c) $\bar{e}$ is monic on $Sq^2 \otimes Y_n \oplus Sq^2 \otimes T_n$. If $\bar{e}(Sq^2y_n+Sq^2t_n) = 0$, then $0 = Sq^2\bar{e}(Sq^2y_n+Sq^2t_n) = Q_1Q_0\bar{e}(y_n+t_n)$ so $\bar{e}(Q_0y_n+Q_0t_n)$ represents a class in $H(M/M^{(n-1)};Q_1)$ and there is an x_{n+1} wirh $\bar{e}(x_{n+1}+Q_0y_n+Q_0t_n) = 0$. Thus the class represented by x_{n+1} in $H(\ ;Q_0)$ is trivial, so $x_{n+1} = 0$, giving $\bar{e}(Q_0y_n+Q_0t_n) = 0$, and by b), $y_n = t_n = 0$.

d) $\bar{e}$ is monic on $Sq^2Sq^1 \otimes Y_n \oplus Sq^2Sq^1 \otimes T_n \oplus Sq^3 \otimes T_n$. If $\bar{e}(Sq^2Sq^1y_n+Sq^2Sq^1t_n+Sq^3t_n') = 0$, apply Sq^1 to get $Q_1Q_0\bar{e}(y_n+t_n) = 0$ and as in c), $y_n = t_n = 0$. Then $\bar{e}(Sq^3t_n') = 0$ so $\bar{e}(Q_1Sq^2t_n') = \bar{e}((Sq^3Sq^2+Sq^2Sq^3)t_n') = 0$. Thus $\bar{e}(Sq^2t_n')$ is a Q_1 cycle and $\bar{e}(Sq^2t_n') = \bar{e}(Sq^2y_n'+x_{n+2})$ for some y_n' and x_{n+2}. Applying Sq^2 to this $\bar{e}(Q_0Q_1t_n'+Q_0Q_1y_n') = 0$ and as in c), $t_n' = y_n' = 0$.

e) $\bar{e}$ is monic on $Q_0Q_1 \otimes Y_n \oplus Q_0Q_1 \otimes T_n$, precisely as in c).

f) $\bar{e}$ is monic on $(Sq^5+Sq^4Sq^1) \otimes T_n$. Note that $Q_1Sq^2 = Sq^5+Sq^4Sq^1$. Thus if $\bar{e}(Q_1Sq^2t_n) = 0$, $\bar{e}(Sq^2t_n)$ represents a class in $H(M/M^{(n-1)};Q_1)$ so $\bar{e}(Sq^2t_n) = \bar{e}(Sq^2y_n+x_{n+2})$ for some y_n, x_{n+2}. Applying Sq^1 gives $\bar{e}(Sq^3t_n) = 0$ and by d), $t_n = 0$.

g) $\bar{e}$ is monic on $Sq^5Sq^1 \otimes T_n$. If $\bar{e}(Q_0Q_1Sq^2t_n) = 0$ then $\bar{e}(Q_0Sq^2t_n)$ is a Q_1 cycle giving $\bar{e}(Q_0Sq^2t_n) = \bar{e}(x_{n+3}+Sq^2y_{n+1}+Q_1(y_n'+t_n'))$. Applying Q_0 gives $Q_0Q_1\bar{e}(y_n'+t_n') = 0$ and by e), $y_n' = t_n' = 0$. Thus the Q_0 homology class of $x_{n+3} + Sq^2y_{n+1}$ is sent to zero, so $x_{n+3} = Sq^2y_{n+1} = 0$ and $\bar{e}(Q_0Sq^2t_n) = 0$. By d) this gives $t_n = 0$.

These computations then contradict the existence of m, and thus, one has $y = z \in (N \oplus (\mathcal{Q}_2 \otimes T))^{(n-1)}$, but $e(y) = e^{(n-1)}(y) = 0$ gives $y = 0$. Thus $e^{(n)}$ is monic, hence an isomorphism. By induction on n, e is then an isomorphism. **

Combining Lemmas 1 through 3 completes the proof of the main Anderson-Brown-Peterson theorem. **

As might be expected after this much effort, the cobordism computation is now a triviality. One has, of course,

$$\Omega_n^{Spin} \cong \lim_{r\to\infty} \pi_{n+r}(TBSpin_r, \infty)$$

and

$$\Omega_n^{Spin^c} \cong \lim_{r\to\infty} \pi_{n+r}(TBSpin_r^c, \infty).$$

Proposition: The groups Ω_n^{Spin} and $\Omega_n^{Spin^c}$ are finitely generated. Further, the maps

$$\pi : BSpin \longrightarrow BSO$$

and

$$\pi' : BSpin^c \longrightarrow BSO \times K(Z,2)$$

are odd primary homotopy equivalences, and induce odd primary homotopy equivalences on the Thom space level. Thus one has isomorphisms

$$\Omega_*^{Spin} \otimes Z[1/2] \cong \Omega_*^{SO} \otimes Z[1/2]$$

and

$$\Omega_*^{Spin^c} \otimes Z[1/2] \cong \Omega_*^{SO}(K(Z,2)) \otimes Z[1/2].$$

In particular, all torsion in Ω_*^{Spin} and $\Omega_*^{Spin^c}$ is two primary.

Proof: Since Spin and $Spin^c$ bundles are naturally oriented for integral cohomology, this all follows from the Thom isomorphism and the fact that π^* and π'^* are isomorphisms on rational and Z_p (p odd) cohomology. **

Turning to the 2-primary structure, one has the 2-primary homotopy equivalences

$$Tf \times g : \underset{\sim}{TBSpin} \longrightarrow \underset{\sim}{BO} \times \Pi \underset{\sim}{K}(Z_2, \dim z_i)$$

and

$$Tf \times g : \underset{\sim}{TBSpin^c} \longrightarrow \underset{\sim}{BU} \times \Pi \underset{\sim}{K}(Z_2, \dim z_i^!).$$

Now $\pi_i(BO) \cong KO^{-i}(pt)$ is given by

i mod 8	0	1	2	3	4	5	6	7
$\pi_i(BO)$	Z	Z_2	Z_2	0	Z	0	0	0

and $\pi_{2i}(BU) \cong Z$, $\pi_{2i+1}(BU) = 0$, so one has easily:

<u>Theorem</u>: All torsion in Ω_*^{Spin} and $\Omega_*^{Spin^c}$ has order 2.

The main structure theorem is:

<u>Theorem</u>: Two Spin manifolds are cobordant if and only if they have the same KO theory and Z_2 cohomology characteristic numbers. Two $Spin^c$ manifolds are cobordant if and only if they have the same rational and Z_2 cohomology characteristic numbers.

<u>Proof</u>: Let $\alpha \in \Omega_n^{Spin}$ with all KO and Z_2 numbers zero. Then $(Tf \times g)_*(\alpha) = 0$ in $\pi_*(\underset{\sim}{BO} \times \Pi \underset{\sim}{K}(Z_2, \dim z_i))$. Since $(Tf \times g)_*$ is an isomorphism mod odd torsion, α is a torsion class of odd order and since all torsion in Ω_*^{Spin} is two primary, $\alpha = 0$.

If $\alpha \in \Omega_*^{Spin^c}$ has all rational and Z_2 characteristic numbers zero, then α is a torsion class. Further, the homotopy homomorphism induced by $g : \underset{\sim}{TBSpin^c} \longrightarrow \Pi \underset{\sim}{K}(Z_2, \dim z_i^!)$ sends α to zero. Since g_* is an isomorphism on the torsion, this makes $\alpha = 0$. **

Proposition: If M and M' are two Spin manifolds with $\pi_R^J[M] = \pi_R^J[M']$ for all sequences J having no 1's, then M and M' have the same KO characteristic numbers.

Proof: Let $Tf : \underset{\sim}{TBSpin} \longrightarrow \underset{\sim}{BO}$ have fiber $\underset{\sim}{G}$. Let $[M - M'] \in \pi_*(\underset{\sim}{TBSpin})$ and then $(Tf)_*[M - M'] = 0$ so $[M - M'] = i_*[g]$ for some $g \in \pi_*(\underset{\sim}{G})$. Let $p : \underset{\sim}{TBSpin} \longrightarrow BO$ be a map realizing some KO characteristic class. Then $pi : \underset{\sim}{G} \longrightarrow BO$, with $p_*i_*[g]$ being the value of the characteristic number defined by p on $M - M'$.

Then $\underset{\sim}{G}$ has mod 2 cohomology a free $\mathcal{A}_2$ module (hence a free E module with homology zero) and trivial rational cohomology since $\underset{\sim}{G}$ has the 2 primary homotopy type of a product of $\underset{\sim}{K}(Z_2)$ spectra. Thus in the stable range, G_n satisfies all the conditions used to analyze filtration on BSO, so that any class in $\widetilde{KO}^*(G_n)$ has filtration at least $2n$. Thus $p_*i_*[g] = 0$, giving the result. **

In order to tell more about the structure of Ω_*^{Spin}, one may analyze the homotopy of $\underset{\approx}{BO} \times \Pi\underset{\sim}{K}(Z_2, \dim z_i)$. From the knowledge of $\pi_*(BO)$ one has:

For each sequence J containing no 1's and having $n(J)$ even, there is a Spin manifold M_J of dimension $4n(J)$, of infinite order in $\Omega_{4n(J)}^{Spin}$, such that $\pi_R^J[M_J]$ is odd (as a multiple of the $KO^*(pt)$ generator), with all other KO theory numbers being zero (and the numbers $z_i[M_J]$ being zero). Applying complexification and the Chern character gives $ch(\pi_R^J[M_J] \otimes \mathbb{C}) = ch(\pi_R^J \otimes \mathbb{C})\hat{A}[M_J] = \mathcal{P}_J[M_J]$, and the mod 2 characteristic number $\mathcal{P}_J[M_J]$ is odd.

For each sequence J containing no 1's and have $n(J)$ odd there are Spin manifolds N_J of dimension $4n(J)-2$ and M_J of dimension $4n(J)$, N_J being of order 2 and M_J having infinite order, such that $\pi_R^J[N_J]$ and $\pi_R^J[M_J]$ are odd (as multiples of the KO*(pt) generator) and having the other KO theory and Z_2 cohomology numbers zero.

One also has classes $\alpha \in \Omega_1^{Spin} \cong Z_2$, $\tau \in \Omega_4^{Spin} \cong Z$ and $\omega \in \Omega_8^{Spin} \cong Z \oplus Z$ for which $1[\alpha]$, $1[\tau]$, and $1[\omega]$ are all odd multiples of the KO*(pt) generator (and with $\pi_R^2[\omega] = 0$). [Note: Applying χ and dualization to $\mathcal{Q}_2/\mathcal{Q}_2 Sq^1 + \mathcal{Q}_2 Sq^2$ gives a polynomial algebra on classes of dimension 4, 6, 7, 15,...; and the generating class for $\underset{\sim}{BO}(4n(2),\ldots,\infty)$ gives another 8 dimensional class. Thus Tf induces isomorphisms on Z_2 cohomology through dimension 9, so no z_i appear in this range.]

Finally one may find manifolds R_i of dimension equal to dim z_i, having order 2, for which all KO numbers vanish and with $z_{i'}[R_i] = 0$ or 1 as $i' \neq i$ or $i' = i$.

One then has the result of Anderson, Brown, and Peterson [7]:

Theorem: A basis for $\Omega_*^{Spin} \otimes Z_2$ is given by

1) $[M_J] \times \omega^k \times \alpha^i$, $k \geq 0$, $0 \leq i \leq 2$, $n(J)$ even,

2) $[M_J] \times \tau \times \omega^k$, $k \geq 0$, $n(J)$ even,

3) $[R_i]$,

4) $[N_J]$, $n(J)$ odd,

5) $[M_J] \times \omega^k$, $k \geq 0$, $n(J)$ odd, and

6) $(([M_J] \times \tau)/4) \times \omega^k \times \alpha^i$, $k \geq 0$, $0 \leq i \leq 2$, $n(J)$ odd.

Proof: This is immediate from the structure of $\pi_*(\underset{\sim}{BO} \times \Pi \underset{\sim}{K}(Z_2, \dim z_i))$, with the classes in cases 1), 2), or 4) through 6) having the number π_R^J (for the same J) an odd multiple of the KO*(pt) generator. One need only note that $\pi_*(BO(k,\ldots,\infty))$ is a $\pi_*(BO)$ module (using the tensor product) and that the image of $\pi_4(BO) \otimes \pi_{8k+4}(BO(8k+4,\ldots,\infty))$ in $\pi_{8k+8}(BO(8k+4,\ldots,\infty))$ is the multiples of 4. **

<u>Theorem</u>: Let $B_n^{SO} = \{x \in H_n(BSO;Q) | s_\omega(e_{\mathcal{O}})\hat{A}[x] \in Z$ for all $\omega\}$, and $B_*^{SO} = \bigoplus_n B_n^{SO} \subset H_*(BSO;Q)$. Then $\tau : \Omega_*^{Spin} \longrightarrow H_*(BSO;Q)$ maps Ω_*^{Spin}/Torsion isomorphically onto the subring of the polynomial ring $B_*^{SO} = Z[x_{4i}]$ consisting of all classes whose dimension is a multiple of 8 and twice every class whose dimension is congruent to 4 mod 8.

In particular, all relations among the Pontrjagin numbers of Spin manifolds are given by the KO theoretic relations:

$$s_\omega(e_{\mathcal{O}})\hat{A}[M] \in Z \quad \text{or} \quad 2Z$$

if $\dim M \equiv 0$ or 4 mod 8, respectively.

<u>Proof</u>: Let M^n be imbedded in S^{8N} for some large N, with Spin normal bundle ν. Then

$$\begin{aligned} s_\omega(e_{\mathcal{O}})\hat{A}[M] &= ch(s_\omega(\pi_R) \otimes \mathbb{C}) \cdot \Phi_H^{-1}(ch(\psi U(\nu)))[M], \\ &= \Phi_H^{-1}(ch(s_\omega(\pi_R) \otimes \mathbb{C})ch(\psi U(\nu)))[M], \\ &= ch(\psi c^*(\pi^*[s_\omega(\pi_R)]U(\nu)))[S^{8N}], \end{aligned}$$

but $c^*(\pi^*[s_\omega(\pi_R)]\cdot U(\nu)) \in \widetilde{KO}^{8N-n}(S^{8N})$ and this Chern character is integral, and is even if $n \equiv 4$ (8).

Thus $\tau : \Omega_*^{Spin} \longrightarrow B_*^{SO}$. From the analysis of B_*^{SO} in the study of SU cobordism, one has $B_*^{SO} = Z[x_{4i}]$ and clearly image τ is contained in the asserted subring $A_* \subset B_*^{SO}$, $A_{8k} = B_{8k}^{SO}$ and $A_{8k+4} = 2B_{8k+4}^{SO}$.

Clearly image $\tau \subset A_*$ has 2 primary index in each dimension, for im $\tau \subset B_*^{SO} \subset$ im Ω_*^{SO} with $\Omega_*^{Spin} \longrightarrow \Omega_*^{SO}$ being an isomorphism mod odd torsion. Also from the study of Ω_*^{SU}, one has $\Omega_{8k+4}^{SU} = \text{im } \partial_{8k+4}$ mapped onto $2B_{8k+4}^{SO}$, and this factors through τ, so $(\text{im } \tau)_{8k+4} = A_{8k+4}$.

For $(\text{im } \tau)_{8k}$, one has $s_\omega(e_{\mathcal{O}})\hat{A}[M^{8k}]\cdot\bar{p}(1)^k = s_\omega(\pi_R)[M^{8k}]$ and for any sequence I belonging to $\pi(2k)$, write $I = (J,1,1,\ldots,1)$, J having no 1's, with p 1's. Let $M_I' = M_J \times \omega^{p/2}$ if p is even, $(M_J \times \tau)/4 \times \omega^{(p-1)/2}$ if p is odd. Then the numbers $\pi_R^K[M_I']$ are 0 if $K \neq J$ and odd if $K = J$. Thus the free group on the $\tau M_I'$ is contained in B_{8k}^{SO} with odd index. Combined with the two primary index property, this gives $B_{8k}^{SO} = (\text{im } \tau)_{8k}$. **

In order to understand a bit more of the structure of Ω_*^{Spin}, suppose M is a Spin $8k + 2$ manifold, and let $\bar{S}^1$ denote the circle with the unusual framing/Spin structure which represents the nonzero class $\alpha \in \Omega_1^{Spin} \cong Z_2$. All Stiefel-Whitney numbers of $\bar{S}^1$ are zero, so the same is true of the numbers of $\bar{S}^1 \times M$. Further, all KO numbers of $\bar{S}^1 \times M$ are zero since the dimension is $8k + 3$. Thus $\bar{S}^1 \times M$ bounds, and let U have boundary $\bar{S}^1 \times M$. Then $2\bar{S}^1 \times M = \partial(2U) = \partial(V \times M)$ where $\partial V = 2\bar{S}^1$, and one may form $T(M) = 2U \cup [-(V \times M)]/\partial(2U) \equiv \partial(V \times M)$ giving a closed Spin manifold of dimension $8k + 4$.

A different choice of cobordism U' of $\bar{S}^1 \times M$ to zero changes $T(M)$ to $T(M) \cup 2[U' \cup (-U)/\partial U' \equiv \partial U]$ in Ω_{8k+4}^{Spin} while a different choice of cobordism V' of $2\bar{S}^1$ to zero replaces $T(M)$ by $T(M) \cup [V' \cup (-V)/\partial V' \equiv \partial V] \times M$ $= T(M) \cup X \times M$, but X is a 2 dimensional Spin manifold so $X \cdot M = a \cdot \alpha^2 \cdot M$ with $a \in Z_2$, and since $\alpha \cdot M = 0$ this is also zero. Thus $T(M)$ gives a well defined class in $\Omega_{8k+4}^{Spin} \otimes Z_2$.

This construction clearly depends only on the cobordism class of M and using disjoint unions is clearly additive, and thus defines a homomorphism

$$T : \Omega_{8k+2}^{Spin} \longrightarrow \Omega_{8k+4}^{Spin} \otimes Z_2.$$

If N is an 8ℓ dimensional Spin manifold, $\partial(U \times N) = \bar{S}^1 \times M \times N$ and so $T(M \times N)$ is represented by $2U \times N \cup [-(V \times M \times N)]/\partial(2U \times N) \equiv \partial(V \times M \times N)$ $= T(M) \times N$. Thus T is an Ω_{8*}^{Spin} module homomorphism.

If one considers Ω_n^{Spin} as $\pi_{n+k}(TBSpin_k)$ for some large k, this may be realized as the following construction:

Let $f : S^{n+k} \longrightarrow TBSpin_k$ represent M, $h : S^{n+k+1} \longrightarrow S^{n+k}$ the suspension of the Hopf map, and $2 : S^{n+k+1} \longrightarrow S^{n+k+1}$ a degree 2 map. One may deform 2 and h to be transverse regular on every point other than the base point, the inverse image of any regular value in S^{n+k} being $\bar{S}^1$. If f is also transverse regular on $BSpin_k$, then $f \circ h \circ 2$ is transverse regular on $BSpin_k$ and defines the Spin manifold $2\bar{S}^1 \times M$. Considering the cobordisms U and V as given by maps $u : D^{n+k+2} \longrightarrow TBSpin_k$ and $v : D^{n+k+2} \longrightarrow S^{n+k}$ (V being a framed cobordism of $2\bar{S}^1$ to zero) extending $f \circ h$ and $h \circ 2$ respectively, $u \circ 2$ and $f \circ v$ fit together along their boundary to define a map of the $n + k + 2$ sphere into $TBSpin_k$ which represents $T(M)$.

Being given any space X, with $\beta \in \pi_n(X)$ such that $\beta \circ \eta \in \pi_{n+1}(X)$ is zero, where $\beta \circ \eta$ is defined by $b \circ h : S^{n+1} \xrightarrow{h} S^n \xrightarrow{b} X$ with b representing β, h being the suspension of the Hopf map, one may deform $b \circ h$ to zero and $h \circ 2$ to zero (if $n \geq 3$) and joining these together define a class $\langle 2,\eta,\beta \rangle$ in $\pi_{n+2}(X)$. A different choice of homotopy for $b \circ h$ changes this by a multiple of 2, while a different homotopy for $h \circ 2$ adds a composition $\beta \circ \eta \circ \eta$ which is zero. Thus the Toda bracket $\langle 2,\eta,\ \rangle : \text{kernel}(\circ\eta)_n \longrightarrow \text{cokernel}(2)_{n+2}$ is well defined. This is just the operation T in the special case of $TBSpin$.

Considering the space Y formed from S^n by attaching an $n + 2$ cell by means of h, $Y = S^n \cup_h e^{n+2}$, the map $b : S^n \longrightarrow X$ extends to $\hat{b} : Y \longrightarrow X$, and clearly the map $i : S^n \longrightarrow Y$ gives $i \circ \eta = 0$. By naturality of the construction, one has $\hat{b}_*(\langle 2,\eta,i \rangle) = \langle 2,\eta,\beta \rangle$. Considering the space Y more closely, it is clear that the cofibration $S^n \xrightarrow{i} Y \xrightarrow{\pi} S^{n+2}$ gives $H_n(Y;Z) \cong H_{n+2}(Y;Z) \cong Z$ and all other positive dimensional groups are zero, and the obvious universal construction of $\langle 2,\eta,i \rangle$ makes the composite $\pi \circ \langle 2,\eta,i \rangle$ of degree 2. Now Y is just the $n + 2$ skeleton of the two stage Postnikov system $K(Z,n+2) \longrightarrow Z \longrightarrow K(Z,n)$ with $\tau\iota_{n+2} = \delta Sq^2\iota_n$, and

$\langle 2,\eta,i \rangle$ represents a generator of $\pi_{n+2}(Y) = Z$ which has Hurewicz image equal to twice a generator of $H_{n+2}(Y;Z)$.

Applying this to BO, with $\gamma \in \pi_{8k+2}(BO)$ the nonzero element ($\gamma \circ \eta \in \pi_{8k+3}(BO) = 0$) one has

$$\begin{array}{ccc} & & BO(8k+2,\ldots,\infty) \xrightarrow{d} BO \\ & & \Big\downarrow c \\ Y & \xrightarrow{b} & BO(8k+2,8k+4) \end{array}$$

with $BO(8k+2,8k+4)$ being the two stage Postnikov system $K(Z,8k+4) \longrightarrow BO(8k+2,8k+4) \longrightarrow K(Z_2,8k+2)$ with $\tau\, i_{8k+4} = \delta Sq^2 i_{8k+2}$, with b, c, and d being isomorphisms on π_{8k+4} and epic on π_{8k+2}. Naturality of $\langle 2,\eta, \ \rangle$ then shows that $\langle 2,\eta,\gamma \rangle$ is the nonzero element in $\pi_{8k+4}(BO) \otimes Z_2$.

This gives:

<u>Proposition</u>: If M is an $8k + 2$ dimensional Spin manifold, then $\pi_R^J[T(M)]$ reduced mod 2 is the same as $\pi_R^J[M]$.

<u>Proof</u>: Let $a : S^{8k+2+8\ell} \longrightarrow TBSpin_{8\ell}$ represent M and $p : TBSpin_{8\ell} \longrightarrow BO$ represent $\pi_R^J \cdot U$. Then $\pi_R^J[M] = p_*[a] \in \pi_{8k+8\ell+2}(BO)$ and $\pi_R^J[T(M)]$ mod 2 is $p_*(\langle 2,\eta,[a] \rangle) = \langle 2,\eta,p_*[a] \rangle \in \pi_{8k+8\ell+4}(BO) \otimes Z_2$, but $\langle 2,\eta, \ \rangle : \pi_{8k+8\ell+2}(BO) \cong Z_2 \longrightarrow \pi_{8k+8\ell+4}(BO) \otimes Z_2 \cong Z_2$ is an isomorphism. **

Now let $I_* \subset \Omega_*^{Spin}$ be the set of classes [M] for which all KO theory' characteristic numbers are zero. Since $\pi_R^i(M \times N) = \sum_{j+k=i} \pi_R^j(M) \cdot \pi_R^k(N)$, I_* is an ideal. Letting $\underset{\sim}{G}$ be the fiber of $Tf : \underset{\sim}{TBSpin} \longrightarrow \underset{\sim}{BO}$, I_* is precisely image$(\pi_*(\underset{\sim}{G}) \longrightarrow \Omega_*^{Spin})$, which is a Z_2 vector space detected by Z_2 cohomology characteristic numbers.

Let R_* be the ring $Z[x_{4i}, Y_{8j+2}, \theta_1 | i \geq 1, j \geq 1]$ modulo the relations $2\theta_1 = 2Y_t = Y_t Y_r = \theta_1 Y_t = \theta_1^3 = 0$, $Y_{8j+2} x_4 = x_{8j+4}\theta_1^2$, $Y_{8j+2} x_{8\ell+4} = Y_{8\ell+2} x_{8j+4}$.

<u>Theorem</u>: Ω_*^{Spin}/I_* is isomorphic as ring with the subring A_* of R_* generated by the Y's, θ_1, the R_{8k} and $2R_{8k+4}$.

<u>Proof</u>: Clearly I_* is the kernel of the homotopy map induced by $Tf : \underset{\sim}{TBSpin} \longrightarrow \underset{\sim}{BO}$, identifying Ω_*^{Spin}/I_* with image($(Tf)_*$). Now $I_* \subset \text{Torsion}(\Omega_*^{Spin})$ so one has $p : \Omega_*^{Spin}/I_* \longrightarrow \Omega_*^{Spin}/\text{Torsion} \subset B_*^{SO} = Z[x_{4i}]$ which completely annihilates the groups of dimension not a multiple of 4, which are Z_2 vector spaces, and since $\pi_{4*}(\underset{\sim}{BO})$ is torsion free, $I_{4*} = \text{Torsion}(\Omega_{4*}^{Spin})$ and thus $p : \Omega_{4*}^{Spin}/I_{4*} \cong \Omega_*^{Spin}/\text{Torsion}$.

Now $(Tf)_* : \Omega_*^{Spin}/I_* \otimes Z_2 \longrightarrow \pi_*(\underset{\sim}{BO}) \otimes Z_2$ is an isomorphism and in fact coincides with the isomorphism $(Tf)_* : (\Omega_*^{Spin}/I_*)_i \longrightarrow \pi_i(\underset{\sim}{BO})$ in dimensions $i \not\equiv 0\ (4)$. Thus one has a class $\theta_1 \in \Omega_1^{Spin}/I_1 \cong \Omega_1^{Spin}$ with $\theta_1^3 = 2\theta_1 = 0$ and the Toda bracket $T = \langle 2, \eta, \ \rangle : \pi_{8k+2}(\underset{\sim}{BO}) \longrightarrow \pi_{8k+4}(\underset{\sim}{BO}) \otimes Z_2$ is an isomorphism, so $T : \Omega_{8k+2}^{Spin}/I_{8k+2} \longrightarrow \Omega_{8k+4}^{Spin}/I_{8k+4} \otimes Z_2$ is an isomorphism. One may choose a unique element $Y_{8k+2} \in \Omega_{8k+2}^{Spin}/I_{8k+2}$, $k \geq 1$, so that $T(Y_{8k+2})$ is the mod 2 reduction of the class which maps by p into $2x_{8k+4}$. Clearly $2Y_t = 0$, $\theta_1 Y_t = 0$ being a class of dimension $8k+3$, and $Y_t Y_r = 0$ being a torsion element in the free group $\Omega_{4*}^{Spin}/I_{4*}$.

$\Omega_{8t}^{Spin}/I_{8t}\cdot\theta_1$ maps onto $\Omega_{8t+1}^{Spin}/I_{8t+1}$, for applying $(Tf)_*$, composition with η sends $\pi_{8k}(\underset{\sim}{BO})$ onto $\pi_{8k+1}(\underset{\sim}{BO})$ with kernel precisely the multiples of 2, making $\Omega_{8t+1}^{Spin}/I_{8t+1} \cong \Omega_{8t}^{Spin}/I_{8t}\cdot\theta_1$. Applying T to $\Omega_{8t}^{Spin}/I_{8t}\cdot\theta_1^2 + \sum_{k=1}^{t} \Omega_{8t-8k}^{Spin}/I_{8t-8k} Y_{8k+2} = B_{8t+2} \subset \Omega_{8t+2}^{Spin}/I_{8t+2}$ gives the subgroup of $\Omega_{8t+4}^{Spin}/I_{8t+4} \otimes Z_2$ spanned by the $\Omega_{8*}^{Spin}/I_{8*}$ multiples of classes mapped under p into $2x_4$ and the $2x_{8k+4}$, and thus B_{8t+2} spans $\Omega_{8t+2}^{Spin}/I_{8t+2}$. The kernel of the map $\Omega_{8t}^{Spin}/I_{8t} \oplus \sum_{k=1}^{t} \Omega_{8t-8k}^{Spin}/I_{8t-8k} \longrightarrow B_{8k+2}$ is precisely the kernel of the composite with T which is precisely the multiples of 2 and the relations given by

$$T(Y_{8\ell+2}x_{8k+4}z_{8s+4}) = 2x_{8\ell+4}x_{8k+4}z_{8s+4} (\mathrm{mod}\ 2) = T(Y_{8k+2}x_{8\ell+4}z_{8s+4})$$

if $k, \ell \geq 1$, $s + k + \ell + 1 = t$, or

$$T(Y_{8k+2}x_4 z_{8s+4}) = 2x_4 x_{8k+4}z_{8s+4} (\mathrm{mod}\ 2) = T(\theta_1^2 \cdot x_{8k+4}z_{8s+4})$$

if $k \geq 1$, $s + k + 1 = t$.

It is then immediate that Ω_*^{Spin}/I_* is isomorphic to the subring A_* of R_*. **

Returning to the $Spin^c$ case one has

$$BU \xrightarrow{t} BSpin^c \xrightarrow{\pi'} BSO \times BU(1)$$

with π'^* an isomorphism on rational cohomology and t^* being monic. One then has Pontrjagin classes $\mathcal{P}_i \in H^{4i}(BSpin^c;Q)$ and a class $c_1 \in H^2(BSpin^c;Q)$ with $H^*(BSpin^c;Q) = Q[c_1, \mathcal{P}_i]$. One may then form the classes $s_{\omega,j}(e) = s_\omega(e_{\mathcal{P}})e^{jc_1}$ and $\mathcal{J} = e^{-c_1/2} \cdot \hat{A}$ in $H^*(BSpin^c;Q)$.

Let $B_n^{Spin^c} = \{x \in H_n(BSpin^c;Q) | s_{\omega,j}(e)\mathcal{J}[x] \in Z$ for all $\omega, j\}$ and $B_*^{Spin^c} = \bigoplus_n B_n^{Spin^c} \subset H_*(BSpin^c;Q)$.

$\Omega_*^{Spin^c}$ is a ring since $BSpin^c$ admits an H-space structure, with both t and π' being H-maps. This ring structure gives a diagonal map in $H^*(BSpin^c;Q)$ given by $\Delta(\mathcal{P}_i) = \sum_{j+k=i} \mathcal{P}_j \otimes \mathcal{P}_k$ and $\Delta(c_1) = c_1 \otimes 1 + 1 \otimes c_1$. Thus $\Delta(s_\omega(e_{\mathcal{P}})) = \sum_{\omega'\omega''=\omega} s_{\omega'}(e_{\mathcal{P}}) \otimes s_{\omega''}(e_{\mathcal{P}})$, $\Delta(\hat{A}) = \hat{A} \otimes \hat{A}$ and $\Delta(e^{vc_1}) = e^{vc_1} \otimes e^{vc_1} (v \in Q)$.

Let $H = Z[\alpha_i] \otimes U$, $i \geq 1$, where U is the free abelian group on elements u_n, $n \in Z$ $(-\infty < n < \infty)$. Define a sum and product in H by

$$(\sum p_i(\alpha)\cdot u_i) \dotplus (\sum q_j(\alpha)\cdot u_j) = \sum (p_k(\alpha) \dotplus q_k(\alpha))\cdot u_k.$$

For any $x \in B_*^{Spin^c}$, let $\mathcal{O}(x) = \sum (s_{\omega,j}(e)\mathcal{S})[x]\cdot\alpha_\omega\cdot u_j \in H$. Using the multiplication in $H_*(BSpin^c;Q)$ induced by the H-structure, the diagonal formulae show that $B_*^{Spin^c}$ is a subring of $H_*(BSpin^c;Q)$ and $\mathcal{O}: B_*^{Spin^c} \longrightarrow H$ is a ring homomorphism.

Letting $\tau : \Omega_*^{Spin^c} \longrightarrow H_*(BSpin^c;Q)$, one has $\text{image}(\tau) \subset B_*^{Spin^c}$ for if M^n is a $Spin^c$ manifold imbedded in S^{2m} with $Spin^c$ normal bundle ν, then

$$\begin{aligned} s_{\omega,j}(e)\mathcal{S}[M] &= ch(s_\omega(\pi_R) \otimes_R \mathbb{C} \otimes_{\mathbb{C}} \xi^j)\cdot\phi_H^{-1}(e^{c_1(\nu)/2}\hat{A}(-\nu))[M], \\ &= chc^*\{\pi^*(s_\omega(\pi_R) \otimes_R \mathbb{C} \otimes_{\mathbb{C}} \xi^j)\cdot U(\nu)\}[S^{2m}], \\ &\in Z \end{aligned}$$

where ξ is the complex line bundle over M with $c_1(\xi) = c_1(\tau M)$.

If M is an almost complex manifold, then M has a $Spin^c$ structure induced by t, with $c_1(\tau M)$ being the first Chern class of M. This defines a ring homomorphism $\Omega_*^U \longrightarrow \Omega_*^{Spin^c} \longrightarrow B_*^{Spin^c}$. In particular, if M is an SU manifold,

$$s_{\omega,j}(e)\mathcal{S}[M] = s_\omega(e_{\mathcal{O}})\mathcal{S}[M],$$

since $c_1(M) = 0$, and hence $\mathcal{O}(\tau M) = \rho'(\tau M)\cdot\sum u_i$, where $\rho'(x) = \sum s_\omega(e_{\mathcal{O}})\mathcal{S}[x]\alpha_\omega$ is the homomorphism defined in the study of SU cobordism.

Let $\bar{\mathcal{O}}_2 : B_*^{Spin^c} \longrightarrow Z_2[\alpha_i]$ be the composition of $\mathcal{O}$ and the homomorphism $H \longrightarrow Z_2[\alpha_i]$ which sends u_j into zero if $j \neq 0$ or into 1 if $j = 0$, and which reduces mod 2.

<u>Proposition</u>: $\bar{\mathcal{O}}_2(\tau(\mathbb{C}P(i)))$ has largest monomial

1) $\alpha_{i/2}$ if i is even, and

2) 1 if $i = 1$.

<u>Proof</u>: If $i = 2j$, one needs only $s_{(j),0}\mathcal{S}[\mathbb{C}P(2j)] = s_{(j)}(e_{\mathcal{P}})\mathcal{S}[\mathbb{C}P(2j)] = s_j(\mathcal{P})[\mathbb{C}P(2j)] = 2j + 1 \not\equiv 0 \bmod 2$. For $\mathbb{C}P(1)$, the total Pontrjagin class is 1, so $s_\omega(e_{\mathcal{P}})$ is zero if $\omega \neq (0)$ and 1 if $\omega = (0)$, and $\hat{A} = 1$. Then $s_{(0),j}\mathcal{S}[\mathbb{C}P(1)] = e^{jc_1+1/2c_1}[\mathbb{C}P(1)] = 2j + 1$, which completes the proof. **

<u>Corollary</u>: $(\Omega_*^{Spin^c}/\text{Torsion}) \otimes Z_2$ is a polynomial algebra over Z_2 on classes y_i, with $i = 2$ or $4k$. In addition, the image of Ω_*^U in $B_*^{Spin^c}$ has odd index in each dimension.

Now let p be an odd prime. Define $\mathcal{O}_p : B_*^{Spin^c} \longrightarrow H \otimes Z_p$ to be the mod p reduction of $\mathcal{O}$. For any integer k and any $\omega \in \pi(2k)$ a partition of $2k$ into integers 2 and $4i$, $\omega = (2,\ldots,2,4i_1,\ldots,4i_r)$ with j 2's, write $j = \lambda_0 + \lambda_1 p + \ldots + \lambda_s p^s$, $0 \leq \lambda_i < p$, and define

$$N_\omega^p = \mathbb{C}P(1)^{\lambda_0} \times \ldots \times \mathbb{C}P(p^s)^{\lambda_s} \times M_{2i_1}^p \times \ldots \times M_{2i_r}^p,$$

where M_{2i}^p denote the SU manifolds whose existence was asserted in Chapter X.

<u>Proposition</u>: The elements $\mathcal{O}_p(\tau N_\omega^p) \in H \otimes Z_p$, for ω belonging to the set of partitions of $2k$ into 2's and $4i$'s, are linearly independent.

<u>Proof</u>: Suppose one has $\sum n_\omega \mathcal{O}_p(\tau N_\omega^p) = 0$, $n_\omega \in Z_p$, with some $n_\omega \neq 0$ ($\omega \in \pi(2k)$ of the given form). Write each $\omega = (2I_j, 4\tilde{\omega})$, where I_j has j 1's. Then among all ω with $n_\omega \neq 0$, let m denote the largest value of j in any I_j, (i.e., $n_\omega \neq 0$ for some $\omega = (2I_m, 4\tilde{\omega})$ and for any $\omega' = (2I_j, 4\tilde{\omega}')$ with $n_{\omega'} \neq 0$, $j \leq m$).

Define $\phi : H \otimes Z_p \longrightarrow Z_p[\alpha_i]$ by $\phi(u_i) = \binom{m}{i}\cdot(-1)^{m-i}$ if $0 \le i \le m$ and zero otherwise. Then for any M, the coefficient of $\alpha_{\tilde\omega}$ in $\phi\mathcal{O}_p(\tau M)$ is

$$\sum_{i=0}^{m}\binom{m}{i}(-1)^{m-i}s_{\tilde\omega}(e_{\mathcal{P}})e^{ic_1}\mathcal{S}[M] = s_{\tilde\omega}(e_{\mathcal{P}})(e^{c_1}-1)^m\mathcal{S}[M],$$

$$= s_{\tilde\omega}(e_{\mathcal{P}})\mathcal{S}\cdot c_1^m[M] + \text{terms with } c_1 \text{ to a higher power.}$$

Since c_1 is zero for an SU manifold, this gives

$$\phi\Big(\sum_\omega n_\omega \mathcal{O}_p(\tau N_\omega^p)\Big) = \sum_\omega\sum_{\omega'} n_\omega s_{\omega'}(e_{\mathcal{P}})\mathcal{S}[M^p_{2\bar\omega}]\cdot c_1^m[V^p_m]\cdot\alpha_{\omega'},$$

the latter sum being over $\omega = (2I_m, 4\bar\omega)$, and V^p_m being $\mathbb{C}P(1)^{\mu_0}\times\ldots\times \mathbb{C}P(p^s)^{\mu_s}$ for $m = \mu_0 + \ldots + \mu_s p^s$, $0 \le \mu_i < p$. Then

$$c_1^m[V_m^p] = \binom{\mu_0+\ldots+\mu_s p^s}{\mu_0}c_1^{\mu_0}[\mathbb{C}P(1)^{\mu_0}]\ldots\binom{\mu_s p^s}{\mu_s p^s}c_1^{\mu_s p^s}[\mathbb{C}P(p^s)^{\mu_s}],$$

and

$$c_1^{\mu p^r}[\mathbb{C}P(p^r)^\mu] = \mu!\cdot(c_1^{p^r}[\mathbb{C}P(p^r)])^\mu = \mu!(p^r+1)^{\mu p^r},$$

so $c_1^m[V_m^p] \not\equiv 0$ modulo p.

Hence $\sum_{\bar\omega}\sum_{\omega'} n_{(2I_m,4\bar\omega)}s_{\omega'}(e_{\mathcal{P}})\mathcal{S}[M^p_{2\bar\omega}]\alpha_{\omega'} = \sum n_{(2I_m,4\bar\omega)}\rho_p'(\tau M^p_{2\bar\omega})$ is zero. As was noted in the SU case, the polynomials $\rho_p'(\tau M^p_{2\bar\omega})$ for $\bar\omega \in \pi((2k-2m)/4)$ are linearly independent, and thus all $n_{(2I_m,4\bar\omega)}$ are zero, contradicting the choice of m. **

This shows that the index of the image of Ω^U_{2k} in $B^{Spin^c}_{2k}$ is not divisible by p for any odd prime p, or:

<u>Theorem</u>: All relations among the Pontrjagin-Chern numbers of $Spin^c$ manifolds are given by K theory; i.e. $\tau\Omega_*^{Spin^c} = B_*^{Spin^c}$. Further, the forgetful homomorphism $\Omega^U_* \longrightarrow \Omega_*^{Spin^c}/\text{Torsion}$ is epic.

Note: Another proof of this may be given by relating $\Omega_*^{Spin^c}$ to $\Omega_*^{SO}(\mathbb{C}P(\infty))$. Since π' is an odd primary homotopy equivalence, one may apply the result that for torsion free spaces all odd primary relations in oriented bordism come from K theory.

For p an odd prime, the generator of $\tau\Omega_2^{Spin^c} \otimes Z_p$ is the image of $\mathbb{C}P(1)$, with $\mathcal{O}_p(\tau\mathbb{C}P(1)) = \sum (2i+1)u_i$, so $\mathcal{O}_p(\tau\mathbb{C}P(1)^p) = \mathcal{O}_p(\tau\mathbb{C}P(1))$. An indecomposable generator of $\tau\Omega_{2(p-1)}^{Spin^c} \otimes Z_p$ is the image of M_{p-1}^p for which $\rho_p'(\tau M_{p-1}^p) = t \not\equiv 0 \bmod p$, so $\mathcal{O}_p(\tau M_{p-1}^p) = t \sum u_i$ and letting $u \cdot t \equiv 1 \pmod p$, $\mathcal{O}_p(\tau(uM_{p-1}^p \times \mathbb{C}P(1))) = \mathcal{O}_p(\tau\mathbb{C}P(1))$. Then $\tau(uM_{p-1}^p) - \tau(\mathbb{C}P(1)^{p-1})$ and $\tau(\mathbb{C}P(1))$ are nonzero in $\tau\Omega_*^{Spin^c} \otimes Z_p$, but their product is zero. Thus:

Proposition: $(\Omega_*^{Spin^c}/\text{Torsion}) \otimes Z_p$, p a prime, is a polynomial algebra over Z_p if and only if $p = 2$.

Since for each prime p (odd) and each integer m, $c_1^m[V_m^p] \not\equiv 0 \bmod p$, there exist manifolds $V_m \in \Omega_{2m}^{Spin^c}$, with $c_1^m[V_m]$ being a power of 2, hence classes $u_{2m} \in \Omega_{2m}^{Spin^c} \otimes Z[1/2]$ with $c_1^m(u_{2m}) = 1$. By the SU-Spin results, $\tau\Omega_*^{Spin} \otimes Z[1/2]$ is a polynomial algebra on classes x_{4i} (x_{4i} being the cobordism class of an SU manifold). Then $\Omega_*^{Spin^c} \otimes Z[1/2]$ is a free module over $\Omega_*^{Spin} \otimes Z[1/2] = Z[1/2][x_{4i}]$ on the classes u_{2m}, and

$$u_{2i} \cdot u_{2j} = \beta_{i,j} u_{2(i+j)} + \sum \beta_{\omega,i,j} x_{4\omega} \cdot u_{2k},$$

$4|\omega| + 2k = 2(i+j)$, $|\omega| > 0$, $\beta_{i,j}, \beta_{\omega,i,j} \in Z[1/2]$, by the module structure. Also $\beta_{i,j} = c_1^{i+j}(u_{2i} \cdot u_{2j}) = \binom{i+j}{i} c_1^i(u_{2k}) c_1^j(u_{2j}) = \binom{i+j}{i}$.

This describes $(\Omega_*^{Spin^c}/\text{Torsion}) \otimes Z[1/2]$ and completes the $Spin^c$ case.

Relation to Framed Cobordism

Theorem: The forgetful homomorphism $F_n : \Omega_n^{fr} \longrightarrow \Omega_n^{Spin}$ is an isomorphism in dimension 0, and for $n > 0$, $\text{image}(F_n) = 0$ for $n \neq 8k + 1$, $8k + 2$, $\text{image}(F_{8k+1}) \cong \text{image}(F_{8k+2}) \cong Z_2$.

The forgetful homomorphism $F_n : \Omega_n^{fr} \longrightarrow \Omega_n^{Spin^c}$ is an isomorphism in dimension zero and is zero if $n > 0$.

Proof: One may factor F_n through Ω_n^{SU} or Ω_n^{U} respectively, showing that these give upper bounds for the image. For the Spin case there are framed manifolds of dimensions $8k + 1$ and $8k + 2$ with the KO theory characteristic number 1 being nonzero. Since KO theory numbers are Spin cobordism invariants, these manifolds have nontrivial image in Ω_*^{Spin}. **

Relation to Unoriented Cobordism

Theorem: The images of the forgetful homomorphisms

$$F_* : \Omega_*^{Spin} \longrightarrow \mathcal{N}_* \quad \text{and} \quad F_* : \Omega_*^{Spin^c} \longrightarrow \mathcal{N}_*$$

are precisely the set of cobordism classes for which all Stiefel-Whitney numbers with a factor w_1 or w_2, and w_1 or w_3 respectively, are zero.

Proof: Since a Spin manifold has w_1 and w_2 zero, while a $Spin^c$ manifold has w_1 and w_3 zero, these sets of cobordism classes contain the images.

Now suppose $v \in H^n(BO;Z_2)$ is zero on all Spin (resp. $Spin^c$) manifolds of dimension n, and let M be any manifold for which all numbers divisible by w_1 and w_2 (resp. w_3) are zero. Then $\hat{p}^* : \tilde{H}^*(\underset{\sim}{TBO};Z_2) \rightarrow \tilde{H}^*(\underset{\sim}{TBSpin^\varepsilon};Z_2)$ is epic and a class in $\tilde{H}^*(\underset{\sim}{TBSpin^\varepsilon};Z_2)$ vanishes on all of the homotopy of

$\underset{\sim}{TBSpin}^{\varepsilon}$ if and only if it belongs to $\bar{\mathcal{A}}_2\tilde{H}^*(\underset{\sim}{TBSpin}^{\varepsilon};Z_2)$. Thus, letting Φ be the Thom isomorphism, $\Phi(v) \equiv \Phi(x) \bmod \bar{\mathcal{A}}_2\tilde{H}^*(\underset{\sim}{TBO};Z_2)$ with $x \in \ker(p^* : H^*(BO;Z_2)) \longrightarrow H^*(BSpin^{\varepsilon};Z_2)$. Since $\Phi^{-1}(\bar{\mathcal{A}}_2\tilde{H}^*(\underset{\sim}{TBO};Z_2))$ vanishes on all manifolds, this gives $v[M] = x[M]$, where $x = \sum \alpha_i\cdot\rho_i(w_1) + \sum \beta_j\cdot\sigma_j(w_2)$ (resp. $\sum \alpha_i\cdot\rho_i(w_1) + \sum \beta_j\cdot\sigma_j(w_3)$) with $\rho_i, \sigma_j \in \mathcal{A}_2$; $\alpha_i, \beta_j \in H^*(BO;Z_2)$. Consider a number $\alpha\cdot\rho(w_1)[M]$. Since $\rho(w_1) = \lambda w_1^k$ with $\lambda \in Z_2$, $\alpha\cdot\rho(w_1)[M] = \alpha\cdot\lambda w_1^k[M] = 0$ since all numbers of M divisible by w_1 are zero. Then consider a number $\beta\cdot\sigma(w_2)[M]$, (resp. $\beta\cdot\sigma(w_3)[M]$). If degree$(\sigma) = 0$, this vanishes since all numbers of M divisible by w_2 (or w_3) are zero. Suppose $\beta'\cdot\sigma'(w_j)[M] = 0$ whenever degree(σ') < degree(σ), and let $\Delta\sigma = \sigma \otimes 1 + \sum \sigma' \otimes \sigma'' + 1 \otimes \sigma$. Then

$$\begin{aligned}\beta\cdot\sigma(w_j)[M] &= \sigma(\beta\cdot w_j)[M] + \sum (\sigma'\beta\cdot\sigma'' w_j)[M] + \sigma\beta\cdot w_j[M],\\ &= \sigma(\beta\cdot w_j)[M]\end{aligned}$$

since the last terms vanish by the inductive assumption. Now $\sigma(\beta\cdot w_j)[M] = v_\sigma\cdot\beta\cdot w_j[M]$ where v_σ is a "Wu class" defined by σ in $H^*(BO;Z_2)$, and this vanishes on M being divisible by w_j. Thus v vanishes on M also.

Since any number which vanishes on image(F_*) also vanishes on M, one must have $[M] \in$ image(F_*). Thus, image(F_*) is as asserted. **

Open question: Can one describe these images nicely as subrings of $\mathcal{N}_*$?

Relation to Oriented Cobordism

Theorem: The forgetful homomorphism $F_* : \Omega_*^{Spin} \longrightarrow \Omega_*^{SO}$ has kernel the ideal generated by $\alpha \in \Omega_1^{Spin}$. In particular, kernel$(F_n) = 0$ if $n \neq 8k + 1$, $8k + 2$, with kernel(F_{8k+i}), $i = 1, 2$, being $U_{8k}\cdot\alpha^i$ where $U_{8k} \subset \Omega_{8k}^{Spin}$ is a subgroup mapping isomorphically onto Ω_{8k}^{Spin}/Torsion.

Proof: Since $\Omega_1^{SO} = 0$, kernel(F_*) contains the ideal generated by α. If $x \in \ker F_*$, then x must be a torsion element, F_* being a rational isomorphism. Thus $\ker F_* \subset \mathrm{Torsion}\,\Omega_*^{Spin}$ mapping monomorphically into $\Omega_*^{Spin} \otimes Z_2$. Since the manifolds M_J, τ, and ω all have infinite order, a basis for $\mathrm{Torsion}\,\Omega_*^{Spin} \subset \Omega_*^{Spin} \otimes Z_2$ is given by the classes $[M_J] \times \omega^k \times \alpha^i$, $i > 0$; $(([M_J] \times \tau)/4) \times \omega^k \times \alpha^i$, $i > 0$; $[R_i]$ and $[N_J]$. (These are the images of torsion classes and modulo these one gets the same rank as $\Omega_*^{Spin}/\mathrm{Torsion} \otimes Z_2$, so this must be the image of the torsion subgroup.) Now the classes $[R_i]$ and $[N_J]$ are detected by Z_2 cohomology, and hence kernel(F_*) is the vector space spanned by the given multiples of α. The classes $[M_J] \times \omega^k$ and $(([M_J] \times \tau)/4) \times \omega^k$ are, of course, a base for $\Omega_{8*}^{Spin}/\mathrm{Torsion} \otimes Z_2$, and hence, choosing a subspace U_{8k} as in the theorem, $\ker F_{8k+i} = U_{8k}\cdot\alpha^i$, $i = 1, 2$. **

Note: The images of Ω_*^{Spin} in $\Omega_*^{SO}/\mathrm{Torsion}$ (recall B_*^{SO}) and in $\mathfrak{N}_*$ have been descirbed, essentially giving image(F_*).

Theorem: The forgetful homomorphism $G_* : \Omega_*^{Spin^c} \to \Omega_*^{SO}(BU(1))$ is monic, while $F_* : \Omega_*^{Spin^c} \to \Omega_*^{SO}$ has torsion free kernel. Further, the composite $\Omega_*^{Spin^c} \xrightarrow{F_*} \Omega_*^{SO} \xrightarrow{\pi} \Omega_*^{SO}/\mathrm{Torsion}$ is epic.

Proof: G_* is induced by π' and is a rational isomorphism, so $\ker G_*$ consists of torsion. Since the torsion is detected by Z_2 cohomology characteristic numbers, all of which come from $H^*(BO;Z_2)$, the torsion of $\Omega_*^{Spin^c}$ maps monomorphically into $\mathfrak{N}_*$. Hence $\ker G_* = 0$. Similarly, $\ker F_*$ must be torsion free. Since $\Omega_*^U \to \Omega_*^{SO}/\mathrm{Torsion}$ is epic and factors through $\Omega_*^{Spin^c}$, $\pi \circ F_*$ is epic. **

Note: $\ker F_*$ may also be described as the ideal generated by the odd projective spaces $\mathbb{C}P(2k+1)$. Certainly all Pontrjagin numbers and Stiefel-Whitney numbers of $\mathbb{C}P(2k+1)$ are zero, so F_* sends this ideal to zero. Let $R_* = Z[y_{4i}] \subset \Omega_*^{Spin^c}$ be a subring mapped isomorphically to $\Omega_*^{SO}/\text{Torsion}$. Then the class y_{4i} may be used to replace $\mathbb{C}P(2i)$ or M_{2i}^p in the mod p calculations proving that $\Omega_*^{Spin^c}/\text{Torsion}$ is generated by R_* and the $\mathbb{C}P(2k+1)$. Let S_* be the subring of $\Omega_*^{Spin^c}$ generated by the $\mathbb{C}P(2k+1)$ and R_*. S_* can contain no torsion classes, for a torsion class in S_* must belong to the ideal generated by the $\mathbb{C}P(2k+1)$ (being in the kernel of $\pi \circ F_*$), hence is in the kernel of F_*, but $\ker F_*$ has no torsion. Thus $\Omega_*^{Spin} = S_* \oplus \text{Torsion}\,(\Omega_*^{Spin^c})$ and so $\ker F_* = \ker F_*|_{S*}$ is the ideal generated by the $\mathbb{C}P(2k+1)$.

Relation to Complex Cobordism

The only reasonable result here is:

Proposition: Ω_*^U maps onto $\Omega_*^{Spin^c}/\text{Torsion}$ under the forgetful homomorphism.

This has, of course, already been proved.

Relation to Special Unitary Cobordism

The interesting result here is:

Proposition: The forgetful homomorphism $F_* : \Omega_*^{SU} \longrightarrow \Omega_*^{Spin}$ is monic on the torsion subgroup.

Proof: If M is an $8k + \varepsilon$ dimensional SU manifold, M is cobordant to $N^{8k} \times \theta^{\varepsilon}$ for some N^{8k}. Then for $\omega \in \pi(2k)$

$$
\begin{aligned}
s_{\omega}(\pi^{s}(\tau \otimes \mathbb{C}))[M] &= s_{\omega}(\pi^{s}(\tau \otimes \mathbb{C}))[N] \bmod 2, \\
&= s_{\omega}(e_{\mathcal{P}})\mathcal{S}[N] \qquad \bmod 2, \\
&= s_{\omega}(\pi_{R})[N] \qquad \bmod 2, \\
&= s_{\omega}(\pi_{R})[M]
\end{aligned}
$$

and these numbers detect $[M]$ in the two cases. **

Relation of Spin and Spin^c

It is readily verified that the map $g : \text{BSpin} \times BU(1) \longrightarrow \text{BSpin}^c$ classifying the sum of the canonical bundles is a homotopy equivalence. Since $TBU(1) = \mathbb{C}P(\infty)$, one has

Proposition: $\Omega_n^{\text{Spin}^c} \cong \tilde{\Omega}_{n-2}^{\text{Spin}}(\mathbb{C}P(\infty))$.

One may relate the pair $(\text{Spin}, \text{Spin}^c)$ through exact sequences in precisely the same way as (SU, U) are related (or as (SO, O) are related). Computationally this is not of much use since one has no way to nicely describe the torsion in $\Omega_*^{\text{Spin}^c}$.

Appendix 1

Advanced Calculus

This appendix collects the results from standard advanced calculus which are needed for geometric arguments in cobordism theory. These results are lifted bodily from the following sources:

1. Milnor, J.: Lectures on Characteristic Classes, mimeographed, Princeton University, Princeton, N. J., 1957.

2. Milnor, J.: Topology from the Differentiable Viewpoint, The University Press of Virginia, Charlottesville, Va., 1965.

3. Spivak, M.: Calculus on Manifolds, W. A. Benjamin, Inc., New York, New York, 1965.

4. Steenrod, N.: The Topology of Fibre Bundles, Princeton University Press, Princeton, N. J. 1951.

5. Sternberg, S.: Lectures on Differential Geometry, Prentice-Hall, New York, 1964.

Definition: A function $f : R^n \longrightarrow R^m$ is differentiable at $a \in R^n$ if there is a linear transformation $\lambda : R^n \longrightarrow R^m$ such that

$$\lim_{h \to 0} \frac{|f(a+h)-f(a)-\lambda(h)|}{|h|} = 0$$

Proposition: If $f : R^n \longrightarrow R^m$ is differentiable at $a \in R^n$, there is a unique linear transformation $\lambda : R^n \longrightarrow R^m$ for which the above holds.

Proof: If $\mu : R^n \longrightarrow R^m$ is another such linear transformation, $x \in R^n$ and $t \in R$, then

$$\frac{|\lambda(x)-\mu(x)|}{|x|} = \lim_{t\to 0} \frac{|\lambda(tx)-\mu(tx)|}{|tx|}$$

$$= \lim_{t\to 0} \frac{|\lambda(tx)-f(a+tx)+f(a)+f(a+tx)-f(a)-\mu(tx)|}{|tx|}$$

$$= \lim_{t\to 0} \frac{|f(a+tx)-f(a)-\lambda(tx)|}{|tx|} + \lim_{t\to 0} \frac{|f(a+tx)-f(a)-\mu(tx)|}{|tx|}$$

$$= 0 + 0$$

so $\lambda(x) = \mu(x)$ for all x. *

<u>Definition</u>: The linear transformation λ satisfying the above conditions is denoted $Df(a)$ and is called the <u>derivative</u> of f at a.

<u>Lemma</u>: If $T : R^n \longrightarrow R^m$ is a linear transformation, there is a number M such that $|T(h)| \le M|h|$ for all $h \in R^n$.

<u>Proof</u>: Let e_i^1, e_j^2 be the usual bases of R^n and R^m respectively and define $t_{ij} \in R$ by $T(e_i^1) = \sum t_{ij} e_j^2$. If $h = \sum h_i e_i^1$, then

$$|T(h)| = \sqrt{\sum_j (\sum_i h_i t_{ij})^2} \le \sum_j |\sum_i h_i t_{ij}|$$

$$\le \sum_j \sum_i |t_{ij}|\cdot|h_i| \le mn \sup_{i,j} |t_{ij}|\cdot|h|.$$

Thus it suffices to take $M = mn \sup_{i,j} |t_{ij}|$. *

<u>Proposition</u>: If $f : R^n \longrightarrow R^m$ is differentiable at $a \in R^n$, then f is continuous at a.

<u>Proof</u>: Let $\varepsilon > 0$. Since $\lim_{x\to a} \frac{|f(x)-f(a)-Df(a)(x-a)|}{|x-a|} = 0$, there is a $\delta_1 > 0$ so that $|x-a| < \delta_1$ implies

$$|f(x) - f(a) - Df(a)(x-a)| < (\varepsilon/2)|x-a|.$$

By the lemma, there is an M such that $|Df(a)(h)| \leq M|h|$. Let δ = minimum of $(\delta_1, \varepsilon/2M, 1)$. Then $|x-a| < \delta$ implies

$$\begin{aligned} |f(x) - f(a)| &\leq |f(x) - f(a) - Df(a)(x-a)| + |Df(a)(x-a)| \\ &< (\varepsilon/2)|x-a| + M|x-a| \\ &\leq (\varepsilon/2) + M(\varepsilon/2M) \\ &= \varepsilon. \end{aligned}$$

Hence f is continuous at a. *

<u>Theorem</u>: (Chain Rule) If $f : R^n \longrightarrow R^m$ is differentiable at $a \in R^n$, and $g : R^m \longrightarrow R^p$ is differentiable at $f(a) = b \in R^m$, then the composition $g\circ f : R^n \longrightarrow R^p$ is differentiable at a, and

$$D(g\circ f)(a) = Dg(f(a)) \circ Df(a).$$

<u>Proof</u>: Define

$$\phi(x) = f(x) - f(a) - \lambda(x-a)$$

$$\psi(y) = g(y) - g(b) - \mu(y-b).$$

where $\lambda = Df(a)$, $\mu = Dg(f(a))$. Then

$$\begin{aligned} g(f(x)) - g(b) - \mu\lambda(x-a) &= g(f(x)) - g(b) - \mu(f(x) - f(a) - \phi(x)) \\ &= [g(f(x)) - g(b) - \mu(f(x) - b)] + \mu(\phi(x)) \\ &= \psi(f(x)) + \mu(\phi(x)). \end{aligned}$$

By the lemma, there is an M_1 such that $|\mu(h)| \leq M_1|h|$, so

$$0 \leq \lim_{x\to a} \frac{|\mu(\phi(x))|}{|x-a|} \leq M_1 \lim_{x\to a} \frac{|\phi(x)|}{|x-a|} = 0$$

Now let $\varepsilon > 0$ and choose an M_2 so that $|\lambda(h)| \leq M_2|h|$. Since $\lim_{y\to b} \frac{|\psi(y)|}{|y-b|} = 0$, there is a $\delta_1 > 0$ so that $|\psi(f(x))| < (\varepsilon/M_2)|f(x) - b|$ if $|f(x) - b| < \delta_1$. Since differentiability implies continuity, there is a $\delta_2 > 0$ so that $|x-a| < \delta_2$ implies $|f(x) - b| < \delta_1$. Thus if $|x-a| < \delta_2$,

$$\begin{aligned} |\psi(f(x))| &< (\varepsilon/M_2)|f(x) - b| \\ &= (\varepsilon/M_2)|\phi(x) + \lambda(x-a)| \\ &\leq (\varepsilon/M_2)|\phi(x)| + \varepsilon|x-a| \end{aligned}$$

and so

$$0 \leq \lim_{x\to a} \frac{|\psi(f(x))|}{|x-a|} \leq \left(\frac{\varepsilon}{M_2}\right) \lim_{x\to a} \frac{|\phi(x)|}{|x-a|} + \varepsilon = \varepsilon,$$

and since this holds for all $\varepsilon > 0$

$$\lim_{x\to a} \frac{|g(f(x))-g(b)-\mu(\lambda(x-a))|}{|x-a|} = 0. \quad *$$

Proposition: 1) If $f : R^n \longrightarrow R^m$ is a constant function, then $Df(a) = 0$.

2) If $f : R^n \longrightarrow R^m$ is a linear transformation, then $Df(a) = f$.

3) If $f : R^n \longrightarrow R^m : x \longrightarrow (f^1(x),\dots,f^m(x))$, then f is differentiable at $a \in R^n$ if and only if each f^i is differentiable at a, and $Df(a) = (Df^1(a),\dots,Df^m(a))$.

4) If $f,g : R^n \longrightarrow R^m$ are differentiable at $a \in R^n$, then $f + g : R^n \longrightarrow R^m$ is differentiable at a, and

$$D(f+g)(a) = Df(a) + Dg(a).$$

5) If $f,g : R^n \longrightarrow R$ are differentiable at $a \in R^n$, then $f \cdot g : R^n \longrightarrow R$ is differentiable at a, and

$$D(f \cdot g)(a) = f(a) \cdot Dg(a) + g(a) \cdot Df(a).$$

<u>Proof</u>: 1) If $f(x) = y$ for all x, then

$$\lim_{h \to 0} \frac{|f(a+h)-f(a)-0|}{|h|} = \lim_{h \to 0} \frac{|y-y-0|}{|h|} = 0.$$

2) $$\lim_{h \to 0} \frac{|f(a+h)-f(a)-f(h)|}{|h|} = \lim_{h \to 0} \frac{|f(a)+f(h)-f(a)-f(h)|}{|h|} = 0.$$

3) If each f^i is differentiable and $\lambda = (Df^1(a), \ldots, Df^m(a))$, then

$$f(a+h)-f(a)-\lambda(h) = (f^1(a+h)-f^1(a)-Df^1(a)(h), \ldots, f^m(a+h)-f^m(a)-Df^m(a)(h))$$

so

$$\lim_{h \to 0} \frac{|f(a+h)-f(a)-\lambda(h)|}{|h|} \leq \sum \lim_{h \to 0} \frac{|f^i(a+h)-f^i(a)-Df^i(a)(h)|}{|h|} = 0.$$

Conversely, f^i is the composition of f and the projection π_i which is linear, so $Df^i(a) = D(\pi_i \circ f)(a) = \pi_i Df(a)$.

4) Let $s : R^m \times R^m \longrightarrow R^m : (x,y) \longrightarrow x + y$, and let $(f,g) : R^n \longrightarrow R^m \times R^m : a \longrightarrow (f(a),g(a))$. Then s is linear, so $Ds = s$ and by 3), $D(f,g) = (Df,Dg)$. By the chain rule,

$$\begin{aligned} D(f+g)(a) &= Ds(f(a),g(a)) \circ D(f,g)(a) \\ &= s(Df(a),D(g)(a)) \\ &= Df(a) + Dg(a). \end{aligned}$$

5) Let $p : R^2 \longrightarrow R : (x,y) \longrightarrow xy$. By the chain rule, it suffices to show that $Dp(a,b)(x,y) = bx + ay$. Letting $\lambda(x,y) = bx + ay$,

$$\lim_{(h,k)\to 0} \frac{|p(a+h,b+k)-p(a,b)-\lambda(h,k)|}{|(h,k)|} = \lim_{(h,k)\to 0} \frac{|hk|}{|(h,k)|}.$$

Since $|hk| \leq \sup(|h|^2, |k|^2) \leq |h|^2 + |k|^2$, one has

$$0 \leq \lim_{(h,k)\to 0} \frac{|hk|}{|(h,k)|} \leq \lim_{(h,k)\to 0} \frac{|(h,k)|^2}{|(h,k)|} = \lim_{(h,k)\to 0} |(h,k)| = 0. \quad *$$

<u>Proposition</u>: If $f : R \longrightarrow R$ is differentiable at $a \in R$ and has either a relative maximum or relative minimum at a, then $Df(a) = 0$.

<u>Proof</u>: Let $Df(a)(h) = th$ with $t \in R$. If a is a relative maximum, then $f(a+h) - f(a) \leq 0$ and so if $th > 0$,

$$0 = \lim_{\substack{h\to 0\\ th\to 0}} \frac{|f(a+h)-f(a)-th|}{|h|} \geq \lim_{h\to 0} \frac{|th|}{|h|} = |t|.$$

If a is a relative minimum, then $f(a+h) - f(a) \geq 0$ so if $th < 0$,

$$0 = \lim_{\substack{h\to 0\\ th\to 0}} \frac{|f(a+h)-f(a)-th|}{|h|} \geq \lim_{h\to 0} \frac{|th|}{|h|} = |t|. \quad *$$

<u>Theorem</u>: (Rolle) Let $[a,b] \subset R$ and $f : [a,b] \longrightarrow R$ a continuous function with $f(a) = f(b) = 0$ and such that $Df(c)$ exists for all $a < c < b$. Then $Df(c) = 0$ for some c with $a < c < b$.

<u>Proof</u>: If f is not identically zero, in which case $Df(c) = 0$ for all $c \in (a,b)$, then f has a positive maximum or a negative minimum which must occur at some $c \in (a,b)$. Thus c is either a relative maximum or relative minimum and so $Df(c) = 0$ by the proposition. *

Theorem: (Mean Value) Let $[a,b] \subset R$ and $f : [a,b] \longrightarrow R$ a continuous function which is differentiable at all points $c \in (a,b)$. Then there is a point $c \in (a,b)$ so that

$$f(b) - f(a) = Df(c)(b-a)$$

Proof: Let $F(x) = f(x) - f(a) - [(f(b) - f(a))/(b-a)](x-a)$. Then F satisfies the conditions of Rolle's theorem, so for some $c \in (a,b)$, $0 = DF(c) = Df(c) - [(f(b) - f(a))/(b-a)]\cdot 1$ where $1 : R \longrightarrow R$ is the identity function. *

Definition: If $f : R^n \longrightarrow R$ and $a \in R^n$, then the limit

$$\lim_{h \to 0} \frac{f(a^1,\ldots,a^{i-1},a^i+h,a^{i+1},\ldots,a^n)-f(a^1,\ldots,a^n)}{h}$$

is called the i-th partial derivative of f at a, denoted $D_i f(a)$, when it exists.

Theorem: If $f : R^n \longrightarrow R^m$ has the property that all of the partial derivatives $D_j f^i(x)$ exist in an open set containing a and are continuous at a, then $Df(a)$ exists.

Proof: It suffices to show $Df^i(a)$ exists, so one may assume $m = 1$. Then

$$f(a+h)-f(a) = \sum_{i=1}^{n} [f(a^1+h^1,\ldots,a^i+h^i,a^{i+1},\ldots,a^n)-f(a^1+h^1,\ldots,a^{i-1}+h^{i-1},a^i,\ldots,a^n)]$$

$$= \sum_{i=1}^{n} h^i D_i f(c_i)$$

for some point $c_i = (a^1+h^1,\ldots,a^{i-1}+h^{i-1},a^i+\theta_i h^i,a^{i+1},\ldots,a^n)$ where $0 < \theta_i < 1$, by the mean value theorem. Hence

$$\lim_{h\to 0} \frac{|f(a+h)-f(a)-\Sigma h^i \cdot D_i f(a)|}{|h|} = \lim_{h\to 0} \frac{|\Sigma h^i [D_i f(c_i)-D_i f(a)]|}{|h|}$$

$$\leq \lim_{h\to 0} \sum |D_i f(c_i)-D_i f(a)| \cdot \frac{|h^i|}{|h|}$$

$$\leq \lim_{h\to 0} \sum |D_i f(c_i)-D_i f(a)|$$

$$= 0$$

by continuity of $D_i f$ at a. Thus $Df(a)(h) = \sum D_i f(a)\cdot h^i$. *

<u>Definition</u>: For $f : R^n \longrightarrow R$, the function $D_{i_1,\ldots,i_r} f$ defined by $D_{i_1,\ldots,i_r} f = D_{i_1}(D_{i_2,\ldots,i_r} f)$ is called an <u>r-th order partial derivative</u> of f. The function f is said to be of <u>class C^∞</u> if all partial derivatives (of all orders) exist.

<u>Theorem</u>: If $f : R^n \longrightarrow R$ and $D_{i,j} f$ and $D_{j,i} f$ exist and are continuous in an open set containing $a \in R^n$, then

$$D_{i,j} f(a) = D_{j,i} f(a).$$

<u>Proof</u>: It suffices to consider the case $n = 2$. Let $a = (c,d)$ and let $(h,k) \in R^2$ be small enough so that both $D_{1,2} f$ and $D_{2,1} f$ are defined on $\{(x,y) \,\big|\, |x-c| \leq h,\ |y-d| \leq k\}$. Let

$$\phi(x) = f(x,d+k) - f(x,d)$$

$$\psi(y) = f(c+h,y) - f(c,y).$$

Then

$$\alpha = f(c+h,d+k) - f(c,d+k) - f(c+h,d) + f(c,d) = \phi(c+h) - \phi(c)$$

$$= \psi(d+k) - \psi(d)$$

There is a $c' \in (c, c+h)$ with

$$\alpha = \phi(c+h) - \phi(c) = D\phi(c')\cdot h$$

$$= [D_1 f(c', d+k) - D_1 f(c', d)]h$$

$$= D_{2,1} f(c', d')hk$$

for some $d' \in (d, d+k)$.

There is a $d'' \in (d, d+k)$ with

$$\alpha = \psi(d+k) - \psi(d) = D\psi(d'')\cdot k$$

$$= [D_2 f(c+h, d'') - D_2 f(c, d'')]k$$

$$= D_{1,2} f(c'', d'')hk.$$

for some $c'' \in (c, c+h)$.

Thus every open set U containing a contains points p', p'' with $D_{1,2}f(p') = D_{2,1}f(p'')$. By continuity of the $D_{i,j}f$ this gives $D_{1,2}f(a) = D_{2,1}f(a)$. *

<u>Proposition</u>: If $f : R^n \longrightarrow R$ is a C^∞ function and $x_0 \in R^n$, there exist C^∞ functions $g_i : R^n \longrightarrow R$, $i = 1, \ldots, n$, with $g_i(x_0) = \frac{\partial f}{\partial x_i}(x_0)$ such that

$$f(x) = f(x_0) + \sum_{i=1}^{n} (x-x_0)_i \cdot g_i(x).$$

<u>Proof</u>: Define $h_x(t) = f(x_0+t(x-x_0))$. Then $h_x(t)$ is a C^∞ function of t and

$$\int_0^1 \frac{dh_x}{dt} \cdot dt = h_x(1) - h_x(0)$$

$$= f(x) - f(x_0).$$

By the chain rule,

$$\frac{dh_x}{dt} = \sum_j \frac{\partial f}{\partial x_j} (x_0+t(x-x_0)) \cdot (x-x_0)_j$$

so

$$f(x) = f(x_0) + \sum_{j=1}^{n} (x-x_0)_j \int_0^1 \frac{\partial f}{\partial x_j} (x_0+t(x-x_0))dt$$

and one may let $g_i(x) = \int_0^1 \frac{\partial f}{\partial x_i} (x_0+t(x-x_0))dt$. Then

$$g_i(x_0) = \int_0^1 \frac{\partial f}{\partial x_i} (x_0)dt = \frac{\partial f}{\partial x_i} (x_0) \int_0^1 dt = \frac{\partial f}{\partial x_i} (x_0). \quad *$$

Lemma: Let $A \subset R^n$ be a rectangle and $f : A \longrightarrow R^n$ continuously differentiable (i.e. each $D_j f^i(x)$ exists and is continuous on A). If there is a number M such that $|D_j f^i(x)| \leq M$ for all x in the interior of A, then

$$|f(x) - f(y)| \leq n^2 M|x-y|$$

for all $x,y \in A$.

Proof: One has

$$\begin{aligned} f^i(y) - f^i(x) &= \sum_{j=1}^{n} [f^i(y^1,\ldots,y^j,x^{j+1},\ldots,x^n) - f^i(y^1,\ldots,y^{j-1},x^j,\ldots,x^n)] \\ &= \sum_{j=1}^{n} |y^j-x^j| \cdot D_j f^i(z_{ij}) \qquad \text{for some } z_{ij} \in \text{interior } A \\ &\leq \sum_{j=1}^{n} |y^j-x^j| \cdot M \\ &\leq nM|y-x| \end{aligned}$$

so

$$|f(y) - f(x)| \leq \sum_{i=1}^{n} |f^i(y) - f^i(x)| \leq n^2 M|y-x|. \quad *$$

<u>Theorem</u>: (Inverse Function) Let $f : R^n \longrightarrow R^n$ be continuously differentiable in an open set containing a, with $Df(a)$ non-singular. Then there is an open set V containing a and an open set W containing $f(a)$ such that $f : V \longrightarrow W$ has a continuous inverse $f^{-1} : W \longrightarrow V$ which is differentiable and $Df^{-1}(y) = [Df(f^{-1}(y))]^{-1}$ for all $y \in W$.

<u>Proof</u>: Let $\lambda = Df(a)$ and then $D(\lambda^{-1} \circ f)(a) = D(\lambda^{-1})(f(a)) \circ Df(a)$ $= \lambda^{-1} \circ Df(a) = 1$. If g is an inverse for $\lambda^{-1} \circ f$, then $g \circ \lambda^{-1}$ is an inverse for f, and hence one may assume $\lambda = 1$. Hence, if $f(a+h) = f(a)$ one has

$$\frac{|f(a+h)-f(a)-\lambda(h)|}{|h|} = \frac{|h|}{|h|} = 1$$

but since

$$\lim_{h\to 0} \frac{|f(a+h)-f(a)-\lambda(h)|}{|h|} = 0$$

this means that $f(x) \neq f(a)$ if x is close to but not equal to a.

Thus there is an closed rectangle U containing a in its interior with

1. $f(x) \neq f(a)$ if $x \in U$, $x \neq a$.

Since f is continuously differentiable in an open set containing a, one may also assume

2. $Df(x)$ is non-singular for all $x \in U$

3. $|D_j f^i(x) - D_j f^i(a)| < 1/2n^2$ for all i,j and $x \in U$.

Since $(D_j f^i(a))$ is the Kronecker delta δ_{ij}, the lemma applies to $g(x) = f(x) - x$ giving that for $x_1, x_2 \in U$

$$|f(x_1) - x_1 - (f(x_2) - x_2)| \leq 1/2\ |x_1 - x_2|$$

so

$$|x_1 - x_2| - |f(x_1) - f(x_2)| \leq |f(x_1) - x_1 - (f(x_2) - x_2)| \leq 1/2\ |x_1 - x_2|.$$

Hence

4. $|x_1 - x_2| \leq 2|f(x_1) - f(x_2)|$ if $x_1, x_2 \in U$.

Since f is continuous, $f(\partial U)$ is compact and by 1. cannot contain $f(a)$, so there is a $d > 0$ such that $|f(x) - f(a)| \geq d$ if $x \in \partial U$. Let $W = \{y \mid |y-f(a)| < d/2\}$. If $y \in W$ and $x \in \partial U$ then

5. $|y-f(a)| < |y-f(x)|$

for $d \leq |f(x) - f(a)| \leq |y - f(x)| + |y - f(a)| < |y - f(x)| + d/2$.

Now let $y \in W$ and let $g : U \longrightarrow R$ by

$$g(x) = |y - f(x)|^2 = \sum_{i=1}^{n} (y^i - f^i(x))^2.$$

Then g is continuous so has a minimum on U, but by 5. $g(a) < g(x)$ for $x \in \partial U$, so the minimum of g must occur at an interior point of U, i.e. is a relative minimum. Thus there is a point $z \in$ interior U with $D_j g(z) = 0$ for all j, or

$$2 \sum_{i=1}^{n} (y^i - f^i(z)) \cdot D_j f^i(z) = 0.$$

Since by 2., $Df(z)$ is non-singular, this gives $y^i - f^i(z) = 0$ or $y = f(z)$ for some $z \in$ interior U. By 4. this z is unique.

Letting $V = (\text{interior } U) \cap f^{-1}(W)$, the function $f : V \longrightarrow W$ has an inverse $f^{-1} : W \longrightarrow V$, and rewriting 4 as $|f^{-1}(y_1) - f^{-1}(y_2)| \leq 2|y_1 - y_2|$ for $y_1, y_2 \in W$ proves continuity of f^{-1}.

To show that f^{-1} is differntiable, let $\mu = Df(x)$ and $y = f(x)$ and for $x_1 \in V$, let us define ϕ by

$$f(x_1) = f(x) + \mu(x_1-x) + \phi(x_1-x)$$

so that

$$\lim_{x_1 \to x} \frac{|\phi(x_1-x)|}{|x_1-x|} = 0.$$

Then

$$\mu^{-1}(f(x_1) - f(x)) = x_1 - x + \mu^{-1}(\phi(x_1-x))$$

and since every $y_1 \varepsilon W$ is of the form $f(x_1)$ with some $x_1 \varepsilon V$, one has

$$f^{-1}(y_1) = f^{-1}(y) + \mu^{-1}(y_1-y) - \mu^{-1}(\phi(f^{-1}(y_1) - f^{-1}(y))).$$

Since μ^{-1} is linear, there is an M with

$$\begin{aligned} \frac{|\mu^{-1}(\phi(f^{-1}(y_1)-f^{-1}(y))|}{|y_1 - y|} &\leq M \frac{|\phi(f^{-1}(y_1)-f^{-1}(y))|}{|y_1 - y|} \\ &= M \frac{|\phi(f^{-1}(y_1)-f^{-1}(y))|}{|f^{-1}(y_1)-f^{-1}(y)|} \cdot \frac{|f^{-1}(y_1)-f^{-1}(y)|}{|y_1 - y|} \\ &\leq 2M \frac{|\phi(f^{-1}(y_1)-f^{-1}(y))|}{|f^{-1}(y_1)-f^{-1}(y)|} \end{aligned}$$

by equation 4. As $y_1 \longrightarrow y$, continuity of f^{-1} gives $f^{-1}(y_1) \longrightarrow f^{-1}(y)$, and by definition of ϕ, this term goes to zero. Thus μ^{-1} is a linear transformation of the form required to show f^{-1} differentiable at y. *

<u>Theorem</u>: (Implicit Function Theorem) Let $f\colon R^n \times R^m \longrightarrow R^m$ be continuously differentiable in an open set containing (a,b) with $f(a,b) = 0$. Let M be the $m \times m$ matrix $(D_{n+j}f^i(a))$ $1 \leq i, j \leq m$.. If M is nonsingular, there is an open set $A \subset R^n$ containing a and an open set $B \subset R^m$ containing b, so that for each $x \varepsilon A$ there is a unique $g(x) \varepsilon B$ such that $f(x,g(x)) = 0$. The function g is differentiable.

<u>Proof</u>: Let $F : R^n \times R^m \longrightarrow R^n \times R^m$ by $F(x,y) = (x,f(x,y))$. Then $DF(a,b)$ is non-singular. There are then open sets $W \subset R^n \times R^m$ containing $F(a,b) = (a,0)$ and $V \subset R^n \times R^m$ containing (a,b), which may be taken to be

of the form $A \times B$, such that $F : V \longrightarrow W$ has a differentiable inverse $h : W \longrightarrow V = A \times B$. Clearly $h(x,y) = (x,k(x,y))$ since F has this form, where k is some differentiable function. Let $\pi : R^n \times R^m \longrightarrow R^m : (x,y) \longrightarrow y$ be the projection. Then

$$f(x,k(x,y)) = f \circ h(x,y) = \pi \circ F \circ h(x,y)$$
$$= \pi(x,y) = y$$

so $f(x,k(x,0)) = 0$ and one may let $g(x) = k(x,0)$. *

__Theorem__: Let $f : R^n \longrightarrow R^p$ be continuously differentiable in an open set containing a, where $p \leq n$. If $f(a) = 0$ and $Df(a)$ is an epimorphism, there is an open set $A \subset R^n$ and a differentiable function $h : A \longrightarrow R^n$ with differentiable inverse so that

$$f \circ h(x_1,\ldots,x_n) = (x_{n-p+1},\ldots,x_n)$$

__Proof__: Since $Df(a)$ has rank p, there are integers $1 \leq i_1 <\ldots< i_p \leq n$ such that the matrix $(D_i f^j(a))$, $1 \leq j \leq p$, $i = i_1,\ldots,i_p$ is non-singular. Let $g : R^n \longrightarrow R^n$ permute the coordinates so that $g(x^1,\ldots,x^n) = (\ldots,x^{i_1},\ldots,x^{i_p})$. Then $f \circ g : R^n = R^{n-p} \times R^p \longrightarrow R^p$ has the matrix $(D_{n-p+j}(f{\circ}g)^j(g^{-1}(a)))$ non-singular $1 \leq i, j \leq p$. As above, there is an $h : A \longrightarrow R^n$, $A \subset R^n$ an open set with $(f{\circ}g) \circ h(x^1,\ldots,x^n) = (x^{n-p+1},\ldots,x^n)$. The function $g \circ h$ satisfies the conditions of the theorem. *

__Lemma__: Let $f : R^n \longrightarrow R^p$ be continuously differentiable in an open set containing a, where $p \geq n$. If $Df(a)$ is monic, there is an open set $U \subset R^p$ containing $f(a)$ and a differentiable function $h : U \longrightarrow R^p$ with differentiable inverse so that

$$h \circ f(x_1,\ldots,x_n) = (x_1,\ldots,x_n,0,\ldots,0)$$

on some neighborhood of a.

<u>Proof</u>: Since $(\partial f_i/\partial x_j)$ has rank n, one may, by reordering coordinates in R^p, assume $(\partial f_i/\partial x_j)_{1\leq i,j\leq n}$ is non-singular. Let $F : R^n \times R^{p-n} \longrightarrow R^p$ by

$$F(x_1,\ldots,x_p) = f(x_1,\ldots,x_n) + (0,\ldots,0,x_{n+1},\ldots,x_p).$$

Since $F(x_1,\ldots,x_n,0,\ldots,0) = f(x_1,\ldots,x_n)$, F extends f. $DF(a,0)$ has

$$\begin{pmatrix} (\partial f_i/\partial x_j) & 0 \\ * & I \end{pmatrix}$$

as matrix so is non-singular. Hence F has an inverse h on a neighborhood of $(a,0)$, so

$$\begin{aligned} hf(x_1,\ldots,x_n) &= hF(x_1,\ldots,x_n,0,\ldots,0) \\ &= (x_1,\ldots,x_n,0,\ldots,0). \quad * \end{aligned}$$

<u>Definition</u>: A <u>rectangle</u> in R^n is a set of the form $\prod_{i=1}^{n} [a_i,b_i]$ with $a_i \leq b_i$, $a_i,b_i \in R$. The <u>volume</u> of the rectangle $S = \prod_{i=1}^{n} [a_i,b_i]$ is $v(S) = \prod_{i=1}^{n} |b_i - a_i|$.

<u>Definition</u>: A subset $A \subset R^n$ has (n-dimensional) <u>measure zero</u> if for every $\varepsilon > 0$ there is a countable collection B_i of rectangles with $A \subset \bigcup B_i$ and $\sum v(B_i) < \varepsilon$.

<u>Theorem</u>: A countable union of sets of measure zero is itself of measure zero.

Proof: If $A = \bigcup A_i$, with each A_i of measure zero, let $\varepsilon > 0$ and choose families $B_{i,j}$ of rectangles with $A_i \subset \bigcup_j B_{i,j}$, $\sum v(B_{i,j}) < \varepsilon/2^i$. Then $A \subset \bigcup_{i,j} B_{i,j}$ and $\sum_{i,j} v(B_{i,j}) < \sum_i \varepsilon/2^i = \varepsilon$. *

Proposition: Let $\mathcal{U}$ be an open cover of the interval $[a,b]$ by intervals of length at most ε. Then there is a finite subcover $\mathcal{U}_0$ of $\mathcal{U}$ so that $\sum_{I_\alpha \in \mathcal{U}_0} v(I_\alpha) \leq 2(|b-a| + \varepsilon)$.

Proof: Let $\mathcal{U}_1$ be a finite cover by elements of $\mathcal{U}$ and let $\mathcal{U}_0$ be a minimal family of elements of $\mathcal{U}_1$ which cover. Order $\mathcal{U}_0$ by writing the elements of $\mathcal{U}_0$ as $I_j = (a_j, b_j)$ where $i < j$ if $a_i < a_j$. Then one has $\mathcal{U}_0 = \{I_j\}$, $j = 1,\ldots,r$ and by minimality of the cover $a_i < a_{i+1} < b_i < b_{i+1}$ for each i and $a_1 < a < a_2$, $b_{r-1} < b < b_r$. The sum of the overlaps is at most

$$(a - a_1) + (b_1 - a_2) + \ldots + (b_i - a_{i+1}) + \ldots + (b_{r-1} - a_r) + (b_r - b) \leq 2\varepsilon + |b-a|$$

since

$$a_1 < a < a_2 < b_1 < a_3 < b_2 < a_4 < b_3 < \ldots < a_{r-1} < b_{r-2} < a_r < b_{r-1} < b < b_r,$$

and this gives the result. *

Theorem: (Fubini) Let $A \subset R^n$ be a compact set such that each set $A \cap (t \times R^{n-1})$ has $(n-1)$-dimensional measure zero. Then A has n-dimensional measure zero.

Proof: Since A is compact $A \subset [a,b] \times R^{n-1}$ for some $a,b \in R$. Let $\varepsilon > 0$ and choose $\varepsilon_1 > 0$ so that $2(|b-a| + 1)\,\varepsilon_1 < \varepsilon$. For each $t \in [a,b]$, $A \cap (t \times R^{n-1})$ has measure 0 so there is a countable collection of rectangles

$B_{t,i} \subset R^{n-1}$ such that $A \cap (t \times R^{n-1}) \subset \bigcup_i t \times B^0_{t,i}$ and $\sum_i v(B_{t,i}) < \varepsilon_1$, where $B^0_{t,i}$ is the interior of $B_{t,i}$. Now $A - R \times \bigcup_i B^0_{t,i}$ is a compact set containing no point of the plane $t \times R^{n-1}$ and hence there is a $1/2 > \delta_t > 0$ so that $A \cap (t-\delta_t, t+\delta_t) \times R^{n-1} \subset (t-\delta_t, t+\delta_t) \times \bigcup_i B^0_{t,i}$. The sets $(t-\delta_t, t+\delta_t)$ cover $[a,b]$ and by the proposition there is a finite family $t_1, \ldots, t_r$ so that the intervals cover $[a,b]$ and have total length at most $2(|b-a| + 1)$. The countable family of all $[t_i-\delta_{t_i}, t_i+\delta_{t_i}] \times B_{ti,j}$ then covers A and has the sum of volumes at most $2(|b-a| + 1)\, \varepsilon_1 < \varepsilon$. *

<u>Definition</u>: Let $f : U \longrightarrow R^p$ be a smooth (C^∞) map, U open in R^n. A point $x \in U$ is a <u>critical point</u> if $Df(x)$ is not epic; it is a <u>regular point</u> if $Df(x)$ is epic. The <u>critical values</u> of f are the images under f of critical points; those points of R^p which are not the image of critical points are called <u>regular values</u>.

<u>Theorem</u>: (Sard) Let $f : U \longrightarrow R^p$ be a C^∞ map, U open in R^n, and let C be the set of critical points of f. Then $f(C) \subset R^p$ has measure zero.

<u>Proof</u>: The statement makes sense for $n \geq 0$, $p \geq 1$, with R^0 being a single point. The proof is by induction on n, being obvious for $n = 0$.

Let $C_i \subset C$ denote the set of $x \in U$ such that all partial derivatives of f of order $\leq i$ are zero at x. For example, $C_1 = \{x \in U | Df(x) = 0\}$.

<u>Step 1</u>: The image $f(C-C_1)$ has measure zero.

One may assume $p \geq 2$ for $C = C_1$ if $p = 1$.

Let $\bar{x} \in C - C_1$. Since $\bar{x} \notin C_1$, there is some partial derivative, say $\partial f^1/\partial x^1$, which is nonzero at $\bar{x}$. Let $h : U \longrightarrow R^n$ by

$$h(x) = (f^1(x), x^2, \ldots, x^n).$$

Since $Dh(\bar{x})$ is non-singular, h maps some neighborhood V of $\bar{x}$ diffeomorphically onto an open set V' of R^n. Then $g = f \circ h^{-1} : V' \longrightarrow R^p$. The set of critical points of g, C', is precisely $h(V \cap C)$, so $g(C') = f(V \cap C)$.

For each $(t, x^2, \ldots, x^n) \in V'$, $g(t, x^2, \ldots, x^n) \in t \times R^{p-1} \subset R^p$ or g takes hyperplanes to hyperplanes. Let

$$g^t : (t \times R^{p-1}) \cap V' \longrightarrow t \times R^{p-1}$$

be the restriction of g. Since

$$(\partial g^i / \partial x^j) = \begin{pmatrix} 1 & 0 \\ * & (\partial g^{ti} / \partial x^j) \end{pmatrix}$$

a point of $t \times R^{p-1}$ is critical for g^t if and only if it is critical for g. By induction, the set of critical values of g^t has measure zero in $t \times R^{p-1}$ and so $g(C')$ intersects each plane $t \times R^{p-1}$ in a set of measure zero, or $f(V \cap C)$ intersects each plane $t \times R^{p-1}$ in a set of measure zero.

Since $C - C_1$ is a countable union of sets of the form $\tilde{V} \cap C$ where $\tilde{V}$ is a compact neighborhood of $\bar{x}$, $\tilde{V} \subset V$, Fubini's theorem shows that $f(C - C_1)$ is a countable union of sets of measure zero, so has measure zero.

<u>Step 2</u>: The image $f(C_i - C_{i+1})$ has measure zero, for $i \geq 1$.

For each $\bar{x} \in C_i - C_{i+1}$ there is some $(i+1)$-st derivative $\partial^{i+1} f_r / \partial x_{s_1} \ldots \partial x_{s_{k+1}}$ which is non-zero. Thus

$$w(x) = \partial^k f_r / \partial x_{s_2} \ldots \partial x_{s_{k+1}}$$

vanishes at $\bar{x}$ but $\partial w / \partial x_{s_1}$ does not. Suppose $s_1 = 1$ for definiteness. Let $h : U \longrightarrow R^n$ by

$$h(x) = (w(x), x^2, \ldots, x^n).$$

Then h carries a neighborhood V of $\bar{x}$ diffeomorphically onto an open set V'. Also h takes $C_i \cap V$ into $0 \times R^{n-1}$. Consider

$$g = f \circ h^{-1} : V' \longrightarrow R^p$$

and let $\bar{g}$ be the restriction of g to $(0 \times R^{n-1}) \cap V'$. By induction, the set of critical values of $\bar{g}$ has measure zero in R^p, but each point of $h(C_i \cap V)$ is a critical point of $\bar{g}$ (since all derivatives of order $\leq i$ vanish). Thus

$$\bar{g}h(C_i \cap V) = f(C_i \cap V)$$

has measure zero. Since $C_i - C_{i+1}$ is covered by countably many such sets V, it follows that $f(C_i - C_{i+1})$ has measure zero.

<u>Step 3</u>: The image $f(C_k)$ has measure 0 for k sufficiently large.

Let $I^n \subset U$ be a cube of edge δ. By Taylor's theorem, the compactness of I^n and the definition of C_k, one has

$$f(x+h) = f(x) + R(x,h)$$

where $|R(x,h)| \leq c|h|^{k+1}$ for $x \in C_k \cap I^n$, $x + h \in I^n$. c being a constant which depends only on f and I^n.

Subdivide I^n into r^n cubes of edge δ/r, and let I_1 be a cube of the subdivision which contains a point $x \in C_k$. Then any point of I_1 is $x + h$ with $|h| \leq \sqrt{n}\,(\delta/r)$. Since $|f(x+h) - f(x)| \leq c|h|^{k+1}$, $f(I_1)$ lies in a cube of edge a/r^{k+1} centered at $f(x)$, where $a = 2c(\sqrt{n}\,\delta)^{k+1}$ is constant. Thus $f(C_k \cap I^n)$ is contained in a union of at most r^n cubes having total volume

$$V \leq r^n(a/r^{k+1})^p = a^p r^{n-(k+1)p}$$

If $k + 1 > n/p$, then $V \longrightarrow 0$ as $r \longrightarrow \infty$, so $f(C_k \cap I^n)$ has measure zero. *

<u>Lemma</u>: Let D, D' be two open rectangles in R^n with $\bar{D} \subset D'$. Then there is a real valued C^∞ function g on R^n such that

a) $0 \leq g(x) \leq 1$ for all x,

b) $g(x) = 1$ for $x \in D$, and

c) $g(x) = 0$ for $x \in R^n - D'$.

<u>Proof</u>: One may write $D = \Pi(a_i, b_i)$, $D' = \Pi(a_i', b_i')$ with $a_i' < a_i < b_i < b_i'$.

For any interval $[c,d] \subset R$, let

$$\psi_{c,d}(x) = \begin{cases} \exp(-\frac{1}{x-c} + \frac{1}{x-d}), & x \in [c,d] \\ 0 & x \notin [c,d]. \end{cases}$$

Then $\psi_{c,d}$ is C^∞ and $\psi_{c,d}(x) \geq 0$. Let

$$\phi_{c,d}(x) = \int_c^x \psi_{c,d}(x)dx \Big/ \int_c^d \psi_{c,d}(x)dx.$$

Then $\phi_{c,d}$ is C^∞, $0 \leq \phi_{c,d}(x) \leq 1$, $\phi_{c,d}(x) = 0$ if $x \leq c$, $\phi_{c,d}(x) = 1$ if $x \geq d$.

For $a' < a < b < b'$, let

$$h_{a',a,b,b'}(x) = \begin{cases} \phi_{a',a}(x) & x \leq b \\ 1 - \phi_{b,b'}(x) & x > b. \end{cases}$$

Then $h_{a',a,b,b'}$ is C^∞, $0 \leq h_{a'abb'}(x) \leq 1$, $h_{a'abb'}(x) = 1$ if $x \in [a,b]$ and $h_{a',a,b,b'}(x) = 0$ if $x \notin [a',b']$.

Let $g(x) = \prod_{i=1}^{n} h_{a_i',a_i,b_i,b_i'}(x_i)$. *

<u>Lemma</u>: Let U be an open set in R^n with $\bar{U}$ compact, and let V be an open set containing $\bar{U}$. Then there is a real valued C^∞ function $g : R^n \longrightarrow [0,1]$ such that $g(x) = 1$ for $x \in \bar{U}$, $g(x) = 0$ for $x \in R^n - V$.

<u>Proof</u>: Since $\bar{U}$ is compact, there are a finite number of open rectangles $D_1, \ldots, D_s$ with $\bar{D}_i \subset V$ covering $\bar{U}$. Let D_i' be an open rectangle containing $\bar{D}_i$ and contained in V. Let g_i be given as in the previous lemma for the pair D_i, D_i'. Then define g by

$$1 - g = (1-g_1)(1-g_2)\ldots(1-g_s).$$

Then g is C^∞, $0 \leq g(x) \leq 1$ for all x. If $x \in \bigcup D_i$ then $g_j(x) = 1$ for some j so $1 - g(x) = 0$. Thus $g(x) = 1$ for $x \in \bar{U} \subset \bigcup D_i$. If $x \notin \bigcup D_i'$ then $g_i(x) = 0$ for all i so $1 - g(x) = 1$. Thus $g(x) = 0$ if $x \in R^n - V \subset R^n - \bigcup D_i'$. *

<u>Lemma</u>: Let $F : W \longrightarrow R$, W open in R^n be a continuous function of class C^∞ in an open set $U \subset W$. Let U', V' be open sets with $\bar{U}' \subset V' \subset \bar{V}' \subset W$, $\bar{U}'$ and $\bar{V}'$ being compact. Let $\delta > 0$. Then there is a continuous function $G : W \longrightarrow R$ with $|G(x) - F(x)| < \delta$ for all $x \in W$, such that G is C^∞ in $U \cup U'$ and $F(x) = G(x)$ if $x \in W - \bar{V}'$.

<u>Proof</u>: By the Weierstrass approximation theorem there is a polynomial $H(x)$ so that $|H(x) - F(x)| < \delta$ for $x \in \bar{V}'$. Let $g : R^n \longrightarrow R$ be C^∞, $0 \leq g \leq 1$ with $g|_{\bar{U}'} = 1$, $g|_{R^n - V'} = 0$. Let

$$G(x) = g(x)\cdot H(x) + (1-g(x))F(x)$$

for all $x \in W$. Then $G(x) = H(x)$ on U' and $G(x) = F(x)$ on $W - V'$. On $\bar{V}'$,

$$|G(x) - F(x)| = |g(x)||H(x) - F(x)| < \delta.$$

Also $G(x)$ is C^∞ when F is, hence on U, so G is C^∞ on $U \cup U'$. *

Proposition: Let $f : W \longrightarrow R^k$ be a C^∞ function, W and open subset of R^n, C a compact subset of W, V a neighborhood of C with $\bar{V} \subset W$, and $\varepsilon > 0$. Then there exists a differentiable $g : W \longrightarrow R^k$ such that

1) $g|_C$ has $0 \in R^k$ as a regular value.

2) $g = f$ on $W - V$

3) $|g_i(x) - f_i(x)| < \varepsilon$, $|\partial g_i/\partial x_j(x) - \partial f_i/\partial x_j(x)| < \varepsilon$

for all $x \in W$, $1 \leq i \leq k$, $1 \leq j \leq n$.

Proof: Let $\lambda : W \longrightarrow R$ be C^∞ with $\lambda|_C = 1$, $\lambda|_{W-V} = 0$ and $0 \leq \lambda(x) \leq 1$ for all x. If y is any regular value of f then

$$g(x) = f(x) - \lambda(x)y$$

satisfies conditions 1) and 2) above. By Sard's theorem, y may be chosen arbitrarily close to 0, and so 3) may be satisfied by taking y small enough. *

Proposition: Let C be a compact subset of W, W open in R^n and $g : W \longrightarrow R^k$ a C^∞ function such that $g|_C$ has 0 as regular value. Then there is an $\varepsilon > 0$ such that if $h : W \longrightarrow R^k$ with

$$|h_i(x) - g_i(x)| < \varepsilon, \quad |\partial h_i/\partial x_j(x) - \partial g_i/\partial x_j(x)| < \varepsilon$$

for all $x \in C$, then $h|_C$ also has 0 as regular value.

Proof: $\{x \in C | x$ is critical for $g\}$ is closed so compact and the set of critical values of g is then closed. Thus there is an $\varepsilon_1 > 0$ so that $|g_i(x)| < \varepsilon_1$ implies x is regular for g. In particular $Dg(x)$ is non-singular and there is an $\varepsilon_2(x) > 0$ such that $|A_{ij} - \partial g_i/\partial x_j(x)| < \varepsilon_2(x)$

implies (A_{ij}) is non-singular. On the set of x for which $|g_i(x)| \leq \varepsilon_{1/2}$, which is compact, there will be an $\varepsilon_3 > 0$ so that $\varepsilon_3 \leq \varepsilon_2(x)$ for all these x. Let $\varepsilon = \min(\varepsilon_1/2, \varepsilon_3)$. If $|h_i(x) - g_i(x)| < \varepsilon$ and $|\partial h_i/\partial x_j\ (x) - \partial g_i/\partial x_j\ (x)| < \varepsilon$ then $h(x) = 0$ implies $|g_i(x)| < \varepsilon \leq \varepsilon_{1/2}$ so $Dg(x)$ is non-singular and since $|\partial h_i/\partial x_j\ (x) - \partial g_i/\partial x_j\ (x)| < \varepsilon \leq \varepsilon_2(x)$ $Dh(x)$ is non-singular. Thus 0 is a regular value for h. *

Appendix 2

Differentiable Manifolds

This appendix covers the basic notions of differentiable manifolds, tangent and normal bundles and proves the transverse regularity theorem which will be basic to the calculation of cobordism groups. In order to get this, one needs basic structure theorems for manifolds such as tubular neighborhoods and imbeddability and these are also proved. Basic references are:

1. Kelley, J. L.: General Topology, D. Van Nostrand Co., Inc., Princeton, N. J., 1955.
2. Milnor, J.: Differential Topology, (mimeographed) Princeton University, 1958.
3. Munkres, J. R.: Elementary Differential Topology, Princeton University Press, Princeton, N. J., 1966.
4. Nomizu, K.: Lie Groups and Differential Geometry, Mathematical Society of Japan, 1956.

__Definition__: $H^n \subset R^n$ is the half space $\{(x_1,\ldots,x_n) \in R^n | x_n \geq 0\}$.

__Definition__: An __n-dimensional differentiable manifold with boundary__ is a pair $(V,\mathcal{F})$ where V is a Hausdorff space with a countable base and $\mathcal{F}$ is a family of real valued continuous functions on V satisfying:

1) $\mathcal{F}$ is local: if $f : V \longrightarrow R$ and for all $p \in V$ there is an open set $U_p \subset V$, $p \in U_p$, and a function $g_p \in \mathcal{F}$ such that $f|_{U_p} = g|_{U_p}$, then $f \in \mathcal{F}$.

2) $\mathcal{F}$ is differentiably complete: if $f_1,\ldots,f_k \in \mathcal{F}$ and $F : R^k \longrightarrow R$ is C^∞, then $F\circ(f_1 \times\ldots\times f_k) : V \longrightarrow R$ belongs to $\mathcal{F}$.

3) For each point $p \in V$ there are n-functions $f_1,\ldots,f_n \in \mathcal{F}$ such that $f_1 \times\ldots\times f_n : V \longrightarrow R^n$ is a homeomorphism of an open neighborhood U of p onto an open subset of H^n. Further, every function $f \in \mathcal{F}$ agrees on U with a function of the form $F\circ(f_1 \times\ldots\times f_n)$ where $F : R^n \longrightarrow R$ is C^∞.

The functions $f \in \mathcal{F}$ are called the <u>differentiable functions</u> on V. A <u>chart</u> at $p \in V$ is a pair (U,h), where U is an open neighborhood of p and $h : V \longrightarrow R^n$ is a function $f_1 \times \ldots \times f_n = h$, with $f_i \in \mathcal{F}$ mapping U homeomorphically onto an open subset of H^n as in 3).

<u>Proposition</u>: V is paracompact.

<u>Proof</u>: Since H^n is locally compact, so is V, and there is a base $U_1, U_2, \ldots$ for V with $\bar{U}_i$ compact for each i. There is a sequence $A_1, A_2, \ldots$ of compact sets with union V and $A_i \subset$ interior A_{i+1}: Let $A_1 = \bar{U}_1$ and if A_i is defined, there is a least integer $k = k(i)$ so that $A_i \subset U_1 \cup \ldots \cup U_k$. Then let $A_{i+1} = \overline{U_1 \cup \ldots \cup U_k}$.

Let $\mathcal{O}$ be any open cover of V. Cover the compact set A_{i+1} - Interior A_i by a finite number of open sets $V_1, \ldots, V_r$ where V_j is contained in an element of $\mathcal{O}$ and in the open set interior $A_{i+2} - A_{i-1}$. Let $\mathcal{P}_i$ denote the family $\{V_1, \ldots, V_r\}$, and $\mathcal{P} = \mathcal{P}_0 \cup \mathcal{P}_1 \cup \ldots$. Then $\mathcal{P}$ refines $\mathcal{O}$, covers V and since any compact set C is contained in some A_i, C can intersect only finitely many elements of $\mathcal{P}$. Thus, for $p \in V$, any compact neighborhood of p meets only a finite number of elements of $\mathcal{P}$. *

<u>Corollary</u>: V is normal.

<u>Proof</u>: a) V is regular. If $a \in V$, $B \subset V$, B closed and $a \notin B$, choose for each $b \in B$ open sets U'_b, V'_b with $a \in U'_b$, $b \in V_b$ and $U'_b \cap V_b = \phi$. Let $U_b = U'_b \cap (V-B)$. Then $a \in U_b$, $b \in V_b$, $U_b \cap V_b = \phi$ and $U_b \subset V - B$. Then $\{V-a-B, U_b, V_b\}_{b \in B}$ is an open cover of V, so has a locally finite refinement $\{C_\alpha\}_{\alpha \in I}$. Let $J = \{\alpha \in I | C_\alpha \cap B\} \neq \phi$, $W = \bigcup_{\alpha \in J} C_\alpha$. Then W is open and contains B. Let N be a neighborhood of a

meeting only a finite number of the sets C_α. There is a finite set $J_0 \subset J$ so that $\alpha \in J$, $N \cap C_\alpha \neq \phi$ implies $\alpha \in J_0$. For each $\alpha \in J_0$, $C_\alpha \cap B \neq \phi$, so there is a $b = b(\alpha) \in B$ with $C_\alpha \subset V_b$. Let $T = N \cap \bigcap_{\alpha \in J_0} U_{b(\alpha)}$. Then T is open, $a \in T$ and $T \cap W = \phi$.

b) V is normal. Let $A, B \subset V$ be closed, $A \cap B = \phi$. For each $a \in A$, there are open sets U'_a, V'_a with $a \in U'_a$, $B \subset V'_a$ and $U'_a \cap V'_a = \phi$. Let $U_a = U'_a \cap (V-B)$, $V_a = V'_a \cap (X-A)$. Then $\{V-A-B, U_a, V_a\}_{a \in A}$ is an open cover of V so has a locally finite refinement $\{C_\alpha\}$. Let $J = \{\alpha | C_\alpha \cap A \neq \phi\}$. For each $b \in B$, there is a neighborhood N_b of b meeting only a finite number of the sets C_α, $\alpha \in J$. Each such C_α is contained in some set U_a and the intersection of N_b with the corresponding sets V_a is a neighborhood T_b of b not meeting any C_α with $\alpha \in J$. Let $T = \bigcup_{b \in B} T_b$, $W = \bigcup_{\alpha \in J} C_\alpha$. Then $B \subset T$, $A \subset W$ and $T \cap W = \phi$. *

Lemma: Let $\mathcal{U}$ be an open cover of V. Then there is a refinement $\mathcal{V}$ of $\mathcal{U}$ so that for each $X \in \mathcal{V}$ there is a set $Y \in \mathcal{U}$ with $\bar{X} \subset Y$.

Proof: Let $\mathcal{U}_0$ be a locally finite refinement of $\mathcal{U}$. Consider the set $\mathfrak{A}$ of all functions F whose domain is a subfamily of $\mathcal{U}_0$, and for each U in the domain of F, $F(U)$ is an open set with closure contained in U, and such that $\bigcup\{F(U) | U \in \text{domain } F\} \cup \bigcup\{W \in \mathcal{U}_0 | W \notin \text{domain } F\} = V$. $\mathfrak{A}$ is non-empty by normality of V. Partially order $\mathfrak{A}$ by $F \leq G$ if G extends F. If F_α is a linearly ordered family, let F be defined on $\bigcup\{\text{domain } F_\alpha\}$ by $F(U) = F_\alpha(U)$ if $U \in \text{domain } F_\alpha$. Let $x \in V$ and suppose $x \notin W$ for any $W \notin \text{domain } F$. Thus if $x \in U$, $U \in \mathcal{U}_0$, then $U \in \text{domain } F$. Since there are only a finite number of sets $U \in \mathcal{U}_0$ with $x \in U$, and each such $U \in \text{domain } F_\alpha$ for some α, there is a β such that $x \in U$, $U \in \mathcal{U}_0$ implies $U \in \text{domain } F_\beta$. Thus $x \in \bigcup\{F_\beta(U) | U \in \text{domain } F_\beta\}$ so

$x \in \bigcup\{F(U) | U \in \text{domain } F\}$. Then $\mathcal{O}\!\mathcal{C}$ has a maximal element F and by normality of V, F must be defined on all of $\mathcal{U}_0$. Thus $\mathcal{V} = \{F(U) | U \in \mathcal{U}_0\}$ suffices. *

<u>Proposition</u>: Let $\mathcal{U}$ be any open cover of V. Then there is a differentiable partition of unity on V subordinate to $\mathcal{U}$. i.e. $\exists$ collection $\Phi \subset \mathcal{F}$ such that:

1) $\phi \in \Phi$ implies $\phi : V \longrightarrow [0,1]$

2) The collection $\mathcal{V} = \{U_\phi | \phi \in \Phi\}$ is a locally finite refinement of $\mathcal{U}$, where $U_\phi = \{x \in V | \phi(x) > 0\}$.

3) For each $x \in V$, $\sum_{\phi \in \Phi} \phi(x) = 1$.

<u>Proof</u>: Let $\mathcal{U}_1$ be the collection of open sets U such that there is a chart (U,h) and such that $U \subset U'$ for some $U' \in \mathcal{U}$. By the lemma, there is a locally finite refinement $\mathcal{U}_2$ of $\mathcal{U}_1$ such that for each $U_2 \in \mathcal{U}_2$ there is a $U_1 \in \mathcal{U}_1$ with $\bar{U}_2 \subset U_1$, and there is a refinement $\mathcal{U}_3$ of $\mathcal{U}_2$ such that for each $U_3 \in \mathcal{U}_3$ there is a $U_2 \in \mathcal{U}_2$ with $\bar{U}_3 \subset U_2$. In particular there is a cover of V by sets U_3 such that $U_3 \in \mathcal{U}_3$, $\bar{U}_3 \subset U_2$, $U_2 \in \mathcal{U}_2$, $\bar{U}_2 \subset U_1$, $U_1 \in \mathcal{U}_1$ and the family of such sets U_1 is a locally finite refinement of $\mathcal{U}$. Let (U_1,h) be a chart and let $\psi_{U_3} : h(U_1) \longrightarrow R$ be C^∞, being 1 on $h(\bar{U}_3)$ and 0 outside $h(U_2)$, with $0 \le \psi_{U_3} \le 1$. Let ϕ'_{U_3} be $\psi_{U_3} \circ h$ on U_1 and 0 on $V - \bar{U}_2$. Then being locally in $\mathcal{F}$, $\phi'_{U_3} \in \mathcal{F}$. Finally let $\phi_{U_3}(x) = \phi'_{U_3}(x) \Big/ \sum_{U_3} \phi'_{U_3}(x)$ and Φ the collection of these ϕ_{U_3}. *

<u>Corollary</u>: Let U and W be open subsets of V with $\bar{U} \subset W$. There is an $f \in \mathcal{F}$ with $f(V) \subset [0,1]$ so that $f|_{\bar{U}} = 1$, $f|_{V-W} = 0$.

Proof: $\{W, V - \bar{U}\}$ is an open cover of V so there is a differentiable partition of unity Φ subordinate to this cover. If $\phi \in \Phi$ and $\phi(x) \neq 0$ for some $x \in \bar{U}$, then $U_\phi \subset W$. Let f be the sum of those $\phi \in \Phi$ which are non-zero on $\bar{U}$. #

The set of points of V may be divided into two classes as follows. For each point $p \in V$, let (U,h) be a chart at p. Then either $h(p) \in R^{n-1} \times 0 \subset H^n$ or $h(p)$ belongs to the interior of H^n. If (U',h') is another chart at p and $h'(p) \notin R^{n-1} \times 0$, then $h \circ h'^{-1} : h'(U \cap U') \longrightarrow h(U \cap U') \subset R^n$ is a C^∞ function with a C^∞ inverse, and by the inverse function theorem, $h \circ h^{-1}$ maps onto an open neighborhood of $h(p)$ in R^n. Thus $h(p) \notin R^{n-1} \times 0$. Hence this property is independent of the choice of (U,h).

Definition: The set of points $p \in V$ for which there is a chart (U,h) with $h(p) \in R^{n-1} \times 0$ is called the boundary of V, and denoted ∂V. The complement of ∂V, $V - \partial V$, is the interior of V.

Proposition: If $(V,\mathcal{F})$ is an n-dimensional differentiable manifold with boundary and $\mathcal{F}|_{\partial V}$ denotes the set of restrictions to ∂V of functions in $\mathcal{F}$, then $(\partial V, \mathcal{F}|_{\partial V})$ is an (n-1)-dimensional differentiable manifold (without boundary; i.e. $\partial(\partial V) = \phi$).

Proof: Clearly ∂V is Hausdorff and has a countable base, and properties 2) and 3) are clear. Suppose $f : \partial V \longrightarrow R$ is any function, and for each $p \in \partial V$ there is an open set $U_p \subset \partial V$ and $g_p \in \mathcal{F}|_{\partial V}$ such that $f|_{U_p} = g_p|_{U_p}$. There is then a function $g'_p \in \mathcal{F}$ and an open neighborhood U'_p of p in V with $U'_p \cap \partial V = U_p$ and $g'_p|_{\partial V} = g_p$. Then $\{V-\partial V, U'_p\}$ is an open cover of V and there is a partition of unity Φ subordinate to this cover. For each $\phi \in \Phi$ such that $U_\phi = \{x \in V | \phi(x) > 0\}$ meets ∂V, there is a set U'_p with $U_\phi \subset U'_p$. Let p_ϕ be one such. Then define $f' : V \longrightarrow R$ by $f'(x) = \sum_{\Phi'} \phi(x) g'_{p_\phi}(x)$ where $\Phi' = \{\phi \in \Phi | U_\phi \cap \partial V \neq \phi\}$. f' is locally a finite sum of elements of $\mathfrak{F}$, so belongs to $\mathfrak{F}$. If $x \in \partial V$ and $\phi(x) \neq 0$ then $x \in U'_{p_\phi}$ so $g'_{p_\phi}(x) = f(x)$. Hence $f'(x) = f(x) \cdot \sum \phi(x) = f(x)$. Thus $f = f'|_{\partial V}$ or $f \in \mathfrak{F}|_{\partial V}$. *

<u>Definition</u>: If $(V, \mathfrak{F}(V))$ and $(W, \mathfrak{F}(W))$ are differentiable manifolds with boundary, a function $f : V \longrightarrow W$ is called a <u>differentiable map</u> if for all $g \in \mathfrak{F}(W)$, $g \circ f \in \mathfrak{F}(V)$. f is a <u>diffeomorphism</u> if f has a differentiable inverse.

<u>Proposition</u>: If $f : (V, \mathfrak{F}(V)) \longrightarrow (W, \mathfrak{F}(W))$ is a differentiable map and $f(\partial V) \subset \partial W$ then $f|_{\partial V} : (\partial V, \mathfrak{F}(V)|_{\partial V}) \longrightarrow (\partial W, \mathfrak{F}(W)|_{\partial W})$ is a differentiable map. The inclusion map $i : (\partial V, \mathfrak{F}(V)|_{\partial V}) \longrightarrow (V, \mathfrak{F})$ is differentiable.

<u>Proposition</u>: If $(V, \mathfrak{F})$ is an n-dimensional manifold with boundary, U is an open subset of V and $\mathfrak{F}|_U$ denotes the set of restrictions to U of functions in $\mathfrak{F}$, then $(U, \mathfrak{F}|_U)$ is an n-dimensional manifold with boundary, and the inclusion map is differentiable.

Let X be a set and suppose there is a countable collection $\mathfrak{A} = \{(X_\alpha, h_\alpha)\}$ where $X_\alpha \subset X$ and $\bigcup_\alpha X_\alpha = X$ and h_α is a bijection of X_α with an n-dimensional manifold with boundary V_α such that for each pair α, β $h_\alpha(X_\alpha \cap X_\beta)$ is an open subset of V_α and

$$h_\beta \circ h_\alpha^{-1} : h_\alpha(X_\alpha \cap X_\beta) \longrightarrow h_\beta(X_\alpha \cap X_\beta)$$

is differentiable. Then X may be given a topology and a differentiable structure so that each set X_α will be open and each function h_α is a diffeomorphism. X is then an n-dimensional differentiable manifold with boundary, and is uniquely determined within diffeomorphism.

For example, let $(V,\mathcal{F})$ and $(W,\mathcal{J})$ be n-dimensional and m-dimensional manifolds with boundary (∂W being empty). Let (U_i,h_i) and (T_j,g_j) be countable families of charts for V and W. Then the collection $\{(U_i \times T_j, h_i \times g_j)\}$ defines a differentiable structure on $V \times W$, giving the product manifold, of dimension $n + m$. Then $\partial(V \times W)$ is diffeomorphic to $\partial V \times W$.

Definition: If $(V,\mathcal{F})$ is a differentiable manifold with boundary a subset $A \subset V$ is called a submanifold of V if for each point $a \in A$ there is a chart (U,h) at a with $h(U \cap A) = h(U) \cap (0 \times R^k)$. The collection $\mathcal{F}|_A$ of restrictions to A of functions of $\mathcal{F}$ is the family of differentiable functions on A.

Note: $\partial A = A \cap \partial V$, is then a submanifold of ∂V.

Definition: A (real) vector bundle $\xi = (E,B,\pi,+,\cdot)$ is a 5-tuple where

1) E and B are topological spaces, called the total space and base space of ξ,

2) $\pi : E \longrightarrow B$ is a continuous map, called the projection,

3) $+ : E + E = \{(e,e') \in E \times E \mid \pi e = \pi e'\} \longrightarrow E$ and

$\cdot : R \times E \longrightarrow E$ are continuous maps such that $\pi\circ +(e,e') = \pi e = \pi e'$, $\pi\circ\cdot(r,e) = \pi e$ and the restrictions to each fiber $\pi^{-1}(b)$ for $b \in B$ make $\pi^{-1}(b)$ into a real vector space.

Definition: A bundle map $f : \xi \longrightarrow \xi'$ is a pair f_E, f_B of continuous maps $f_E : E \longrightarrow E'$, $f_B : B \longrightarrow B'$ such that $\pi' \circ f_E = f_B \circ \pi$ and $f_E \circ + = +' \circ (f_E + f_E)$, $f_E \circ \cdot = \cdot' \circ f_E$, where $f_E + f_E$ is the restriction to $E + E$ of $f_E \times f_E$. f is an isomorphism if there is a bundle map $g : \xi' \longrightarrow \xi$ which is inverse to f.

For example one has the product bundle $(B \times R^n, B, \pi, +, \cdot)$ where π is the projection of the product space.

Definition: The bundle $\xi = (E, B, \pi, +, \cdot)$ is locally trivial if for each point $b \in B$ there is an open set U in B containing b and a bundle isomorphism $h_U : \xi|_U \longrightarrow (U \times R^n, U, \pi, +, \cdot)$ where $\xi|_U$ is the bundle $(\pi^{-1}(U), U, \pi|_{\pi^{-1}(U)}, +, \cdot)$ with induced operations with the induced map of base spaces being the identity map of U.

Definition: A differentiable vector bundle is a vector bundle ξ for which the total space and base space are differentiable manifolds with boundary, the projection is a differentiable map and such that for each point $b \in B$ the open set U and map h_U may be chosen to give a diffeomorphism of total spaces.

Note: $+$ and $\cdot$ are forced to be differentiable by the local triviality.

Definition: Let $(V, \mathcal{F})$ be an n-dimensional manifold with boundary, and $v \in V$. A tangent vector X and v is a function $X : \mathcal{F} \longrightarrow R$ such that:

1) If $f, g \in \mathcal{F}$ and there is an open neighborhood U of v with $f|_U = g|_U$, then $X(f) = X(g)$,

2) For $f, g \in \mathcal{F}$, $a, b \in R$, $X(af + bg) = aX(f) + bX(g)$,

3) If $f, g \in \mathcal{F}$, then $X(f \cdot g) = X(f) \cdot g(v) + f(v) \cdot X(g)$.

The set of tangent vectors at v forms a vector space induced from the additive structure in R, called the tangent space to V at v and denoted τ_v.

Denote by $\tau(V)$ the union over all $v \in V$ of the sets τ_v and let $\pi : \tau(V) \longrightarrow V$ be the function which sends each subset τ_v into the point v.

Proposition: Let $v \in V$ and let (U,h) be a chart at v, with $h = f_1 \times \dots \times f_n$. Then

$$\lambda_U : \pi^{-1}(U) \longrightarrow U \times R^n : X \longrightarrow (\pi(X),(X(f_i)))$$

is a bijection. If (U',h') is another chart at v, then $\lambda_U \circ \lambda_{U'}^{-1} : (U \cap U') \times R^n \longrightarrow (U \cap U') \times R^n$ is given by $\lambda_U \circ \lambda_{U'}^{-1}(u,\alpha) = (u,D(h \circ h'^{-1})(h'(u))(\alpha))$.

Proof: First note that if $X \in \tau_v$ then X annihilates constant functions. To see this, one has $X(c) = cX(1) = cX(1 \cdot 1) = c\{1X(1) + X(1) \cdot 1\} = 2cX(1)$. Thus $cX(1) = 2cX(1)$ must be zero, so $X(c) = 0$. Then let $f \in \mathcal{F}$ be any function. There is a C^∞ function $F : R^n \longrightarrow R$ with $f = F \circ h$ and one may write $F(x) = F(h(v)) + \sum_{i=1}^{n} (x-h(v))_i g_i(x)$ with g_i being C^∞ and

$$g_i(h(v)) = \frac{\partial F}{\partial x_i}(h(v)).$$

Thus

$$f = f(v) + \sum_{i=1}^{n} (f_i - f_i(v)) \cdot (g_i \circ h)$$

so

$$X(f) = X(f(v)) + \sum_{i=1}^{n} \{X(f_i - f_i(v)) \cdot g_i \circ h(v) + (f_i(v) - f_i(v))X(g_i \circ h)\}$$

$$= \sum_{i=1}^{n} X(f_i) \cdot \frac{\partial F}{\partial x_j}(h(v)).$$

Thus λ_U is one-to-one, and letting $X_\alpha(f) = \sum \alpha_i \frac{\partial F}{\partial x_j}(h(v))$ for $\alpha \in R^n$ one has λ_U onto. Thus λ_U is a bijection.

Then $\lambda_U \circ \lambda_{U'}^{-1}(u,\alpha) = (u,(\lambda_{U'}^{-1}(u,\alpha)(f_i)))$ and

$$\lambda_{U'}^{-1}(u,\alpha)(f_i) = \sum_{j=1}^{n} \alpha_j \frac{\partial(\pi_i \circ h \circ h'^{-1})}{\partial x_j}(h'(v)),$$

$$= [D(h \circ h'^{-1})(h'(v))(\alpha)]_i. \quad *$$

<u>Corollary</u>: $\tau = (\tau(V), V, \pi, +, \cdot)$ may be given the structure of a differentiable fiber bundle so that if (U,h) is a chart in V, λ_U is a local trivialization of τ and $(\pi^{-1}(U), (h \times 1)\circ\lambda_U)$ is a chart of $\tau(V)$. The boundary of $\tau(V)$ is $\pi^{-1}(\partial V)$.

<u>Proposition</u>: If $\phi : (V, \mathcal{F}(V)) \longrightarrow (W, \mathcal{F}(W))$ is a differentiable map, $v \in V$ and $X \in \tau_v$, let $\phi_*(X)$ be defined by

$$\phi_*(X)(f) = X(f\circ\phi)$$

if $f \in \mathcal{F}(W)$. Then $\phi_* : \tau(V) \longrightarrow \tau(W)$ is a differentiable map covering ϕ and (ϕ_*, ϕ) is a differentiable bundle map.

<u>Definition</u>: Let $M(p,n)$ denote the set of $p \times n$ matrices with differentiable manifold structure given by identification with R^{pn}. Let $M(p,n;k)$ denote the subset consisting of matrices of rank k.

<u>Lemma</u>: $M(p,n;k)$ is a differentiable manifold of dimension $k(p+n-k)$ if $k \leq \min(p,n)$.

<u>Proof</u>: Let $E_0 \in M(p,n;k)$ and by reordering coordinates write

$$E_0 = \begin{pmatrix} A_0 & B_0 \\ C_0 & D_0 \end{pmatrix}$$

where A_0 is non-singular and $k \times k$. There is an $\varepsilon > 0$ so that if all entries of $A - A_0$ are less than ε, then A is also non-singular. Let $U \subset M(p,n)$ consist of all

$$E = \begin{pmatrix} A & B \\ C & D \end{pmatrix}$$

with entries of $A - A_0$ less than ε.

Then $E \in M(p,n;k)$ if and only if $D = CA^{-1}B$. To see this, note that

$$\begin{pmatrix} A & B \\ XA+c & XB+D \end{pmatrix} = \begin{pmatrix} I_k & 0 \\ X & I_{p-k} \end{pmatrix} \begin{pmatrix} A & B \\ C & D \end{pmatrix}$$

has the same rank as E. If $X = -CA^{-1}$, this is

$$\begin{pmatrix} A & B \\ 0 & -CA^{-1}B+D \end{pmatrix}$$

so if $D = CA^{-1}B$ this has rank k, while if any element of $-CA^{-1}B + D$ is non-zero the rank is greater than k.

Let W be the open set in R^m, $m = k(p+n-k) = pn - (p-k)(n-k)$, consisting of matrices

$$\begin{pmatrix} A & B \\ C & 0 \end{pmatrix}$$

with all entires of $A - A_0$ less than ε. Then

$$\begin{pmatrix} A & B \\ C & 0 \end{pmatrix} \longrightarrow \begin{pmatrix} A & B \\ C & CA^{-1}B \end{pmatrix}$$

maps W homeomorphically onto the neighborhood $U \cap M(p,n;k)$ of E_0. *

Definition: A differentiable map $\phi : (V, \mathcal{F}(V)) \longrightarrow (W, \mathcal{F}(W))$ is an immersion if ϕ_* is a monomorphism on each fiber of $\tau(V)$. It is an imbedding if it is also a homeomorphism into.

Proposition: Let U be an open subset in R^n and $f : U \longrightarrow R^p$ a differentiable map with $p \geq 2n$. Given $\varepsilon > 0$, there is a $p \times n$ matrix A with all entries less than ε such that $g(x) - f(x) + Ax$ is an immersion.

Proof: For any $p \times n$ matrix A, $Dg = Df + A$ and one wants to choose A so that Dg has rank n at all points of U, or equivalently, $A = Q - Df$ where Q has rank n.

Define $F_k : M(p,n;k) \times U \longrightarrow M(p,n) : (Q,x) \longrightarrow Q - Df(x)$. Then F_k is differentiable and domain F_k has dimension $k(p+n-k) + n < pn = \dim M(n,p)$. [Taking partials one has $p+n-2k$ so the dimension is a monotone function of k and for $k < n$ this is at most $(n-1)(p+n-(n-1)) + n = (2n-p) + pn - 1 < pn$]. Thus for any chart (W,h) of $M(p,n;k) \times U$, $F_k oh^{-1}$ has no regular values. By Sard's theorem, $F_k(W) = F_k oh^{-1}(h(W))$ has measure zero but image F_k is a countable union of such sets so has measure zero. Hence there is an A arbitrarily near zero which is not in $\bigcup_{k<n}$ image F_k. This A suffices. *

Remark: If U were an open subset of H^n the same argument suffices since f is the restriction of a differentiable map from R^n into R^p.

Theorem: Given a differentiable map $f : (V,\mathcal{F}(V)) \longrightarrow R^p$ where $p \geq 2n$ and a continuous positive function δ on V, there is an immersion $g : (V,\mathcal{F}(V)) \longrightarrow R^p$ such that $|g(v) - f(v)| < \delta(v)$. If f_* is monic on τ_v for all $v \in N$, N a closed subset of V, then one may let $g|_N = f|_N$.

Proof: Since $f_*|_{\tau_v}$ is monic for all $v \in N$, it is monic for all $v \in U$ where U is an open neighborhood of N. One may then find a refinement of the open cover $\{V - N,U\}$ by a locally finite countable family of sets V_i such that each set $\bar{V}_i$ is compact and such that each V_i is the underlying set of a chart (V_i,h_i). [There is a countable base consisting of sets W with $\bar{W}$ compact and (W,h) a chart. The proof that V is paracompact shows that one may find a countable locally finite refinement]. Index the sets V_i so that the V_i contained in U have $i \leq 0$, while the remainder have $i > 0$, with $i \in Z$. Applying the proof of the shrinking lemma twice constructs open sets

$W_i \subset \bar{W}_i \subset U_i \subset \bar{U}_i \subset V_i$ with $\{W_i\}$ being a cover of V.

Let $f_0 = f$ and suppose $f_{k-1} : V \longrightarrow R^p$ is defined such that $(f_{k-1})_*|_{\tau_v}$ is monic for all $v \in N_{k-1} = \bigcup_{j<k} \bar{W}_j$. For any $p \times n$ matrix A let $F_A : h_k(V_k) \longrightarrow R^p$ be given by

$$F_A(x) = f_{k-1} \circ h_k^{-1}(x) + \phi(x)\cdot A(x)$$

where ϕ is a C^∞ function from $R^n \longrightarrow [0,1]$ with $\phi|_{h_k(\bar{W}_k)} = 1$, $\phi_k(R^n - U_k) = 0$.

First, one wants $DF_A(x)$ to have rank n on $K = h_k(N_{k-1} \cap \bar{U}_k)$. [$\bar{U}_k$ has a finite cover by open sets each meeting only finitely many $\bar{W}_j$ so $N_{k-1} \cap \bar{U}_k$ is compact] and $D(F_A)(x) = D(f_{k-1}h_k^{-1}(x)) + A(x)\cdot D\phi(x) + \phi(x)\cdot A$ with $D(f_{k-1} \circ h_k^{-1})(x)$ having rank n on K. This is a continuous function from $K \times M(p,n)$ to $M(p,n)$ sending $K \times 0$ into $M(p,n;n)$, so if A is sufficiently small one has $K \times A$ mapped into $M(p,n;n)$. Assume A is this small.

Next, choose A small enough so that $|A(x)| < \varepsilon_k/2^k$ where $\varepsilon_\mu = \inf\{\delta(x) | x \in \bar{U}_k\}$ for all $x \in h_k(V_k)$.

Finally, as above A may be chosen arbitrarily small so that $f_{k-1}h_k^{-1}(x) + A(x)$ has rank n on $h_k(U_k)$.

Let A satisfy all these requirements.

Then define $f_k : V \longrightarrow R^p$ by

$$\begin{aligned} f_k(y) &= f_{k-1}(y) + \phi(h_k(y))A(h_k(y)) && \text{if } y \in V_k \\ &= f_{k-1}(y) && \text{if } y \in V - \bar{U}_k. \end{aligned}$$

These agree on the overlap $V_k - \bar{U}_k$ so f_k is differentiable. By the first condition on A, DF_A has rank n on N_{k-1}, and by the third it has rank n on $\bar{W}_k$, hence $f_*|_{\tau_v}$ is injective for each $v \in N_k$. By the second condition, f_k is a $\delta/2^k$ approximation to f_{k-1}.

Since the cover V_i is locally finite, the f_k agree on any given compact set if k is sufficiently large, so $g(x) = \lim_{k\to\infty} f_k(x)$ exists and g is differentiable, $g_*|_{\tau_v}$ is monic for all $v \in V$, and g is a δ approximation to f. *

Lemma: Let $(V,\mathcal{F})$ be a differentiable manifold with boundary and $f : V \longrightarrow R^p$ an immersion. Then for each point $a \in V$ there is an open set U containing a such that $f|_U$ is an imbedding.

Proof: Let (W,h) be a chart at a. Then $f\circ h^{-1} : h(W) \longrightarrow R^p$ extends to a differentiable map $k : R^n \longrightarrow R^p$ with $Dk(h(a))$ monic. Thus there is an open set $T \subset R^p$ containing $k(h(a))$ and differentiable map $g : T \longrightarrow R^p$ with differentiable inverse so that $gk(x) = (x,0)$ on a neighborhood S of $h(a)$. Then

$$(h^{-1}\circ\pi\circ g)\circ f(y) = h^{-1}\circ\pi\circ g f\circ h^{-1}(h(y)) = h^{-1}\circ\pi\circ g\circ k(h(y)) = h^{-1}\circ\pi(h(y),0) = h^{-1}h(y) = y$$

for all y in a neighborhood of a, $h^{-1}(S) = U$, where π projects R^p on the first n coordinates. *

Lemma: If $p > 2n$ any immersion $f : (V,\mathcal{F}) \longrightarrow R^p$ may be δ-approximated by a 1-1 immersion g. If f is 1-1 on a neighborhood U of the closed set N, one may choose $g|_N = f|_N$.

Proof: Choose a covering of V by sets $\{U_\alpha\}$ such that $f|_{U_\alpha}$ is an imbedding for each α, refining the cover $\{U,V-N\}$. Construct a countable locally finite refinement by sets V_i, of the cover $\{U_\alpha\}$, indexed so that the $V_i \subset U$ have $i \leq 0$. Apply the shrinking lemma twice to get $W_i \subset \bar{W}_i \subset U_i \subset \bar{U}_i \subset V_i$ and let $\phi_i : V \longrightarrow R$ be a function of $\mathcal{F}$ so that $0 \leq \phi_i \leq 1$, $\phi_i(\bar{W}_i) = 1$, $\phi_i(V-U_i) = 0$.

Let $f_0 = f$ and suppose the immersion $f_{k-1} : V \longrightarrow R^p$ is defined. Then f_k is defined by

$$f_k(x) = f_{k-1}(x) + \phi_k(x)b_k$$

where $b_k \in R^p$ is yet to be chosen. As above, for small b_k, f_k will be an immersion, so let b_k be this small. b_k may also be made small enough so that f_k is a $\delta/2^k$ approximation to f_{k-1}.

Finally, let N be the open subset of $V \times V$ consisting of pairs (x,x') with $\phi_k(x) \neq \phi_k(x')$, and let $\sigma : N \longrightarrow R^p$ by $\sigma(x,x') = -[f_{k-1}(x)-f_{k-1}(x')]/[\phi_k(x)-\phi_k(x')]$. N is the union of the manifolds $(V-\partial V) \times (V-\partial V) \cap N$, $\partial V \times (V-\partial V) \cap N$, $(V-\partial V) \times \partial V \cap N$, and $\partial V \times \partial V \cap N$ on each of which σ is differentiable, and since each of these have dimension at most $2n < p$, $\sigma(N)$ has measure zero. Thus b_k may be chosen arbitrarily small and not in this image.

Then $f_k(x) = f_k(x')$ if and only if $\phi_k(x) = \phi_k(x')$ and $f_{k-1}(x) = f_{k-1}(x')$ for $k > 0$.

Let $g(x) = \lim_{k\to\infty} f_k(x)$. If $g(x) = g(x_0)$ and $x \neq x_0$ it follows that $f_{k-1}(x) = f_{k-1}(x_0)$ and $\phi_k(x) = \phi_k(x_0)$ for all $k > 0$. Thus $f(x) = f(x_0)$ so x and x_0 cannot belong to the same set V_i, and since $\phi_k(x) = \phi_k(x_0)$ for $k > 0$ neither can belong to a set W_i with $i > 0$. Thus x and x_0 must lie in U, contradicting the fact that f is 1-1 on U. *

Definition: Let $f : (V, \mathcal{F}) \longrightarrow R^p$. The limit set $L(f)$ of f is the set of $y \in R^p$ such that $y = \lim f(x_i)$ for some sequence $\{x_1, x_2, \ldots\}$ in V which has no limit point in V.

Proposition: $f(V)$ is a closed subset of R^p if and only if $L(f) \subset f(V)$.

Proof: Let $y \in \overline{f(V)}$. Then there is a sequence of points $y_i \in f(V)$ with $\lim y_i = y$. Let $x_i \in V$ with $f(x_i) = y_i$. If the sequence x_i has a limit point $x \in V$, then $f(x) = y$ by continuity of f. If the sequence x_i has no limit point in V, then $y \in L(f)$ so $y \in f(V)$. Thus $y \in f(V)$, so $f(V)$ is closed. *

Proposition: f is a topological imbedding if and only if f is 1-1 and $L(f) \cap f(V)$ is empty.

Proof: Let $T \subset V$ be closed and $y \in \overline{f(T)} \cap f(V)$. Then there is a sequence $y_i \in f(T)$ with $\lim y_i = y$. Let $x_i = f^{-1}(y_i) \in T$. If the sequence x_i has no limit point then $y \in L(f)$, but $L(f) \cap f(V) = \phi$. Thus there is a limit point x of the sequence x_i, and since T is closed, $x \in T$. By continuity of f, $f(x)$ is a limit point of the sequence y_i, and since y is the limit of the sequence y_i and R^p is Hausdorff, $y = f(x)$. Thus $\overline{f(T)} \cap f(V) = f(T)$ so $f(T)$ is closed in $f(V)$. Hence $f^{-1} : f(V) \longrightarrow V$ is continuous, or f is a topological imbedding. *

Lemma: There is a differentiable map $f : (V, \mathcal{F}) \longrightarrow R$ with $L(f) = \phi$.

Proof: Let V_i be a countable, locally finite cover of V by sets V_i with compact closure. Apply the shrinking lemma twice to give $W_i \subset \bar{W}_i \subset U_i \subset \bar{U}_i \subset V_i$, with $\{W_i\}$ a cover of V, and let $\phi_i \in \mathcal{F}$ with $0 \leq \phi_i \leq 1$, $\phi_i(\bar{W}_i) = 1$, $\phi_i(V-U_i) = 0$. Let $f(x) = \sum j\phi_j(x)$. This sum is finite for each x since V_i is locally finite. If x_i is a sequence in V having no limit point, then only finitely many x_i can lie in any compact subset of V. Given any integer m, there is an integer $N(m)$ such that $i \geq N(m)$ implies $x_i \notin \bar{W}_1 \cup \ldots \cup \bar{W}_m$. Thus if $i \geq N(m)$, there is a $j > m$ with $x_i \in \bar{W}_j$, so $f(x_i) \geq j > m$.. Hence the sequence $f(x_i)$ can have no limit point. *

Corollary: Every n-dimensional differentiable manifold with boundary can be imbedded in R^{2n+1} as a closed subset.

Proof: Let $f : (V, \mathcal{F}) \longrightarrow R \subset R^{2n+1}$ be a differentiable map with $L(f) = \phi$ constructed as above. Let $\delta(x) = 1$ for all $x \in V$ and let g be a 1-1 immersion of $(V, \mathcal{F})$ in R^{2n+1} with $|g(x) - f(x)| < \delta(x)$ for all $x \in V$. Let x_i be any sequence in V having no limit point. Given any integer m there is an integer $P(m) = N(m+1)$ so that if $i \geq P(m)$, then $|g(x_i)| > m$, for $|g(x_i)| \geq |f(x_i)| - 1 > m + 1 - 1$. Thus the sequence $g(x_i)$ cannot have a limit point. Hence $L(g) = \phi$ and g is a topological imbedding as a closed subset. *

Definition: Let V_1, V_2 be differentiable manifolds, $f : V_1 \longrightarrow V_2$ an immersion. The normal bundle of f, ν_f is defined as follows. Let τ_1 and τ_2 denote the tangent bundles of V_1 and V_2. Then $f_* : \tau_1 \longrightarrow \tau_2$ induces a monomorphism into the bundle $f^!\tau_2$ over V_1, where $f^!\tau_2$ is the pull-back. The quotient bundle of $f^!\tau_2$ by τ_1 is a differentiable vector bundle over V_1 which is ν_f.

Now let $(V, \mathcal{F})$ be a differentiable manifold and let $g : V \longrightarrow R^p$ be an imbedding. Since the tangent bundle of R^p is trivial, i.e. the total space is $R^p \times R^p$ one may use the usual inner product in R^p to give an inner product in each fiber of $\tau(R^p)$ and hence in $g^!(\tau(R^p))$. The orthogonal complement of the image of each fiber of $\tau(V)$ in each fiber of $g^!(\tau(R^p))$ is a subspace mapped isomorphically to the fiber of ν_g. The orthogonal complements fit together to form the total space of a differentiable vector bundle $\tau(V)^{\perp}$ over V isomorphic to ν_g, via the quotient map $\alpha : g^!(\tau(R^p)) \longrightarrow \nu_g$. The bundle map $\beta : g^!(\tau(R^p)) \longrightarrow \tau(R^p)$ then gives a differentiable map $\gamma = \beta \circ (\alpha|_{\tau(V)^{\perp}})^{-1}$ mapping $E(\nu_g)$ diffeomorphically onto the submanifold of $R^p \times R^p = E(\tau(R^p))$ given by $\{(x,y) \in R^p \times R^p | x = g(v)$ and $y \perp g_*(\tau_v),\ v \in V\}$.

Let $e : R^p \times R^p \longrightarrow R^p : (x,y) \longrightarrow (x+y)$.

<u>Theorem</u>: If $(V,\mathcal{F})$ in an n-dimensional differentiable manifold with $\partial V = \phi$ and $g : V \longrightarrow R^p$ is an imbedding, then the differentiable function $e\circ\gamma : E(\nu_g) \longrightarrow R^p$ maps an open neighborhood of the zero section of ν_g diffeomorphically onto an open neighborhood of $g(V)$ in R^p.

<u>Proof</u>: First $e\circ\gamma$ is differentiable and has rank p at all points of the zero section. [To see this, let (U,h) be a chart on V so that ν_g is trivial over U. One then has a local trivialization
$k : h(U) \times R^{p-n} \longrightarrow E(\nu_g)$ with $(\pi^{-1}(U),k^{-1})$ a chart of $E(\nu_g)$. Then the function $\delta = e\circ\gamma\circ k : h(U) \times R^{p-n} \longrightarrow R^p$ is given by
$\delta(x,\alpha) = g\circ h^{-1}(x) + \sum \alpha_i y_i(x)$ where for each $x \in h(U)$, $\{y_i(x)\}$ form a base for the orthogonal complement to $D(g\circ h^{-1})(x)[R^n] = g_*(\tau_{h(x)})$. Then
$D\delta(x,\alpha)\{y,\beta\} = D(g\circ h^{-1})(x)(y) + \sum \beta_i y_i(x) + \sum \alpha_i Dy_i(x)(y)$ where
$\{y,\beta\} \in R^n \times R^{p-n} = R^p$. For $\alpha = 0$, this gives
$D\delta(x,0)\{y,\beta\} = D(g\circ h^{-1})(x)(y) + \sum \beta_i y_i(x)$ which spans R^p as $\{y,\beta\}$ runs through R^p because of the choice of y_i]. Hence $e\circ\gamma$ has rank p in some neighborhood of the zero section of $E(\nu_g)$, so that it is a local diffeomorphism at points of the zero section: i.e. it maps an open neighborhood of each point x in the zero section diffeomorphically onto an open neighborhood of $e\circ\gamma(x)$ in R^p.

To complete the proof it suffices to show:

<u>Lemma</u>: Let X and Y be Hausdorff space with countable bases and X locally compact. If $f : X \longrightarrow Y$ is a local homeomorphism and the restriction

of f to a closed subset A is a homeomorphism, then f is a homeomorphism on some neighborhood V of A.

Proof: 1) If A is compact, the lemma is true. If not, then every neighborhood N of A contains a pair x,y of points for which $f(x) = f(y)$. One may then find a countable family N_i of compact neighborhoods of A with $N_{i+1} \subset N_i^0$ and $\bigcap N_i = A$. For each i, let $x_i, y_i \in N_i^0$ with $f(x_i) = f(y_i)$. Since N_1 is compact, the sequence x_i and y_i have limit points x and y. Since $V-N_{i+1}$ contains only a finite number of points x_j and y_j, one must have $x,y \in \bigcap N_i = A$. But $f(x) = \lim f(x_i) = \lim f(y_i) = f(y)$ contradicting the fact that $f|_A$ is a homeomorphism.

2) Let A_0 be a compact subset of A. Then there is a neighborhood U_0 of A_0 such that $\bar{U}_0$ is compact and f is a homeomorphism on $\bar{U}_0 \cup A$. To see this, let V_0 be a neighborhood of A_0 with $\bar{V}_0$ compact and $f|_{\bar{V}_0}$ 1-1, which is possible by 1). If no neighborhood of A_0 in V_0 satisfies the requirements for U_0, there is a sequence of points $x_n \in X-A$ converging to $x \in A_0$ with $f(x_n) \in f(A)$. Let $y_n \in A$ with $f(y_n) = f(x_n)$. Since f is continuous, $f(y_n)$ converges to $f(x)$; and since $f|_A$ is a homeomorphism, y_n converges to x. Since $x_n \neq y_n$ this contradicts the assumption that f is a local homeomorphism at x.

3) Express A as the union of an ascending sequence of compact sets $A_1 \subset A_2 \subset \ldots$. Suppose V_i is a neighborhood of A_i with $\bar{V}_i$ compact and f is a homeomorphism on $\bar{V}_i \cup A$. Then $\bar{V}_i \cup A_{i+1}$ is a compact subset of $\bar{V}_i \cup A$ on which f is a homeomorphism and by 2) there is a neighborhood V_{i+1} of $\bar{V}_i \cup A_{i+1}$ with $\bar{V}_{i+1}$ compact and f a homeomorphism on $\bar{V}_{i+1} \cup A$. Let $V = \bigcup V_i$. The sets V_i are an ascending sequence of open sets so if $x,y \in V$ with $f(x) = f(y)$ then there is an i with $x,y \in V_i$, but $f|_{V_i}$ is 1-1 so $x = y$. Thus f is 1-1 on V and being a local homeomorphism, $f|_V$ is a homeomorphism. *

<u>Lemma</u>: Let $(V,\mathcal{F})$ be a manifold with boundary. Then there is a differentiable function $g : V \longrightarrow [0,\infty)$ such that $g(\partial V) = 0$ and $g_*|_{\tau_v}$ is non-zero for each $v \in \partial V$.

<u>Proof</u>: Let (V_i,h_i) be a countable locally finite cover of V by charts and apply the shrinking lemma twice to get $W_i \subset \bar{W}_i \subset U_i \subset \bar{U}_i \subset V_i$. Let $\phi_i \in \mathcal{F}$ with $0 \leq \phi_i \leq 1$, $\phi_i(\bar{W}_i) = 1$, $\phi_i(V-U_i) = 0$. Let K be the set of i such that $W_i \cap \partial V \neq \phi$. For each $i \in K$, $h_i : V_i \longrightarrow H^n$ is of the form $f_1^i \times \ldots \times f_n^i$ and $V_i \cap \partial V = h_i^{-1}(R^{n-1} \times 0) = (f_n^i)^{-1}(0)$. Let $g(x) = \sum_{i\in K} \phi_i(x)\cdot f_n^i(x)$. Then $g \in \mathcal{F}$ and $g_i : V \longrightarrow [0,\infty)$ with $g(\partial V) = 0$.

Let $v \in \partial V$. There is then an $i \in K$ with $v \in W_i$. Let $\lambda : [0,\infty) \longrightarrow H^n$ by $\lambda(t) = h(v) + (0,\ldots,0,t)$. Then there is an $\varepsilon > 0$ with $\lambda([0,\varepsilon)) \subset h(V_i)$. Then $h^{-1}\circ\lambda : [0,\varepsilon) \longrightarrow V$ is a differentiable map and to show $g_*|_{\tau_v} \neq 0$ it suffices to prove that $\frac{d}{dt}(g\circ h^{-1}\circ\lambda) \neq 0$ at $t = 0$.

For the $i \in K$ used to define λ, we have $\phi_i\circ h^{-1}\circ\lambda(t) = 1$ for all $t \in \lambda^{-1}(W_i)$ and $f_n^i\circ h^{-1}\circ\lambda(t) = t$ for all $t \in [0,\varepsilon)$. Thus $\frac{d}{dt}((\phi_i\circ f_n^i)\circ h^{-1}\circ\lambda) = 1$.

For any $i' \neq i$, $i' \in K$ with $v \in V_{i'}$, one has

$$\frac{d}{dt}((\phi_{i'}\cdot f_n^{i'})\circ h^{-1}\circ\lambda) = (\phi_{i'}\circ h^{-1}\circ\lambda)\cdot\frac{d}{dt}(f_n^{i'}\circ h^{-1}\circ\lambda) + f_n^{i'}\circ h^{-1}\circ\lambda\cdot\frac{d}{dt}(\phi_{i'}\circ h^{-1}\circ\lambda)$$

Now $\phi_{i'}\circ h^{-1}\circ\lambda(0) \geq 0$, $f_n^{i'}\circ h^{-1}\circ\lambda(0) = 0$ and for $t > 0$ $f_n^{i'}\circ h^{-1}\circ\lambda(t) > 0$ in a neighborhood of $t = 0$, and hence $\frac{d}{dt}(f_n^{i'}\circ h^{-1}\circ\lambda) \geq 0$.

Adding these up, one has $\frac{d}{dt}(g\circ h^{-1}\circ\lambda) \geq 1$. *

<u>Theorem</u>: Let $(V,\mathcal{F})$ be a differentiable manifold with boundary. There is an open neighborhood U of ∂V in V such that $(U,\mathcal{F}|_U)$ is diffeomorphic to $\partial V \times [0,1)$.

Proof: Let $\bar{g} : V \longrightarrow R^p$ be an imbedding. Then $\bar{g}|_{\partial V} : \partial V \longrightarrow R^p$ is an imbedding so there is an open neighborhood N of ∂V in R^p diffeomorphic to a neighborhood of the zero section in $E(\nu_{\bar{g}|_{\partial V}})$, with $\alpha : N \longrightarrow E(\nu_{\bar{g}|_{\partial V}})$ the diffeomorphism into. Then $\pi\circ\alpha\circ\bar{g} : \bar{g}^{-1}(N) \longrightarrow \partial V$ is a differentiable retraction of the open neighborhood $\bar{g}^{-1}(N)$ of ∂V onto ∂V. Let $g : V \longrightarrow [0,\infty)$ be as given previously. Then $r = (\pi\circ\alpha\circ\bar{g}) \times g : \bar{g}^{-1}(N) \longrightarrow \partial V \times [0,\infty)$ is a differentiable map. For any $v \in \partial V$, the kernel of $g_*|_{\tau_v}$ contains the image of $\tau(\partial V)_v$, hence by dimension this is precisely the kernel. $(\pi\circ\alpha\circ\bar{g})_*$ maps the image of $\tau(\partial V)_v$ isomorphically. Thus $r_*|_{\tau_v}$ is an isomorphism. Thus $r_*|_{\tau_w}$ is monic for all w in some open neighborhood W of ∂V, and so is a local diffeomorphism of W with an open neighborhood of $\partial V \times 0$ in $\partial V \times [0,\infty)$, and is a homeomorphism on ∂V. Thus there is an open neighborhood Q of ∂V in V diffeomorphic to an open neighborhood of ∂V in $\partial V \times [0,\infty)$. By means of a countable locally finite cover of ∂V by charts, with compact closure, one may take a neighborhood of ∂V of the form $\{(x,y) \in \partial V \times [0,\infty) | y < \beta(x)\}$ for some $\beta \in \mathcal{F}(\partial V)$ with $\beta > 0$, within this neighborhood. Sending $(x,y) \longrightarrow (x,y/\beta(x))$ maps this diffeomorphically onto $\partial V \times [0,1)$. *

Theorem: Let $(V,\mathcal{D})$ and $(W,\mathcal{D})$ be differentiable manifolds with boundary such that V is a submanifold of W with inclusion $i : V \hookrightarrow W$ and suppose there is a neighborhood U of ∂W in W and a diffeomorphism $f : (U, U \cap V) \longrightarrow (\partial W \times [0,1), \partial V \times [0,1))$. Then there is an open neighborhood of V in W diffeomorphic to an open neighborhood of the zero section in ν_i.

Proof: Let $\alpha = \pi_1 \circ f : U \longrightarrow \partial W$, $\beta = \pi_2 \circ f : U \longrightarrow [0,1)$. There is a function $\mu \in \mathcal{F}(W)$ with $0 \leq \mu \leq 1$, $\mu(\beta^{-1}([0,3/4))) = 1$, $\mu(W-U) = 0$ and a function $\nu \in \mathcal{F}(W)$ with $0 \leq \nu \leq 1$, $\nu(\beta^{-1}([0,5/8])) = 0$,

$\nu(W-\beta^{-1}([0,3/4))) = 1$ and so $\sigma = \mu\cdot\beta + \nu : W \longrightarrow [0,\infty)$ is in $\mathcal{F}(W)$ and $\sigma|_{U'} = \beta|_{U'}$ where $U' = \beta^{-1}([0,1/2))$.

Let $\phi : [0,1/2) \longrightarrow [0,1]$ be the C^∞ function with $\phi[0,1/4] = 0$, $\phi[3/8,1/2) = 1$ given by $\phi_{1/4,3/8}$. Let $q : W \longrightarrow W$ be $f^{-1}\circ(1\times u)\circ f$ on U' where $u(s) = \phi(s)\cdot s$ and the identity on $W - f^{-1}([0,3/8))$.

Let $g : W \longrightarrow R^p$ be any imbedding and define $h = (g\circ q) \times \sigma : W \longrightarrow H^{p+1}$ h is easily seen to be an imbedding and $h\circ f^{-1} : \partial W \times [0,1/2) \longrightarrow H^{p+1}$ agrees with $g|_{\partial W} \times 1$.

The inner product on R^{p+1} gives inner products on $\tau(W)_w$ and $\tau(V)_v$ so that one may identify ν_i with

$$\{(x,y) \in H^{p+1} \times R^{p+1} | x = hi(v), y \in h_*\tau(W)_{i(v)}, y \perp h_* i_* \tau(V)_v\}.$$

The evaluation map e sends this subset into R^{p+1}, and by the agreement of $h\circ f^{-1}$ with $g|_{\partial W} \times 1$ on $\partial W \times [0,1/2)$ will send $\{(x,y)| x = h(u'), u' \in U'\}$ into H^{p+1} (since y can have no component orthogonal to $R^p \times 0$) and hence sends a neighborhood of $h(i(V)) \times 0$ into H^{p+1}. Since W is imbedded nicely by h, there is a retraction of a neighborhood of W into W (as in the tubular neighborhood theorem for closed manifolds in Euclidean space). The composite map of a neighborhood of the zero section in $E(\nu_i)$ into W is of maximal rank at the zero section, and checking along the tubular neighborhood of ∂V shows that this is a diffeomorphism of smaller neighborhoods. *

Note: Such a nice tubular neighborhood U seems to always exist if one has sufficient regularity at the intersection of V and ∂W. In particular, our definition of submanifold appears sufficiently restrictive to give this. No simple proof seems possible, and hoping that we won't need this existence, we will avoid the argument.

Definition: Let $f : M^m \longrightarrow N^n$ be a differentiable map, N'^k a closed submanifold of N. f is said to be transverse regular to N'^k at $x \in M^m$ if

1) $f(x) \notin N'^k$, or

2) $f(x) \in N'^k$ and the composite

$$\tau(M)_x \xrightarrow{f_*} \tau(N)_{f(x)} \longrightarrow \tau(N)_{f(x)} / i_* \tau(N')_{f(x)}$$

is epic, where $i : N' \hookrightarrow N$ is the inclusion.

f is said to be transverse regular on N' if f is transverse regular at each point of M.

Proposition: The set of points $x \in M$ at which f is transverse regular to N' is open.

Proof: $f^{-1}(N')$ is closed so the set of points of type 1 is open. Suppose x is of the second type and choose a chart at $f(x)$, (U,h), with $h(U \cap N') = h(U) \cap (0 \times R^k)$. Let (V,k) be a chart at x with $V \subset f^{-1}(U)$. With coordinates u_i in $h(U)$, v_j in $k(V)$, one has $h \circ f \circ k^{-1} : k(V) \longrightarrow h(U)$ and the transversality condition at x is the assertion that

$$(\partial u_i / \partial v_j)_{\substack{i=1,\dots,n-k \\ j=1,\dots,m}}$$

has rank $n - k$ at $k(x)$. This matrix has rank $n - k$ in a neighborhood of $k(x)$, so f is transverse regular on a neighborhood of x. *

Proposition: If $f : M^n \longrightarrow N^n$ is transverse regular to N'^k and the restriction of f to ∂M is also transverse regular to N'^k then $f^{-1}(N')$ is a submanifold of M of dimension $m - (n-k)$. Further, the normal bundle of $f^{-1}(N')$ in M is induced from the normal bundle to N' in N.

Proof: Let f be transverse regular at $x \in f^{-1}(N')$ and choose charts (U,h) and (V,k) as above. By reordering coordinates in V, one may assume $(\partial u_i/\partial v_j)$ $i,j = 1,\ldots,n-k$ is non-singular at $k(x)$. Hence by the inverse function theorem the functions $(u_1,\ldots,u_{n-k},v_{n-k+1},\ldots,v_m)$ give a chart at $k(x)$ in $k(V)$ and hence a chart at x, (V',k') so that $k'(V' \cap f^{-1}(N')) = k'(V') \cap (0 \times R^{m-(n-k)})$. If $x \in M - \partial M$ this is a chart of the required type. If $x \in \partial M$, then the condition on $f|_{\partial M}$ implies that in the reordering the function v_m is not replaced by any u_i, and hence that $k'(V') \subset H^m$. Thus, the chart (V',k') is as required. The normal bundle condition is clear since the induced map is epic on fibers.

Theorem: Let $f : M \longrightarrow N$ be a differentiable map; let N' be a closed differentiable submanifold of N'. Let A be a closed subset of M such that the transverse regularity condition for f on N' is satisfied at all points of $A \cap f^{-1}(N')$. There exists a differentiable map $g : M \longrightarrow N$ such that

1) g is homotopic to f,

2) g is transverse regular on N'

3) $g|_A = f|_A$.

Proof: There is a neighborhood U of A in M such that f satisfies the transverse regularity condition on U. Cover N by $N - N' = Y_0$ and coordinate systems (Y_i,k_i) for $i > 0$ with coordinate functions $(v_1,\ldots,v_n)$ such that $N' \cap Y_i$ is mapped precisely to the set for which $v_1 =\ldots= v_{n-k} = 0$. The sets $f^{-1}(Y_i)$ cover M, as do the sets U and $M - A$. Let (V_j,h_j) be a refinement of both coverings which is countable and locally finite, indexed so that $j \leq 0$ if $V_j \subset U$ and the others have $j > 0$. Apply the shrinking lemma twice to get $W_j \subset \bar{W}_j \subset U_j \subset \bar{U}_j \subset V_j$ and let $\phi_j \in \mathcal{F}(M)$,

$0 \leq \phi_j \leq 1$, $\phi_j(\bar{W}_j) = 1$, $\phi_j(M-U_j) = 0$. For each j choose an $i(j) \geq 0$ with $f(V_j) \subset Y_{i(j)}$.

Let $f_0 = f$ and suppose f_{k-1} has been defined, satisfies transverse regularity on $\bigcup_{j<k} \bar{W}_j$ with $f_{k-1}(\bar{U}_j) \subset Y_{i(j)}$ for each j. In particular, letting $i = i(k)$, $f_{k-1}(\bar{U}_k) \subset Y_i$.

Consider the function $\pi k_i f_{k-1} \circ h_k^{-1} : h_k(U_k) \longrightarrow R^{n-k}$ where π projects on the first $n - k$ coordinates. By the approximation of regular values theorem, there are arbitrarily small vectors $y \in R^{n-k}$ such that $\pi k_i^{-1} f_{k-1} \circ h_k^{-1} - (\phi_k \circ h_k^{-1})y$ has the origin as regular value. We then define f_k by

$$f_k(x) = k_i^{-1}\{k_i f_{k-1}(x) - \phi_k(x)(y,0)\} \quad \text{for } x \text{ in a neighborhood of } \bar{U}_k$$

$$= f_{k-1} \quad \text{if } x \in M - U_k.$$

where $y \in R^{n-k}$ is yet to be chosen.

First one needs y small enough that $k_i f_{k-1} - \phi_k(y,0)$ lies in $k_i(Y_i)$ for all $x \in \bar{U}_k$. If Y_i is a neighborhood meeting ∂N then $(y,0)$ is "parallel" to ∂N and one is not translated out of $k_i(Y_i)$ across ∂N. Hence for small y this holds and thus k_i^{-1} may be applied. Next y is chosen to give a $\delta/2^k$ approximation to f_{k-1}. Also y is chosen small enough that $f_k(\bar{U}_j) \subset Y_{i(j)}$ for each j. This is possible since only a finite number of sets $\bar{U}_j$ meet $\bar{U}_k$. Under these conditions f_k will be transverse regular on N' at each point of $f_k^{-1}(N') \cap \bar{W}_k$.

Now f_{k-1} is transverse regular on N' at each point of the compact set $\bar{U}_k \cap (\bigcup_{j<k} \bar{W}_j)$ and since small changes preserve regularity, for sufficiently small y, f_k will also be transverse regular on this set, hence on $\bigcup_{j\leq k} \bar{W}_j$.

After all these limitations, we have such a y and hence an f_k. Let $g(x) = \lim f_k(x)$. A homotopy from f_{k-1} to f_k is given by contracting y and a limit of these homotopies defines a homotopy from f to g. *

BIBLIOGRAPHY

1. J. F. Adams: On the non-existence of elements of Hopf invariant one, Ann. of Math., 72 (1960), 20-104.

2. ____: On formulae of Thom and Wu, Proc. London Math. Soc., 11 (1961), 741-752.

3. ____: On the groups J(X),II, Topology, 3 (1965), 137-173.

4. ____: On the groups J(X),IV, Topology, 5 (1966), 21-71.

5. D. W. Anderson: Thesis, Univ. of Calif. Berkeley (not yet published).

6. D. W. Anderson, E. H. Brown, Jr., and F. P. Peterson: SU cobordism, KO-characteristic numbers, and the Kervaire invariant, Ann. of Math., 83 (1966), 54-67.

7. ____: Spin cobordism, Bull. Amer. Math. Soc., 72 (1966), 256-260.

8. ____: The structure of the spin cobordism ring, Ann. of Math., 86 (1967), 271-298.

9. P. G. Anderson: Cobordism classes of squares of orientable manifolds, Ann. of Math., 83 (1966), 47-53.

10. ____: Note on a problem of Conner and Floyd, (to appear).

11. S. Araki and H. Toda: Multiplicative structures in mod q cohomology theories, I, Osaka J. Math., 2 (1965), 71-115.

12. M. F. Atiyah: Immersions and embeddings of manifolds, Topology, 1 (1961), 125-132.

13. ____: Bordism and cobordism, Proc. Camb. Phil. Soc., 57 (1961), 200-208.

14. ____: Thom complexes, Proc. London Math. Soc., 11 (1961), 291-310.

15. ____: K-theory, mimeographed notes, Harvard University, Cambridge, Mass., 1964.

16. M. F. Atiyah, R. Bott, and A. Shapiro: Clifford modules, Topology, 3, suppl. 1 (1964), 3-38.

17. M. F. Atiyah and F. Hirzebruch: Riemann-Roch theorems for differentiable manifolds, Bull. Amer. Math. Soc., 65 (1959), 276-281.

18. ____: Vector bundles and homogeneous spaces, Proc. Symp. Pure Math. vol. III, Amer. Math. Soc., Providence, R. I., 1961, 7-38.

19. ____: Cohomologie Operationen und charakteristische Klassen, Math. Zeit., 77 (1961), 149-187.

20. M. F. Atiyah and I. M. Singer: The index of elliptic operators on compact manifolds, Bull. Amer. Math. Soc., 69 (1963), 422-433.

21. B. G. Averbuh: Algebraic structure of intrinsic homology groups, Dokl. Akad. Nauk SSSR., 125 (1959), 11-14.

22. J. M. Boardman: On manifolds with involution, Bull. Amer. Math. Soc., 73 (1967), 136-138.

23. _____: Unoriented bordism and cobordism, (to appear).

24. R. Bott: Lectures on K(X), mimeographed notes, Harvard University, Cambridge, Mass., 1962.

25. W. Browder: The Kervaire invariant of framed manifolds and its generalization, (to appear).

26. W. Browder and J. Levine: Fibering manifolds over a circle, Comm. Math. Helv. 40 (1966), 153-160.

27. W. Browder, A. Liulevicius, and F. P. Peterson: Cobordism theories, Ann. of Math., 84 (1966), 91-101.

28. E. H. Brown, Jr. and F. P. Peterson: Algebraic bordism groups, Ann. of Math., 79 (1964), 616-622.

29. _____: Relations among characteristic classes-I, Topology, 3, suppl. 1 (1964), 39-52; -II, Ann. of Math., 81 (1965), 356-363.

30. _____: A spectrum whose Z_p cohomology is the algebra of reduced p^{th} powers, Topology, 5 (1966), 149-154.

31. R. O. Burdick: Oriented manifolds fibered over the circle, Proc. Amer. Math. Soc., 17 (1966), 449-452.

32. J. M. Cohen: The Hurewicz homomorphism on MU, (unpublished).

33. P. E. Conner: A bordism theory for actions of an abelian group, Bull. Amer. Math. Soc., 69 (1963), 244-247.

34. P. E. Conner and E. E. Floyd: Fixed point free involutions and equivariant maps, Bull. Amer. Math. Soc., 66 (1960), 416-441.

35. _____: Cobordism theories, Seattle conference on Differential and Algebraic Topology (mimeographed), 1963.

36. _____: "Differentiable Periodic Maps", Springer, Berlin, 1964.

37. _____: Periodic maps which preserve a complex structure, Bull. Amer. Math. Soc., 70 (1964), 574-579.

38. _____: Fibring within a cobordism class, Michigan Math. J., 12 (1965), 33-47.

39. _____: Torsion in SU-bordism, Memoirs Amer. Math. Soc., no. 60, 1966.

40. P. E. Conner and E. E. Floyd: *Maps of odd period*, Ann. of Math., 84 (1966), 132-156.

41. ____: The relation of cobordism to K-theories, Springer, Berlin, 1966.

42. P. E. Conner and P. S. Landweber: The bordism class of an SU manifold, Topology, 6 (1967), 415-421.

43. A. Dold: Erzeugende du Thomschen Algebra $\mathfrak{N}$., Math. Z., 65 (1956), 25-35.

44. ____: Vollständigkeit der Wuschen Relationen zwischen den Stiefel-Whitneyschen Zahlen differenzierbarer Mannigfaltigkeiten., Math. Z., 65 (1956), 200-206.

45. ____: Démonstration elémentaire de deux résultats du cobordisme, Ehresmann seminar notes, Paris, 1958-59.

46. ____: Structure de l'anneau de cobordisme Ω, Bourbaki seminar notes, Paris, 1959-60.

47. ____: Relations between ordinary and extraordinary cohomology, Notes, Aarhus Colloquium on Algebraic Topology, Aarhus, 1962.

48. S. Eilenberg: On the problems of topology, Ann. of Math., 50 (1949), 247-260.

49. F. T. Farrell: The obstruction to fibering a manifold over a circle, (mimeographed) Yale University, New Haven, Conn., 1967.

50. S. Gitler and J. D. Stasheff: The first exotic class of BF, Topology, 4 (1965), 257-266.

51. P. S. Green: A cohomology theory based upon self-conjugacies of complex vector bundles, Bull. Amer. Math. Soc., 70 (1964), 522-524.

52. A. Hattori: Integral characteristic numbers for weakly almost complex manifolds, Topology, 5 (1966), 259-280.

53. M. Hirsch: Immersions of manifolds, Trans. Amer. Math. Soc., 93 (1959), 242-276.

54. F. Hirzebruch: Komplexe Mannigfaltigkeiten, Proc. Int. Cong. Math., 1958, 119-136.

55. ____: Topological Methods in Algebraic Geometry, Springer, Berlin, 1966.

56. L. Hodgkin: K-theory of Eilenberg-MacLane complexes I-II, (multilithed notes), Institute for Advanced Study, Princeton, N. J., (about 1965).

57. C. S. Hoo: Remarks on the bordism algebra of involutions, Proc. Amer. Math. Soc., 17 (1966), 1083-1086.

58. W. C. Hsiang and C. T. C. Wall: Orientability of manifolds for generalized homology theories, Trans. Amer. Math. Soc., 118 (1965), 352-359.

59. D. Husemoller: "Fibre bundles", McGraw-Hill Book Company, New York, N. Y., 1966.

60. M. A. Kervaire: A manifold which does not admit any differentiable structure, Comm. Math. Helv., 34 (1960), 257-270.

61. M. A. Kervaire and J. W. Milnor: Groups of homotopy spheres I, Ann. of Math., 77 (1963), 504-537.

62. V. Y. Kraines: Topology of quaternionic manifolds, Trans. Amer. Math. Soc., 122 (1966), 357-367.

63. P. S. Landweber: Cobordism operations, mimeographed, University of Virginia (about 1965).

64. _____: Cobordism operations, Notices Amer. Math. Soc., 12 (1965), 578.

65. _____: Künneth formulas for bordism theories, Trans. Amer. Math. Soc., 121 (1966), 242-256.

66. _____: Steenrod representability of stable homology, Proc. Amer. Math. Soc., 18 (1967), 523-529.

67. _____: The bordism class of a quasi-symplectic manifold, (to appear).

68. _____: On the symplectic bordism groups of the spaces Sp(n), HP(n), and BSp(n), (to appear).

69. _____: Fixed point free conjugations on complex manifolds, Ann. of Math., 86 (1967), 491-502.

70. _____: Conjugations on complex manifolds and equivariant homotopy of MU, Bull. Amer. Math. Soc., 74 (1968), 271-274.

71. R. Lashof: Poincaré duality and cobordism, Trans. Amer. Math. Soc., 109 (1963), 257-277.

72. A. Liulevicius: The factorization of cyclic reduced powers by secondary cohomology operations, Memoirs of the Amer. Math. Soc., no. 42, 1962.

73. _____: A proof of Thom's theorem, Comm. Math. Helv., 37 (1962), 121-131.

74: _____: A theorem in homological algebra and stable homotopy of projective spaces, Trans. Amer. Math. Soc., 109 (1963), 540-552.

75. _____: Notes on homotopy of Thom spectra, Amer. J. Math., 86 (1964), 1-16.

76. A. Markov: The insolubility of the problem of homeomorphy, Dokl. Akad. Nauk SSSR, 121 (1958), 218-220 (Russian).

77. J. W. Milnor: On manifolds homeomorphic to the 7-sphere, Ann. of Math., 64 (1956), 399-405.

78. _____: Lectures on characteristic classes, mimeographed, Princeton University, Princeton, N. J., 1957.

79. J. W. Milnor: The Steenrod algebra and its dual, Ann. of Math., 67 (1958), 150-171.

80. _____: Differential topology, mimeographed, Princeton University, Princeton N. J., 1958.

81. _____: On the cobordism ring Ω^*., Notices Amer. Math. Soc., 5 (1958), 457.

82. _____: On the cobordism ring Ω^* and a complex analogue, Part I., Amer. J. Math., 82 (1960), 505-521.

83. _____: A procedure for killing homotopy groups of differentiable manifolds, Proc. Symp. Pure Math. vol. III, Amer. Math. Soc., Providence, R. I., 1961, 39-55.

84. _____: A survey of cobordism theory, Enseignement Mathematique, 8 (1962), 16-23.

85. _____: Spin structures on manifolds, L'Enseignment Mathematique, 9 (1963), 198-203.

86. _____: Microbundles I, Topology, 3, suppl. 1 (1964), 53-81.

87. _____: On the Stiefel-Whitney numbers of complex manifolds and of spin manifolds, Topology, 3 (1965), 223-230.

88. _____: Lectures on the h-cobordism theorem, Princeton University Press, Princeton, N. J., 1965.

89. _____: Characteristic classes for spherical fibre spaces, (mimeographed), Princeton University, 1965.

90. J. W. Milnor and J. C. Moore: On the structure of Hopf algebras, Ann. of Math., 81 (1965), 211-264.

91. J. R. Munkres: "Elementary Differential Topology", Princeton University Press, Princeton, N. J., 1966.

92. S. P. Novikov: Some problems in the topology of manifolds connected with the theory of Thom spaces, Dokl. Akad. Nauk SSSR, 132 (1960), 1031-1034, (Soviet Math Doklady 1, 717-720).

93. _____: Homotopy properties of Thom complexes, Mat. Sb. (N.S.), 57 (1962), 407-442, (Russian).

94. _____: The Topology Summer Institute, Seattle 1963, Russian Mathematical Surveys, 20 (1965), 145-167.

95. _____: Topological invariance of rational Pontrjagin classes, Dokl. Akad. Nauk SSSR, 163 (1965), 298-300 (Soviet Math Doklady, 6 (1965), 921-923).

96. _____: Operation rings and spectral sequences of the Adams type in extraordinary cohomology theories. U-cobordisms and K-theory, Dokl. Akad. Nauk SSSR, 172 (1967), 33-36 (Soviet Math Doklady, 8 (1967), 27-31).

97. R. S. Palais: Seminar on the Atiyah-Singer index theorem, Princeton University Press, Princeton, N. J., 1965.

98. F. P. Peterson: Relations among Stiefel-Whitney classes of manifolds, Notes, Aarhus Colloquium on Algebraic Topology, Aarhus, 1962.

99. H. Poincaré: Analysis Situs, Journal de l'Ecole Polytechnique, 1 (1895), 1-121.

100. L. S. Pontrjagin: Characteristic cycles on differentiable manifolds, Math. Sbor. (N.S.), 21 (63)(1947), 233-284 (AMS translations, no. 32).

101. _____: Smooth manifolds and their applications in homotopy theory, Trudy Mat. Inst. im Steklov, no. 45, Izdat. Akad. Nauk SSSR, Moscow, 1955 (AMS translations, series 2, vol. 11, 1959).

102. B. L. Reinhart: Cobordism and the Euler number, Topology, 2 (1963), 173-178.

103. V. A. Rohlin: A 3 dimensional manifold is the boundary of a 4 dimensional manifold, Dokl. Akad. Nauk SSSR, 81 (1951), 355.

104. _____: Intrinsic homologies, Dokl. Akad. Nauk SSSR, 89 (1953), 789-792.

105. _____: Intrinsic homology theories, Uspekhi Mat. Nauk, 14 (1959), 3-20 (AMS translations, series 2, 30 (1963), 255-271).

106. V. A. Rohlin and A. S. Svarc: Combinatorial invariance of the Pontrjagin classes, Dokl. Akad. Nauk SSSR, 114 (1957), 490-493.

107. J. P. Serre: Groupes d'homotopie et classes de groupes abéliens, Ann. of Math., 58 (1953), 258-294.

108. L. Smith and R. E. Stong: The structure of BSC, (to appear).

109. _____: Exotic cobordism theories associated with classical groups, (to appear).

110. E. H. Spanier: "Algebraic Topology", McGraw-Hill Book Company, New York, 1966.

111. M. Spivak: Spaces satisfying Poincaré duality, Topology, 6 (1967), 77-101.

112. J. Stasheff: A classification theorem for fiber spaces, Topology, 2 (1963), 239-246.

113. N. E. Steenrod: Cohomology invariants of mappings, Ann. of Math., 50 (1949), 954-988.

114. N. E. Steenrod and D. B. A. Epstein: Cohomology operations, Princeton University Press, Princeton, N. J., 1962.

115. R. E. Stong: Determination of $H^*(BO(k,\ldots,\infty);Z_2)$ and $H^*(BU(k,\ldots,\infty);Z_2)$ Trans. Amer. Math. Soc., 107 (1963), 526-544.

116. R. E. Stong: Cobordism and Stiefel-Whitney numbers, Topology, 4 (1965), 241-256.

117. ____: Relations among characteristic numbers-I, Topology, 4 (1965), 267-281; -II, Topology, 5 (1966), 133-148.

118. ____: On the squares of oriented manifolds, Proc. Amer. Math. Soc., 17 (1966), 706-708.

119. ____: Cobordism of maps, Topology, 5 (1966), 245-258.

120. ____: Involutions fixing projective spaces, Michigan Math. J., 13 (1966), 445-447.

121. ____: On complex-spin manifolds, Ann. of Math., 85 (1967), 526-536.

122. ____: Some remarks on symplectic cobordism, Ann. of Math., 86 (1967), 425-433.

123. __: Stationary point free group actions, Proc. Amer. Math. Soc., 18 (1967), 1089-1092.

124. J. C. Su: A note on the bordism algebra of involutions, Michigan Math. J., 12 (1965), 25-31.

125. D. Sullivan: The Hauptvermutung for manifolds, Bull. Amer. Math. Soc., 73 (1967), 598-600.

126. R. Thom: Espaces fibrés en spheres et carrés de Steenrod, Ann. Sci. Ecole Norm. Sup., 69 (1952), 109-181.

127. ____: Quelques propriétés globales des variétés differentiables, Comm. Math. Helv., 28 (1954), 17-86.

128. ____: Les classes caractéristiques de Pontrjagin des variétés trianguleés, Symposium Internacional de Topologia Algebraica, Mexico, 1958.

129. ____: Travaux de Milnor sur le cobordisme, Séminaire Bourbaki, 1958/59, Paris.

130. C. T. C. Wall: Determination of the cobordism ring, Ann. of Math., 72 (1960), 292-311.

131. ____: Cobordism of pairs, Comm. Math. Helv., 35 (1961), 136-145.

132. ____: A characterization of simple modules over the Steenrod algebra mod 2, Topology, 1 (1962), 249-254.

133. ____: Cobordism exact sequences for combinatorial and differentiable manifolds, Ann. of Math., 77 (1963), 1-15.

134. ___: Cobordism of combinatorial n-manifolds for $n \leq 8$, Proc. Camb. Phil. Soc., 60 (1964), 807-811.

135. C. T. C. Wall: Topology of smooth manifolds, Journal London Math. Soc., 40 (1965), 1-20.

136. _____: Addendum to a paper of Conner and Floyd, Proc. Camb. Phil. Soc., 62 (1966), 171-175.

137. A. H. Wallace: Modifications and cobounding manifolds, Canadian J. Math., 12 (1960), 503-528.

138. R. Wells: Cobordism groups of immersions, Topology, 5 (1966), 281-294.

139. G. W. Whitehead: Generalized homology theories, Trans. Amer. Math. Soc., 102 (1962), 227-283.

140. J. H. C. Whitehead: On C^1-complexes, Ann. of Math., 41 (1940), 809-824.

141. R. E. Williamson, Jr.: Cobordism of combinatorial manifolds, Ann. of Math., 83 (1966), 1-33.

142. Wu, Wen-Tsün: Classes caractéristiques et i-carrés d'une variété, C. R. Acad. Sci. Paris, 230 (1950), 508-511.

Milton Keynes UK
Ingram Content Group UK Ltd.
UKHW010023170724
445591UK00005B/140